Frontiers in Protein and Peptide Sciences
(Volume 1)

Editor

Ben M. Dunn
Department of Biochemistry and Molecular Biology
University of Florida
College of Medicine
USA

CONTENTS

FOREWORD

This year, *Protein and Peptide Letters* is in its 20th anniversary. Over the years, this journal played an important role not only in the illumination of protein and peptide science, but also in the development of several generations of protein and peptide scientists around the globe. As of June 12, 2014, WEB of Science listed 2,265 papers published in *Protein and Peptide Letters* that were cited 12,405 times. Curiously, although the number of annually published papers reached a plateau in 2005-2006, the number of annual citations of the papers published in *Protein and Peptide Letters* keeps its steady growth.

The first volume of "Frontiers in Protein and Peptide Science" edited by Prof. Ben Dunn represents an exciting collection of 11 papers which are the revised and updated versions of the manuscripts originally published in *Protein and Peptide Letters*. Papers for this volume were selected based on the number of citations that the original articles received. Therefore, the resulting volume represents a set of "the best of the best of the best" studies extracted from the *Protein and Peptide Letters* and covers a broad variety of topics ranging from protein function, to protein folding, and to computational methods for prediction of protein structural and functional properties.

Curiously, some authors of the best performing publications are frequent guests in *Protein and Peptide Letters*, whereas others published just one "hit". Papers included in this volume talk about structure and functional properties of hemoglobins from snakes, discuss major enzymes of the endocannabinoid metabolism, represent anti-apoptotic and oncogenic functions of Hsp27, show the results of the molecular modeling of CYP2C19 SNPs, represent new updates related to understanding of the folding rates and folding nuclei of globular proteins, discuss the roles of heat shock response agents in diseases, talk about prediction of protein secondary structure content, provide analysis of sequence divergence in metabolic proteins of *Plasmodium falciparum*, and show how computational tools can be used to predict protein subcellular locations, types of membrane proteins, and types of Golgi-resident protein. The volume will be particularly useful to basic investigators, applied scientists, and clinicians interested in the latest advances in the field of protein and peptide research.

Vladimir N. Uversky
University of South Florida
USA

PREFACE

The first issue of Protein and Peptide Letters (Vol. 1 Issue 1) was published in July 1994. Since then, this international journal has grown steadily to become widely recognized for publishing excellent work in the field of proteins and peptide research. The Aims and Scope of the journal include a range of topics, but the journal is firmly rooted in studies of the properties of the title molecules and their interactions. Over the years we have welcomed a truly international group of authors and provided a valuable service to many new scientists publishing their first papers in English.

As we have moved through the years, the popularity of the journal and its content has increased. Recently, an analysis was done of the citation record for the past few years for all published papers. From that listing we selected the the top 100 papers receiving the highest citation numbers. Those authors were invited to submit revised and updated manuscripts; the ones received were subjected to a new round of peer review, revisions were requested, and the final manuscripts were edited by the editor. The resulting top papers were selected to be included in this first volume of "Frontiers in Protein and Peptide Science". Future volumes will follow based on a continuing analysis of citation history.

I would like to thank all contributors to this volume for their efforts to update their papers with new information and references and for their efforts to prepare the new manuscripts according to the format specified in the invitation to submit. I would also like to thank two managers at Bentham Science Publishers, Dur-e-Shahnaz Shafi, who worked with me to initiate this book and Sara Yaser, who took over the book and saw it through to completion.

Ben M. Dunn

Department of Biochemistry and Molecular Biology
University of Florida
College of Medicine
USA

List of Contributors

Alexei V. Finkelstein

Institute of Protein Research, Russian Academy of Sciences, Pushchino, Moscow Region, 142290, Russian Federation

Aykut Özgür

Gaziosmanpasa University, Faculty of Natural Sciences and Engineering, Department of Bioengineering, Tokat, Turkey

Balapal S. Basavarajappa

Division of Analytical Psychopharmacology, New York State Psychiatric Institute, New York, USA; Department of Psychiatry, College of Physicians & Surgeons, Columbia University, New York, NY 10032, USA and Nathan Kline Institute for Psychiatric Research, Orangeburg, NY 10962, USA

Bing Niu

College of Life Science, Shanghai University, 99 Shang-Da Road, Shanghai, 200444, PRChina

Bi-Qing Li

Institute of Systems Biology, Shanghai University, Shanghai, People's Republic of China

Chao Chen

School of Traditional Chinese Medicine, Guangdong Pharmaceutical University, Guangzhou 510006, PR China

Dmitry N. Ivankov

Institute of Protein Research, Russian Academy of Sciences, Pushchino, Moscow Region, 142290, Russian Federation

Dong-Qing Wei

National Micro-measurement Center, Xibei University, Xian 710069, China

Esen Tutar

Kahramanmaraş, Sütçü İmam University, ÜSKİM Research Center, Central Laboratory, Kahramanmaraş, Turkey

Fábio Renato Lombardi

Centro Universitário de Jales (UNIJales), Jales – S.P., Brazil

Gayatri Ramakrishnan

Indian Institute of Science Mathematics Initiative and Department of Microbiology and Cell Biology, Indian Institute of Science, Bangalore-560012, India

Giovanni César Santos

Depto. de Bioquímica, UNICAMP, Campinas, S.P., Brazil

Guo-Zheng Li

School of Computer Science & Engineering, Shanghai University, Shanghai, 200072, China

Gustavo Orlando Bonilla-Rodriguez

Depto. de Química e Ciências Ambientais, IBILCE-UNESP, São José do Rio Preto, S.P., CEP: 15054-, Brazil

Hao Lin

Key Laboratory for NeuroInformation of Ministry of Education, Center of Bioinformatics, School of Life Science and Technology, University of Electronic Science and Technology of China, Chengdu, 610054, China

Hui Ding

Department of Physics, School of Sciences, Center for Genomics and Computational Biology, Hebei United University, Tangshan, China

Jing-Fang Wang

College of Life Science and Biotechnology, Shanghai Jiaotong University, Shanghai 200240, China

Johnny Rizzieri Olivieri

Depto. de Física, IBILCE-UNESP, São José do Rio Preto, S.P., CEP: 15054-, Brazil

Kai-Yan Feng

Division of Imaging Science & Biomedical Engineering, Room G424 Stopford Building, The University of Manchester, M13 9PT, UK

Kübra Açıkalın Coşkun

Kahramanmaraş, Sütçü İmam University, Faculty of Science and Letters, Department of Biology, Kahramanmaraş, Turkey

Kuo-Chen Chou

Bioinformatics Center, Key Lab of Systems Biology, Shanghai Institutes for Biological Sciences, Graduate School of the Chinese Academy of Sciences, Chinese Academy of Sciences, Shanghai 200031, China

Lei Chen

Department of Genetics and Genomic Sciences, Mount Sinai School of Medicine, New York, United States of America

Liang Liu

Department of Chemistry, College of Sciences, Shanghai University, 99 Shang-Da Road, Shanghai 200444, People's Republic of China

Li-Na Ma

Bioinformatics Center, Key Lab of Systems Biology, Shanghai Institutes for Biological Sciences, Graduate School of the Chinese Academy of Sciences, Chinese Academy of Sciences, Shanghai 200031, China

Lixuan Chen

School of Traditional Chinese Medicine, Guangdong Pharmaceutical University, Guangzhou 510006, PR China

Lütfi Tutar

Kahramanmaraş, Sütçü İmam University, Faculty of Science and Letters, Department of Biology, Kahramanmaraş, Turkey

Maristella Conte Anazetti

Centro Universitário de Jales (UNIJales), Jales – S.P., Brazil

Narayanaswamy Srinivasan

Department of Microbiology and Cell Biology, Indian Institute of Science, Bangalore-560012, India

Oxana V. Galzitskaya

Department of Genome Oriented Bioinformatics, Technische Universitaet Muenchen, Wissenschaftszentrum Weihenstephan, 85354 Freising, Germany

Patrícia Peres Polizelli

Depto. de Física, IBILCE-UNESP, São José do Rio Preto, S.P., CEP: 15054-, Brazil

Peixiang Cai

Guangzhou Institute of Standardization, Guangzhou 510170, PR China and School of Chemistry and Chemical Engineering, Sun Yat-Sen University, Guangzhou 510275, PR China

Pratik Dave

Department of Microbiology and Cell Biology, Indian Institute of Science, Bangalore-560012, India

Sergiy O. Garbuzynskiy

Pushchino Branch of the Moscow State Lomonosov University; 142290 Pushchino, Moscow Region, Russian Federation

Smita Mohanty

Molecular Biophysics Unit, Indian Institute of Science, Banglore, India

Tao Huang

Key Laboratory of Systems biology, Shanghai Institutes for Biological Sciences, Chinese Academy of Sciences, Shanghai, People's Republic of China

Walter Filgueira de Azevedo Jr.

Depto. de Química e Ciências Ambientais, IBILCE-UNESP, São José do Rio Preto, S.P., CEP: 15054-, Brazil

Wei Chen

Department of Physics, School of Sciences, Center for Genomics and Computational Biology, Hebei United University, Tangshan, China

Wen-Cong Lu

Department of Mathematics, University of Manchester, Manchester M60, 1QD, UK

Xiaoyong Zou

Guangzhou Institute of Standardization, Guangzhou 510170, PR China and School of Chemistry and Chemical Engineering, Sun Yat-Sen University, Guangzhou 510275, PR China

Yixue Li

Bioinformatics Center, Key Lab of Systems Biology, Shanghai Institutes for Biological Sciences, Graduate School of the Chinese Academy of Sciences, Chinese Academy of Sciences, Shanghai 200031, China

Yu-Dong Cai

Department of Combinatorics and Geometry, CAS-MPG Partner Institute for Computational Biology, Shanghai Institutes for Biological Sciences, Chinese Academy of Sciences, Shanghai 200031, People's Republic of China and Department of Mathematics, University of Manchester, Manchester M60, 1QD, UK

Yu-Huan Jin

School of Materials Science and Engineering, Shanghai University, 149 Yan-Chang Road, Shanghai 200072, People's Republic of China

Yusuf Tutar

Kahramanmaraş, Sütçü İmam University, ÜSKİM Research Center, Central Laboratory, Kahramanmaraş, Turkey

Send Orders for Reprints to reprints@benthamscience.net
Frontiers in Protein and Peptide Sciences, Vol. 1, **2014, 3-30**

CHAPTER 1

Oxygen Binding Properties and Tetrameric Stability of Hemoglobins From the Snakes *Crotalus Durissus Terrificus* and *Liophis Miliaris*

Fábio Renato Lombardi[1], Maristella Conte Anazetti[2], Giovanni César Santos[3], Patrícia Peres Polizelli[3], Johnny Rizzieri Olivieri[4], Walter Filgueira de Azevedo Jr.[4] and Gustavo Orlando Bonilla-Rodriguez[5,*]

[1]Centro Universitário de Jales (UNIJales), Jales – S.P., Brazil; [2]Depto. de Bioquímica, UNICAMP, Campinas, S.P., Brazil; [3]Depto. de Física, IBILCE-UNESP; [4]Depto. de Física, IBILCE-UNESP and [5]Depto. de Química e Ciências Ambientais, IBILCE-UNESP, São José do Rio Preto, S.P., CEP: 15054-, Brazil

Abstract: Hemoglobins (Hbs) from the South American snake *Liophis miliaris* undergo a reversible oxygen-linked dissociation phenomenon $\alpha_2\beta_2 + 4O_2 \leftrightarrow 2\alpha\beta(O_2)$, due to substituted glutamate residues (43β and 101β) at the $\alpha_1\beta_2$- interface, according to Matsuura *et al.*, 1989. This work reports the oxygen-binding properties and analysis of tetrameric stability from the major hemoglobins of two snakes: a terrestrial species (*Crotalus durissus terrificus),* and a semiaquatic one, *L. miliaris*. The major hemoglobins from both species were analyzed by classic functional experimental approaches and other analytical techniques: dimer-tetramer association constant by analytical gel filtration chromatography and small angle x-ray scattering (SAXS). In addition, we performed molecular modeling of the hemoglobin from *L. miliaris*. The functional results display cooperative oxygen binding, alkaline Bohr Effect, and conformational transitions, a behavior which is compatible with tetrameric Hbs. Also, the estimated dimer-tetramer association constant $^4K_2$, radius of gyration, and maximum dimension exhibit similar compatibility. A molecular modeling of the structure of *L. miliaris* Hb, on the other hand, allows us to conclude that the residues Glu43β(CD2) and Glu101β(G3) are not essential in the maintenance of the tetrameric form, supporting our findings and allowing us to rule out the hypothesis proposed in the literature for the Hb of *Liophis miliaris* and other species of snakes. Therefore, the results reported in the paper suggest that the major Hbs of *Liophis miliaris* and *Crotalus durissus terrificus* are stable tetramers *in vivo* and *in vitro*.

Keywords: *Crotalus durissus terrificus*, Dimer-tetramer association constant, Hemoglobin, *Liophis miliaris*, Oxygen binding properties and SAXS.

*Address correspondence to Gustavo Orlando Bonilla-Rodriguez:** Departamento de Química e Ciências Ambientais, IBILCE-UNESP, Cristovão Colombo, 2265, São José do Rio Preto – S.P., Brazil, CEP: 15054-000; Tel: 55-17-3221-2361; Fax: 55-17-3221-2356; E-mail: bonilla@ibilce.unesp.br

Ben M. Dunn (Ed)

1. INTRODUCTION

Several papers concerning snake hemoglobins [1-4] reported that those proteins present distinct functional properties among vertebrates: high oxygen affinity in the stripped form, low or absent Bohr effect, strong phosphate binding, and a pH dependent cooperativity, which decreases as pH rises. ATP binding also raises cooperativity, mainly around neutral pH [5-7].

For the hemoglobin from a particular South American water snake, *Liophis miliaris*, an enhanced tetramer dissociation dependent on oxygen binding was proposed [8, 9], which would be as follows: $\alpha_2\beta_2$ (tetramer) + $4O_2 \leftrightarrow {}_2\alpha\beta(O_2)_2$. Although most hemoglobins show increased dissociation for the oxygenated form due to broken interdimeric bonds produced by the conformational change that follows O_2 binding, there is, in fact, a mixture of tetrameric and dimeric forms, with prevalence of the former. For *Liophis miliaris* hemoglobin, the proposal [1-4] is that the tetramer splits as oxygen binds, having only (or mostly) the dimeric oxygenated form. The explanation for such a drastic mechanism would be the substitution of two negative charges of β glutamate residues in the $\alpha_1\beta_2$ interface: Glu43β (CD2) → Thr and Glu101β (G3) → Val [2]. It would affect the electrostatic balance between the dimers [10] and probably its functional behavior would be sensitive to anion concentration.

Motivated to quantify and analyze that mechanism, we analyzed this hemoglobin by several experimental approaches: Analytical Gel Filtration Chromatography (AGFC), Small Angle X-ray Scattering (SAXS), osmotic stress and molecular modeling. In the AGFC, SAXS, and osmotic stress experiments, we used purified human adult hemoglobin A_0 as a control.

2. MATERIALS AND METHODS

Blood Collection – Adult rattlesnake (*Crotalus durissus terrificus*) specimens weighing 300 to 500g and *Liophis miliaris* specimens weighing 100 to 150g were supplied by Instituto Butantan (São Paulo) and the Central Animal Facility of our campus.

The animals were anesthetized with ethyl ether, and blood was collected by cardiac puncture, using a syringe containing buffered saline (0.05M Tris-HCl pH

8.0 containing 0.9% NaCl and 1mM EDTA) solution. Red blood cells were washed four times by centrifugation (2,000xg for 10 minutes each) with buffered saline using a refrigerated *Jouan* CR3i centrifuge. Hemolysis was accomplished with 30mM Tris-HCl buffer pH 9.0 containing 1mM EDTA, followed by centrifugation (9,000 x g for 60 minutes).

The stripped form (Hb without phosphates) was obtained by dialysis against the same buffer and concentrated by centrifugation using Centriprep-30 (*Amicon*), followed by gel filtration on Sephadex G-25 (*Pharmacia*) on a 2.6 x 30cm and subsequently Sephacryl S-100 HR (*Pharmacia*), to remove contaminants. Both columns were equilibrated with the same buffer, with different NaCl concentration: 0.02 and 0.2M for Sephadex and Sephacryl, respectively.

Oxygen Binding Properties - Oxygen equilibrium determinations were performed at 20°C by the tonometric-spectrophotometric method [11] using stripped Hb form and in the presence of anions (ATP and chloride). Protein concentration were about 50-60μM (heme) in suitable 0.05 M buffers (Hepes and Tris) covering the pH range from 7.0 to 9.0. The experiments were performed adding catalase and superoxide dismutase in order to decrease methemoglobin formation [12]. Samples containing more than 5% methemoglobin were discarded. The spectra of hemoglobin solutions were collected in the range from 500 to 700nm using a Cary 100 Scan spectrophotometer *(Varian)*.

The calculation of the binding constants of ATP for both the oxyhemoglobin (K_O) and deoxyhemoglobin (K_D) forms of *Crotalus* Hb-II was performed using Equation 1 as a model by non-linear regression using the Levenberg-Marquardt algorithm:

$$Log(P_{50})_p = \log(P_{50})_a + \frac{1}{4}\log\left(\frac{1+K_D[X]}{1+K_o[X]}\right) \qquad \textbf{(1)}$$

where log $(P_{50})_p$ and $(P_{50})_a$ refer to the measurements in the presence and in the absence of ATP, respectively, K_D and K_O are the ATP association constants for the deoxy and oxyhemoglobins and X is the molar concentration of free ATP [7, 13].

The number of ATP molecules bound per subunit was calculated using Wyman's linkage equation [14]:

$$-\Delta x = \Delta \log P50/\Delta \log[ATP] \qquad \textbf{(2)}$$

These experiments were done using 0.05 M Tris-HCl buffer, pH 7.5.

Analytical Gel Filtration Chromatography (AGFC) - The dimer to tetramer association constant ($^4K_2$) was determined by analytical gel filtration chromatography using the method developed by Ackers [15]. The experiments were performed on a Sephadex G-75 column (1 x 20cm) at room temperature, previously equilibrated with 0.05M Tris-HCl buffer, pH 8.0. The hemoglobin samples, stabilized with carbon monoxide to avoid oxidation, were applied as large volumes in the concentration range from 1 to 60mM (heme).

The column flow rate was maintained at about 12 mL/hr with a peristaltic pump P-1 (Pharmacia). Continuous absorbance readings of the elution profile were collected using a quartz flow cell attached to a Cary 100 spectrophotometer (*Varian*), at 541 or 412 nm depending on hemoglobin concentration.

The void volume V_0 was determined with blue-dextran solution, and the column was calibrated with Myoglobin (*Sigma*), Carbonic Anhydrase (*Sigma*) and deoxygenated Human Hemoglobin A. The internal volume Vi was measured with ATP *(Sigma),* reading the absorbance at 260nm. The usual approach involves transforming the elution volumes to partition coefficients using the equation 3:

$$\sigma = (Ve - V_0)/Vi \qquad \textbf{(3)}$$

where Ve is the elution volume, V0 the exclusion volume (dextran blue), and Vi the internal volume (ATP). V0 and Vi correspond to the limits 0 and 1, respectively. Since we did not find differences between using the partition coefficients or the elution volumes for the fitting, we chose to use these for the estimation of $^4K_2$ Dimers' association can be represented by the expression 4 below:

$$Hb_d + Hb_d = Hb_t \qquad \textbf{(4)}$$

where the dimer's fraction fd will be described by equation 5:

$$f_d = \frac{Hb_d}{(Hb_d + Hb_t)} \quad \textbf{(5)}$$

The association constant for the bound state ($^4K_2$) is represented by the fraction of tetramers divided by the fraction of dimers, as shown by:

$$^4K_2 = \frac{f_t}{f_d^2 x Hb} = \frac{(1-f_d)}{f_d^2 x Hb} \quad \textbf{(6)}$$

The total Hb concentration is represented by fd+ft, and we can rearrange the previous expression as shown below:

$$^4K_2 x f_d^2 Hb + f_d - 1 = 0 \quad \textbf{(7)}$$

obtaining a quadratic equation with fd=X. Accordingly, we can calculate the root fd as shown by equation 8:

$$f_d = \frac{\left(-1+\sqrt{1+{}^4K_2 x Hb}\right)}{2x\,{}^4K_2 x Hb} \quad \textbf{(8)}$$

The association constant $^4K_2$ was estimated using nonlinear regression, applied to the equation 9:

$$V_w=(f_d(V_d-V_t))+V_t \quad \textbf{(9)}$$

where Vw refers to the Hb solution, fd is the fraction of dimers, and Vd and Vt represent the elution volumes for the dimers and tetramers, respectively [16], constrained within the upper and lower limits obtained independently as described below. In the model fd was substituted by equation 8. Vt was obtained by passing deoxygenated human hemoglobin, and estimates of Vd were obtained based on the approach used by [17] and [18]:

$$V^d=(V_{Mb}+V_t)/2 \quad \textbf{(10)}$$

Osmotic Stress Experiments - The oxygen-binding curves with the hemolysate from *Liophis miliaris* and Hb-II from *Crotalus durissus terrificus* were determined by the tonometric-spectrophotometric method [19], using pure sucrose (Acros Organics) to set water activity [20,21]. In the osmotic stress method changes in the Hb-oxygen affinity are related to changes in water activity that can be converted to changes in protein hydration by use of linkage equations [21].

Oxygen binding experiments were performed with 60μM/heme Hb solution in 30mM TAPS, pH 8.0 in the absence (stripped form) and presence of anions (ATP and chloride) [20-22]. All equilibrium measurements were carried out at 20°C. The functional parameters P_{50} (O_2 partial pressure at half saturation) and cooperativity (n_H) were calculated from Hill plots by linear regression around half saturation. Hemoglobin and methemoglobin concentrations were estimated using the extinction coefficients for human Hb [23]. Data obtained from samples containing more than 5% methemoglobin (final concentration) were discarded.

The Hb solution osmolality (Osm) were determined after binding experiments from freezing point depression measurements using an Osmette A model 5002 osmometer (Precision Systems Inc.). The osmolality was transformed to the natural logarithm of water activity through the following relationship [20]:

$$\ln a_w = \frac{\Delta}{K_f M_w} = \frac{-Osm}{M_w} \tag{11}$$

where Δ is the freezing point depression, $K_f = 1.86\ K.kg.mol^{-1}$ is the cryoscopic constant, and M_w is the molarity of pure water ($55.56\ mol.L^{-1}$).

The effect of water as a single heterotropic ligand on oxygenation is typically analyzed with the following linkage equation [24,25]:

$$\frac{d\ln(P_{50}/P_{50}^0)}{d\ln(a_w)} = \frac{\Delta n_w}{4} = -\left(n_w^{oxy} - n_w^{deoxy}\right) \tag{12}$$

where a_w is the water activity and P_{50}^0 values are those obtained in the absence of sucrose. The slope of the linkage plot $\ln(P_{50}/P_{50}^0)$ *vs.* $\ln(a_w)$, gives the differential number of water molecules bound in the conformational transition from the deoxy to the oxy structures, Δn_w.

Small Angle X-ray Scattering (SAXS) - The equipment utilized in the data collected SAXS was a generator of X-ray with rotating anode *Rigaku* RU300. SAXS experiments utilized a target of copper and radiation $K_\alpha = 1.54Å$. The X-ray beam was monochromatic by using a Ni filter (20mm) and collimated with a block slit system [26]. The equipment potency was 4.5kW (50kV – 90mA).

The contributions to the scattering intensity from the solvent, capillary and air were subtracted from the total intensity. The SAXS measurements were carried out using hemolysate from *Liophis miliaris* (concentration 10 mg/mL) and collecting time between 6 to 8h, at room temperature (20°C). The buffer used in the experiments was 30mM HEPES pH 7.0 and 8.0.

Information of the quaternary molecular structure was obtained from the distance distribution function p(r). This function is proportional to the number of pairs of electrons separated by the distance r which is encountered by combinations between all the elements of the macromolecule, as illustrated by equation 13:

$$p(r) = \frac{1}{2\pi^2} \int_0^\infty I(h)(hr) \sin(hr)\, dh \qquad \textbf{(13)}$$

Where, I(h) is a function that represents the intensity of scattering to a given angle of scattering h.

The radius of gyration (R_g) of macromolecule in solution was determined either by Guinier approximation or by p(r) function. The advantage to utilize p(r) function is that the whole scattering curve is used in the calculation and not only a small fraction of curve, as when Guinier approximation is used. The radius of gyration (R_g) of the molecule is related to p(r) by equation 14:

$$R_g^2 = \frac{\int_0^\infty p(r) r^2 dr}{2\int_0^\infty p(r) dr} \qquad \textbf{(14)}$$

The p(r) function has been determined by indirect Fourier transformation using the ITP program [27]. This method was shown to be more efficient than the direct Fourier transformation for which termination and background effects are strong [28]. This program was also used to determine the desmeared and smoothed scattered intensity I(h), free from smearing collimation effects.

The theoretical p(r) functions were calculated using the program Multibody [29], modified in order to make molecular model building easier [30]. The models used in this program were generated from an arrangement of small spheres which reproduce the desired molecular shape and size.

We utilized the atomic coordinates of human hemoglobin (1HHO) to obtain the theoretical p(r) function of the dimeric form and used the coordinates of an alpha and a beta globins to generate a p(r) function of the dimeric form. With these coordinates, we built a dimer utilizing the Multibody program [29], in order to calculate its radius of gyration and maximum dimension. The same atomic coordinates and procedures were utilized to beget a p(r) function to tetrameric form.

Molecular Modeling – Molecular modeling was used for the purpose of investigating whether the residues Glu 43(β) and Glu 101(β) are essential in the maintaining of the tetramer, so that their replacements generate the destabilization of the tetramer in dimer.

The literature [32, 38-41] reports that residues such as Trp 37 and Arg 102 are essential in the maintenance of the tetramer; however, we did not find any reference for the residues Glu 43 and Glu 101. Since the primary sequences from *Liophis miliaris* α and β chains are available at the Protein Data Bank with access codes: P16417 and P16418, respectively, we used them to perform molecular modeling.

The program MODELLER [31] was used with *Gallus gallus* oxy-Hb [32] as a template. The atomic coordinates of all waters were removed from the templates. Next, the spatial restraints and CHARMM energy terms enforcing proper stereochemistry [33] were combined into an objective function. Finally, the model was obtained by optimizing the objective function in the Cartesian space.

The optimization was carried out by the use of the variable target function method [34] employing the method of conjugated gradient. A total of 1,000 models were generated for *Liophis miliaris* Hb structure. The final model was selected based on stereochemical quality. Utilizing the program LSQKAB [35], we assembled three-dimensional coordinates of α and β chains in the same file, generating the dimer template. Subsequently, using the same program, we rotated the dimers, and using a rotation matrix, generated the tetramers. Further optimization of the tetrameric structure was performed by employing the method of conjugated gradient.

Analysis of the Model - The overall stereo chemical quality of the final model for hemoglobin from *Liophis miliaris* was assessed by the program PROCHECK [36].

The following were residues that we considered as part of the dimeric interface: 1- α Chain**:** Pro37 (C2), Thr38 (C3), Lys40 (C5), Thr41 (C6), Tyr42 (C7), Pro44 (CD2), Ala88 (F9), Leu91 (FG3), Arg92 (FG4), Asp94 (G1), Pro95 (G2), Val96 (G3), Asn97(G4), Leu100(G7). 2- β Chain**:** Val34 (B16), Tyr35 (C1), Pro36 (C2), Trp37 (C3), Gln39(C5), Arg40 (C6), Phe41(C7), Glu43b(CD2), His97 (FG4), Val98 (FG5), Asp99 (G1), Pro100 (G2), Glu101(G3), Asn102 (G4), Leu105 (G7) [37, 38].

The position of residues Glu43(β) and Glu101(β) was analyzed in the modeled Hb from *Liophis miliaris* and also in the following hemoglobins: 1. human oxyHb crystallized in high salt concentration, PDB code: 1HHO [39]; 2. human oxyHb determined in low salt concentration, PDB code: 1BBB [40]; 3. Hb D from *Gallus gallus,* PDB code: 1HBR [32] and 4. CO-Hb from trout (*Oncorhyncus mykiss*), PDB code: 1OUU [41].

For analysis of the structure we used the program Swiss Pdb-Viewer version 3.7 [42]. The criteria used to determine side chain interactions were as follows: for hydrogen bonds, a distance range between 3.0 and 3.6Å [34]. For van der Waals contacts, we considered atoms whenever the addition of their van der Waals radius was similar or smaller than the addition of their individual values. For each atom, the following radius was considered: carbonyl carbon (backbone): 1.80Å, carbonyl oxygen (backbone): 1.52Å, and all the atoms of chains and groups: 1.8Å [43].

3. RESULTS

The hemolysate of *Crotalus durissus terrificus* showed six components. The analysis of several specimens (not shown here) did not show differences with that pattern.

The second component (denominated Hb-II) was the major fraction (about 35%), as shown by Fig. **1**. The hemolysate was submitted to ion-exchange chromatography using DEAE-Sepharose in order to isolate the major (second) component, which was denominated Hb-II. The present work was performed with the isolated Hb-II (Fig. **1**).

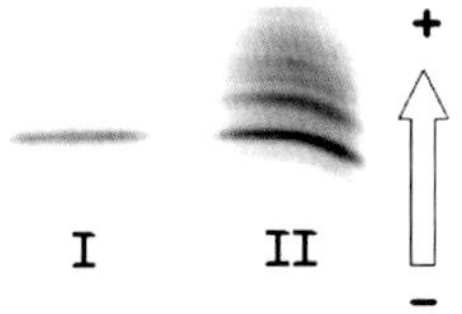

Figure 1: Hb pattern obtained by agarose isoelectric focusing after ion exchange chromatography showing I) purified Hb-II and II) hemolysate. The agarose gel was performed at 4°C, utilizing ampholines covering pH ranges 3-10 and 7-9 (*Pharmacia*).

The hemolysate from *Liophis miliaris* displayed, when analyzed by isoelectric focusing (not shown), a major Hb and two negligible components. Accordingly, we decided to perform a partial purification as previously described, obtaining a stripped hemolysate.

Fig. **2a** shows the Bohr effect of *Crotalus* Hb-II at 20°C. The stripped form showed high oxygen affinity values: $\log P_{50}$=-0.15 (P_{50}=0.7 mmHg) at pH 7.7. The Bohr effect for the stripped form was –0.39±0.08 H^+/heme. Analysis of the cooperativity (n_H), as observed in Fig. **2b**, shows a small dependence with pH change. At pH 7.0 n_H=1.8 and decreases to 1.4 at pH 9.0.

In the presence of 1mM ATP, the oxygen affinity of Hb-II decreased sharply to log P_{50}=0.64 (P_{50}=4.4 mmHg) at pH 7.7, and the Bohr effect had a slight increase (-0.44±0.06 H^+/heme). Under the same experimental conditions, cooperativity did not show dependence with varying pH, remaining higher than for the stripped form, around 2.0.

In the presence of 0.1M NaCl, we verified a small decrease in the affinity of Hb-II in comparison to the stripped form, above at pH 8.0. However, below pH 7.5 there is no difference with the stripped form. The values of cooperativity show n_H=1.8 at pH 7.0 dropping slightly to 1.5 at pH 9.0, similar to the behavior reported for the stripped form. Proton binding was –0.25±0.13 H^+/heme.

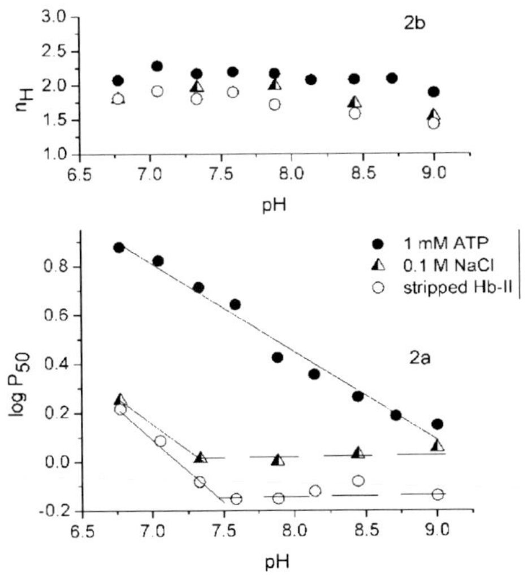

Figure 2: Oxygen affinity (Log P_{50}) and cooperativity (n_H) of Hb-II from *Crotalus durissus terrificus* in different pH values at 20°C, in the absence and in the presence of anions (1mM ATP and 0.1M Chloride). Hemoglobin concentration was 60μM (heme).

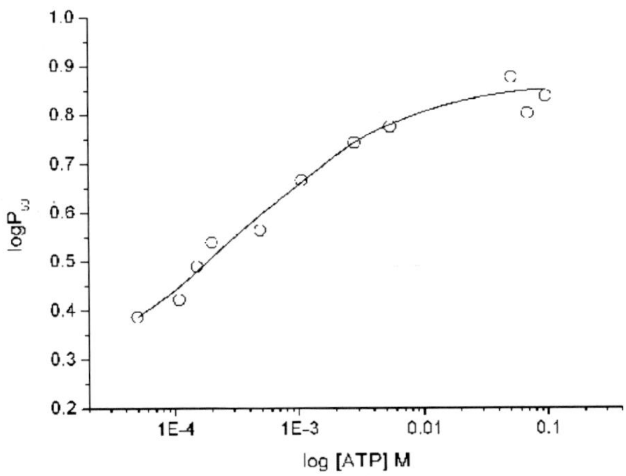

Figure 3: Effect of ATP concentration on the oxygen affinity. The slope of the straight portion of the line indicates the amount of ATP bound per subunit. The experiments were performed with 0.05M Tris-HCl buffer pH 7.5, at 20°C. Hemoglobin concentration was 60μM/heme.

Fig. **3** shows the effect of ATP concentration on oxygen-affinity in the range from 5mM to 5mM, to Hb-II from *Crotalus*. The results showed a differential binding of ATP between the deoxygenated and oxygenated forms (-Δx) of 0.25 per heme, as expected for a 1:1 relationship (ATP:Hb tetramer). The values for the ATP association constants for deoxyhemoglobin were K_D=5.13±0.85x10^{5}M^{-1} and K_O=1.92±0.63x10^{2}M^{-1}.

Concerning the conformational changes assessed by the osmotic stress approach, Fig. **4** shows the data gathered for *Crotalus* Hb-II. The results show that the number of water molecules differentially bound to the oxygenated form (Δn_w), was 36±7 water molecules in the stripped form, increasing to 52±10 water molecules in the presence of chloride. In the presence of 1mM ATP, Δn_w decreased to 45±11 water molecules, from Hb-II from *Crotalus*.

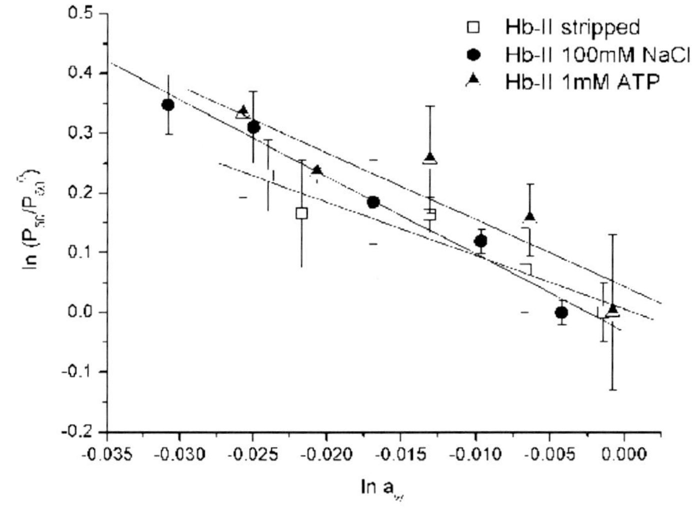

Figure 4: Dependence of ln P_{50} of Hb-II of *Crotalus* on water activity (ln a_w) at different sucrose concentrations. The buffer utilized was 0.03M Hepes-NaOH pH 7.5, at 20°C, in the presence and absence of anions. The sample concentration was 60μM/heme.

For *Liophis miliaris* Hb the osmotic stress results (not shown) the stripped hemolysate form binds 62±4 water molecules during the T to R transition,. In the presence of 0.1M chloride and 1 mM ATP, we found 50±6 and 57±7 water molecules differentially bound to oxy-hemoglobin, respectively. It must be stressed here that oxygen binding was always cooperative, with the stripped form displaying lower values, around 1.3.

As verified for hemoglobins from other snake species, in the presence of ATP, n_H for *Liophis*' Hb reached around 2.5, close to those values reported for human hemoglobin. Clearly there is a dependence on the presence of anionic species, as chloride also raised cooperativity around 1.6, although it is less than in the presence of ATP. For the stripped form and in the presence of ATP, n_H tends to increase as a_w decreases. In contrast, in the presence of ATP, n_H raises as a_w increases.

For Hb-II from *Crotalus* the direct measurement of the Dimer-Tetramer equilibrium Association Constant ($^4K_2$) point out a high tetrameric stability. Fig. **5** shows the relationship between hemoglobin concentration and the centroid elution volume for *Crotalus* Hb-II.

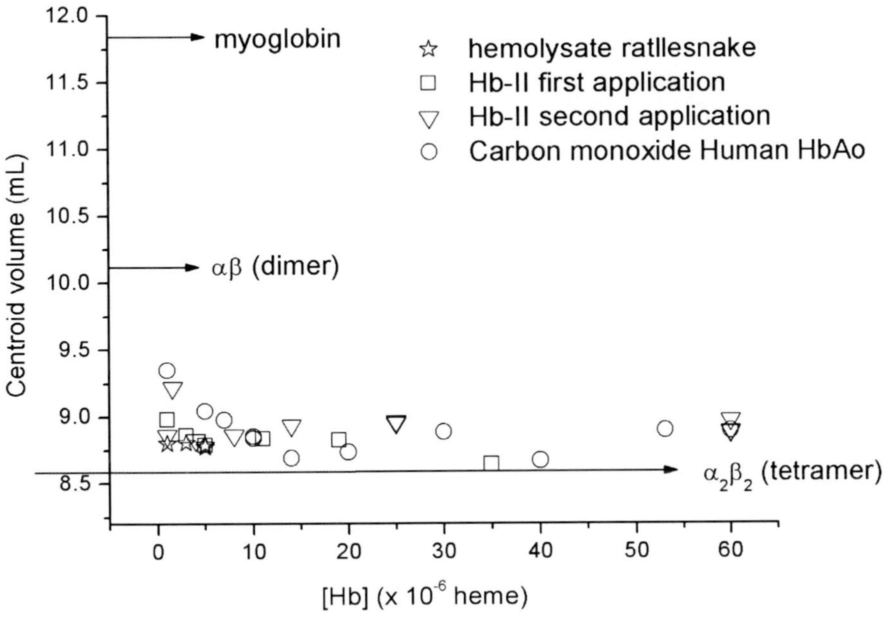

Figure 5: Comparison between the centroid elution volume of *Crotalus* Hb-II and human hemoglobin (HbA_0) (both stabilized with CO), as a function of hemoglobin concentration. The analytical gel filtration chromatography was done in Sephadex G-75, using 0.05M Tris-HCl pH 8.0, at room temperature.

The results obtained after the elution of the stripped form, by analytical gel filtration chromatography, had an elution volume of 8.7±0.08mL, which is very similar to human hemoglobin (Hb A_0): 8.67±0.01 mL between 30-40 mM (heme).

At lower concentration (between 5-10 mM) the elution volume was also very similar to human hemoglobin (8.85±0.01mL): 8.84±0.05 mL for Hb-II.

We also analyzed the behavior of the hemolysate from *Crotalus*. We did not observe significant differences between the hemolysate (stars) and purified Hb-II (down triangles and squares). The estimated Dimer-tetramer association constant of Hb-II was $1.5 \pm 0.24 \times 10^{7} M^{-1}$.

Fig. **6** shows a representative profile from the AGFC. Since there is no conformational difference between oxy and CO-Hb, we stabilized the samples with carbon monoxide. The centroid elution volume from the hemolysate (Fig. **7**) displays close values to deoxygenated human hemoglobin (HbA_0; down triangles) and is far from the expected values for dimeric species, even at the lower tested concentrations. The centroid elution volume for the hemolysate (1μM heme) from *Liophis miliaris* was 12.68mL (circles), whereas for deoxygenated human Hb, it was 12.28mL at the same concentration. Carbonic anhydrase, with a molecular mass close to that from dimers (squares), eluted at 14.22mL. Based on the data given, we obtained a good fit, and estimated dimer-tetramer equilibrium association constant as $2.89 \pm 0.35 \times 10^{7} M^{-1}$ (Fig. **7**).

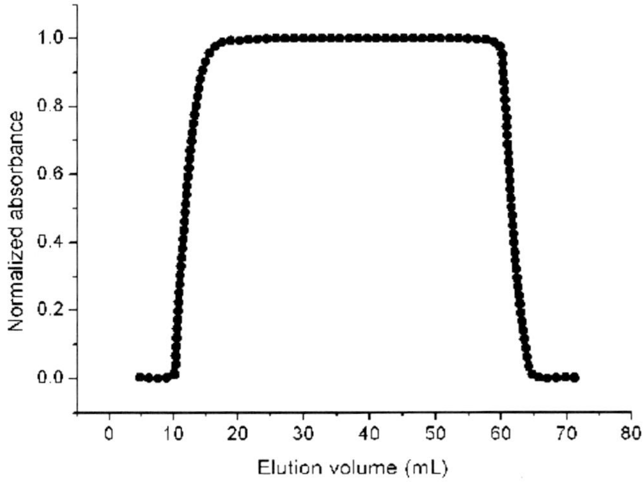

Figure 6: Representative elution profile of *Liophis miliaris'* hemoglobin analyzed by analytical gel filtration chromatography.

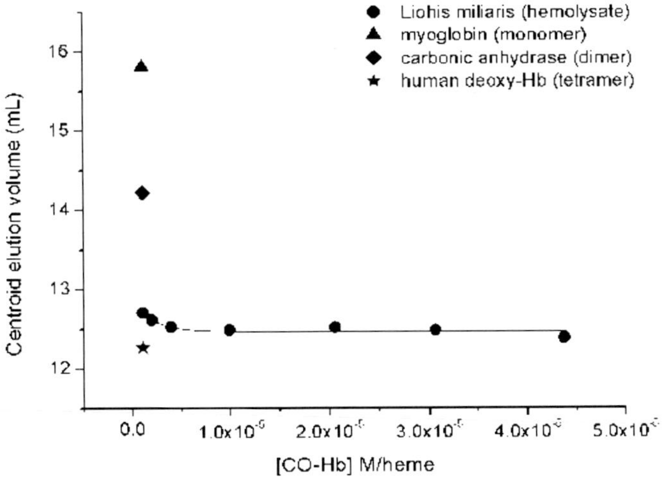

Figure 7: Centroid elution volume of the hemolysate from *Liophis miliaris* as a function of hemoglobin concentration (filled circles). The CO-H of *Liophis* was submitted to analytical gel filtration chromatography in Sephadex G-75, utilizing 30mM TAPS buffer pH 8.0, at room temperature.

Concerning SAXS experiments, Fig. **8** exhibits a plot of the distance distribution function p(r) for Hb-II compared to human hemoglobin.

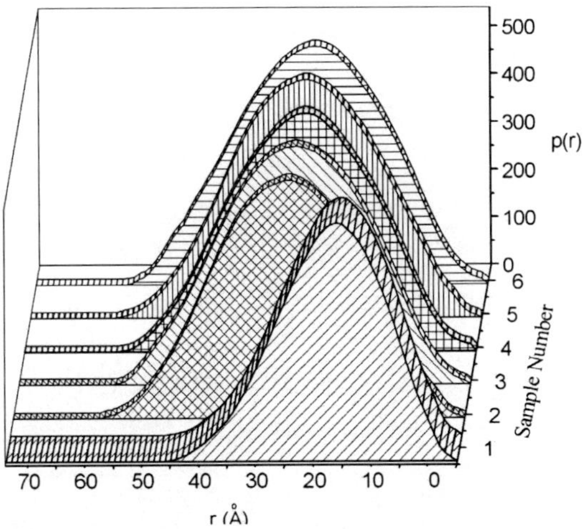

Figure 8: Comparison between the distance distribution functions, p(r), of human hemoglobin (theoretical data: left diagonal line (dimeric form) and chess line (tetrameric form) and experimental data: right diagonal line (pH 7.0) and checked line (pH 8.0)) and Hb-II from *Crotalus durissus terrificus* in pH 7.0 (vertical line) and pH 8.0 (horizontal line). All experiments were performed in 0.03M Hepes buffer pH 7.0 and 8.0, at 20°C. The sample concentration was 0.6mM/heme.

Initially the data concerning the radius of gyration and maximum dimension obtained for human hemoglobin were determined from atomic coordinates (1HHO) of this protein deposited in the Protein Data Bank. The radius of gyration and maximum dimension obtained for this first data set were 23Å and 63Å, respectively.

Subsequently, we performed a data collection for a second data set of Hb human, at pH 7.0 and 8.0, in order to confirm the theoretical data obtained in the first data set. The radius of gyration measurements, at pH 7.0 and 8.0, were 23.2Å and 22.9Å respectively. The maximum dimension was 63Å, confirming the theoretical data. These results validated our theoretical calculations. The data show that Hb-II has a radius of gyration of 22.6Å and maximum dimension of 63Å. These data are very similar to the values observed for human hemoglobin, as discussed above.

The plot of the distance distribution function p(r) for the hemolysate from *Liophis* (not shown) was compared to human hemoglobin (theoretical and experimental curves) as used for *Crotalus*. The data show that at pH 7.0 and 8.0, the gyration radius and the maximum dimension for *Liophis* Hb are 24.0Å and 64.0Å, respectively. These data are very similar to those found for human oxyhemoglobin, which possesses a radius of gyration of 23.2Å (at pH 7.0) and 22.9Å (at pH 8.0), and maximum dimension of 63.0Å. For comparison, dimeric human oxyhemoglobin radius of gyration was 17.0Å (theoretical data).

After analysis of *Liophis'* and *Crotalus*' Hbs by AGFC, oxygen binding properties, SAXS and osmotic stress, we used molecular modeling, searching for structural explanations for the absence of significant dissociation. We used the BLAST (Basic Local Alignment Search Tool) program [43, 44] to find the hemoglobin sequence with the highest identity relative to *Liophis miliaris'* Hb sequences.

As hemoglobin is a tetramer constituted by two α and two β chains, it was necessary to perform the search for each globin individually. The results showed that Hb D from *Gallus gallus* is the protein that possesses the highest identify with *Liophis'* Hb, having 52% and 73% identity with the α and β chains, respectively.

Accordingly, this structure was resolved at 2.3Å, and was used it as a template, following recommendations from Sanchez and Sali [45]. The second sequence with high identity was human Hb, followed by *Oncorhyncus mykiss* (trout) Hb. This procedure was not adopted for Hb-II from *Crotalus* because this Hb does not have its sequence determined.

Analysis of the Ramachandran Plot for the modeled *Liophis miliaris*' Hb (not shown) indicates that 91.5% of the residues lie in the most favorable regions, 8.1% are in the additional allowed regions, and only 0.8% is in generously allowed regions. We can emphasize that no residue was found in the non-allowed regions, except glycine and proline, due to the freedom and the restriction to rotation of their side chains, respectively. These results indicate that the generated model for *Liophis miliaris'* Hb possesses high stereochemical quality.

Since the previous methods point out that the major Hb from *Liophis miliaris* is a tetramer, we performed a molecular modeling, in order to examine whether the residues Glu 43(CD2) and Glu 101(G3) are or are not important for stabilizing the dimeric interface [2], and for that purpose we used the generated model (Fig. **9**).

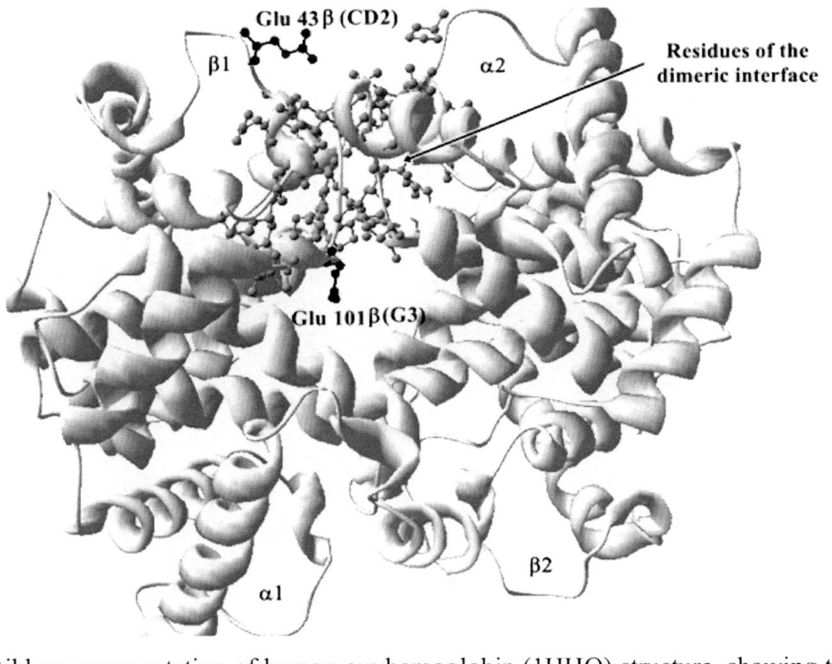

Figure 9: Ribbon representation of human oxyhemoglobin (1HHO) structure, showing the overall folding of this protein, as well as the location of residues Glu43β(CD2) and Glu101β(G3) (black) and other residues from the dimeric interface (gray). The figure was generated with the program Swiss Pdb-Viewer version 3.7.

First, we analyzed that the residue Glu43β(CD2) of human hemoglobin (PDB code: 1BBB); OE2 (epsilon oxygen 2) forms a hydrogen bond with NH_2 (amine group) from Arg92α(FG4), separated by 3.09Å. Nevertheless, in the other human Hb structure (PDB code: 1HHO), the distance between OE2 and NH_2 atoms is 5.38Å (Fig. **10**). This fact suggests that this interaction is not important to maintain the Hb tetramer.

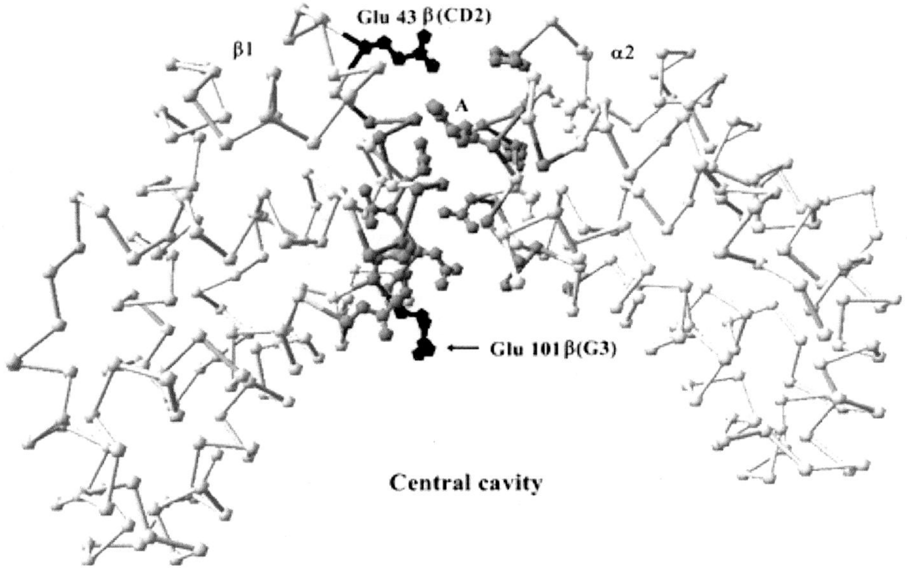

Figure 10: Sticks and balls representation of human oxyhemoglobin (1HHO), showing the residue Glu101β(G3) facing the central cavity of hemoglobin. This also happens in human oxyhemoglobin structure 1BBB (R2 state). The figure also shows Glu43β(CD2); this residue is separated by 5,38Å of NH1 of Arg92α(FG4) (letter A). The figure was generated with the program Swiss Pdb-Viewer version 3.7.

Concerning the analysis of Glu101β(G3), its side chain is not located at the $\alpha_1\beta_2$ interface, but in the central cavity, for both human Hb structures, 1BBB and 1HHO, and, accordingly, it would not contribute to Hb stability either.

The used template, Hb D from *Gallus gallus*, has been studied functionally and structurally [32]. Analyzing the glutamate residues, we can observe that Glu43β(CD2) is substituted by Ala. Looking for possible contacts, the distance from Cβ (carbon beta) from Ala43β(CD2) to NH_2 (amine group) from Arg92α(FG4) is 9.4Å. The position of Glu101β(G3) is the same reported for human Hb.

For *Oncorhyncus mykiss* (trout) Hb, both residues are substituted: Glu43β(CD2)→ Gly and Glu101β(G3) → Asp. For Glu101β(G3) → Asp is a conservative substitution, and Asp101 keeps the same orientation towards the central cavity. However, introduction of a glycine at position β43 could bring greater implications due to the freedom of rotation given by its short side chain. The distance between glycine and Arg92α was 7.76 Å.

For modeled *Liophis miliaris* Hb residue, Thr43 does not form any contact at the dimeric interface; its OG1 (oxygen gama 1) is 11.48Å away from OG (oxygen gama) (Ser44α CD2) and at 7.75Å from CD (carbon delta) Arg92α(FG4). This residue forms hydrogen bonds with the carbonyl oxygen (backbone) from residues Pro36β(C2) and Trp37β(C3).

On the other side, the residueVal101β(G3) establishes a van der Waals contact with CD2 (carbon delta 2) from Val96α(G3) at the dimeric interface, separated by 3.29Å (Fig. **11**), an interaction not observed for the other hemoglobins cited previously, which, in fact, contributes for the tetramer stability.

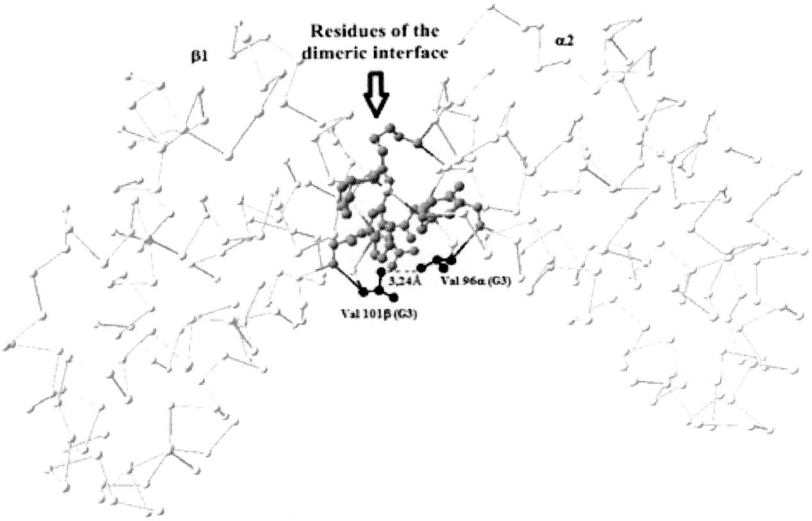

Figure 11: Sticks and balls representation of the interdimeric interface of modeled oxyhemoglobin from *Liophis miliaris*. We observe that Val101β(G3) is facing the dimeric interface establishing a van der Waals contact with Val96α(G3), at a distance of 3.24Å. Therefore, substitution Glu101β(G3) to Val, actually contributes for the tetramer maintenance of *Liophis'* hemoglobin, instead of promoting dissociation, as reported in the literature. The figure was generated with the program Swiss Pdb-Viewer version 3.7.

4. DISCUSSION

The number of hemoglobins found for the rattlesnake blood is higher than the reported number for other snakes like *Liophis miliaris*, *Bothrops alternatus*, *Mastigodryas bifossatus*, and *Laticauda laticaudata*, having only two components [1,3,7,46]. Intraspecific variations in the hemolysate pattern could be ascribed to sex, age, season, *etc.* Although we analyzed blood from several specimens, we did not find any variations of that pattern.

Several postulates try to explain the presence of various hemoglobin isoforms in the red blood cell. For example, they could prevent precipitation of the deoxygenated form, and, since it is less soluble than oxy-hemoglobin [47], it could be an adaptive advantage to sustain different physiological demands [48], or even just evolutionary remnants, harmless to the animal.

The results of the functional properties showed that *Crotalus* Hb-II has high oxygen affinity in the stripped form. However, when ATP is adding to the solution, affinity decreases significantly, as observed in other snake hemoglobins [1,8,3,7].

Bonafé *et al.* [4] studied the hemolysate from *Helicops modestus,* finding that the stripped hemoglobin has high oxygen affinity, absence of cooperativity, and no proton binding. These functional properties would agree with the existence of a dimeric form of that hemoglobin, as proposed for the semiaquatic snake *Liophis miliaris*. According to the authors, in the presence of ATP, there is a large decrease of O_2-affinity, the alkaline Bohr effect becomes evident, and oxygen binding, at pH 7.0 is cooperative (n_H=2.0). However, above this pH, oxygen binding becomes noncooperative (n_H=1.0).

In other words, at pH 7.0, as proposed by the authors, ATP-induced tetramer stabilization would occur and also an increase of the cooperativity. At pH 8.0, oxygen binding to hemoglobin would be noncooperative, and hemoglobin would assume the properties of the stripped form, suggesting a weakening of the electrostatic interaction between ATP and hemoglobin [4].

Our results concerning the Bohr effect for Crotalus Hb-II (Fig. **2a**) contrast with the behavior described by Bonafé *et al.* [4] for *Helicops modestus* hemoglobin.

The stripped form of Hb-II showed between pH 7 to 8, an alkaline Bohr effect of ΔH^+=-0.38 H^+/heme, and in the presence of ATP ΔH^+=-0.44 H^+/heme.

Fig. **2b** shows that the cooperativity in the stripped form, never reached values close to 1.0. The lowest value was obtained for Hb-II at pH 9.0, where n_H was 1.4, characterizing cooperative oxygen binding. In the presence of ATP, the values of n_H remained about 2.0, in the entire pH range studied.

As suggested by Fig. **3**, ATP exerts a strong binding, suggested by the significant decrease on O_2-affinity, as described for other snake hemoglobins [7,49]. The titration with that phosphate indicated the presence of one ATP binding site per tetramer. The value obtained for the ATP association constant for the deoxygenated form (K_D) was $5.13 \pm 0.85 x 10^5 M^{-1}$. This value is one order of magnitude higher than the respective value for human deoxyhemoglobin ($K_D = 10^4 M^{-1}$), whereas the value found for the association constant for the oxygenated form (K_O) was $1.92 \pm 0.63 x 10^2 M^{-1}$, similar to the value reported for human hemoglobin ($K_O = 10^2 M^{-1}$) [50].

The value of K_O for Hb-II is in agreement with results obtained for *Liophis miliaris*, *Bothrops alternatus* and *Boa constrictor*, which are about 10^2-$10^3 M^{-1}$ [7]. The value for $K_D = 10^5 M^{-1}$ is three orders of magnitude higher than the value for $K_O = 10^2 M^{-1}$, showing preferential ATP binding to the deoxygenated state.

Concerning the "osmotic stress" experiments, the hydration change between the oxy and deoxygenated forms of hemoglobin, reflects quantitative differences on the conformational states of oxy and deoxyhemoglobin [21].

The data obtained for Hb-II suggest the presence of an intermediate allosteric state, which agrees with data reported for human and animal hemoglobins [22, 51, 52], and reported, also, for other hemoglobins. As proposed by those authors, deoxyhemoglobin would exist in equilibrium in two unliganded states, named T_X and T_0. In the absence of anions, this equilibrium is displaced to the T_0 state, whereas in the presence of anions the equilibrium it is displaced to the T_X state. In the absence of anions, 25 extra molecules of water were bound to the oxygenated state, but increasing to 70 in the presence of anions.

We found that the number of extra water molecules bound to oxy Hb-II increases in the presence of chloride, with this anion driving the equilibrium to the T_X state. In the presence of 1mM ATP Δn_w decreases, probably due to phosphate binding to the oxygenated form.

These results showed that, in the absence of a chloride ion, Hb-II from *Crotalus* assumes a new conformational state T_0, as Colombo and Seixas [51] reported on human hemoglobin (Hb A_0). A recent work from Kodkjaer *et al.* [53] pointed out that erythrocytes from this snake species can undergo significant changes in their water content, so there is a potential effect on O_2-binding due to water and anionic activity.

Concerning *Liophis*' Hb, using the osmotic stress method we attempted to verify whether drastic changes in the differential number of water molecules bound between the oxy and deoxyHb would occur. Following the proposal, if Hb splitting occurs during oxygen binding, Δn_w would be significantly larger than the reported values for human Hb. If, conversely, the stripped form stays dimeric, we would expect no or just a few water molecules bound during oxygenation, as supported in the literature [54,55].

Our data showed that water binding for *Liophis*' Hb closely resembles the values reported for human Hb [51]: for the stripped form, 25 water molecules, increasing to 72 in the presence of chloride. Similar numbers were observed for hemoglobins from *Hoplosternum littorale* [22] and trout [52]. On the other hand, for the dimeric hemoglobin from the gastropod mollusc *Scapharca inaequivalvis*, only 17 water molecules bind at the dimer interface in the deoxy conformation; upon ligation of CO or O_2, this number is reduced to 11 water molecules [56].

Matsuura *et al.* [2] sequenced the Hb from *Liophis miliaris* and proposed that *Liophis*' Hb dissociation should occur due to substituted Glu residues at the $\alpha_1\beta_2$ interface. The lack of these two negative charges should favor the dissociation into dimers. Most of the background for the proposal arose from O_2-binding properties, redox potential measurements, and molecular weight determinations by ordinary gel filtration chromatography, using small sample volumes [1,3,4,9].

For the authors, Hb dissociation would ensure its oxygenation during the periods the snake stays under water, being accordingly a physiological advantage for the animal. ATP binding would guarantee restoration of the tetramer and a decrease of O_2-affinity. Nevertheless, although the dissociation mechanism may be considered very beneficial, the percentage of dimers and/or tetramers in solution was never measured through suitable methods, such as measuring the dimer-tetramer association constant determination of the oxygenated state.

Analytical gel filtration chromatography, also denominated "large zone gel filtration", is based on the application of large sample volumes, ensuring the equilibrium between dimers and tetramers in solution, since they depend on Hb concentration.

We measured the dimer-tetramer association constant of human hemoglobin as a control, obtaining $^4K_2$=2.0±0.2x10^{6}M^{-1}, which is in close agreement with those values reported in the literature by Valdes *et al.* [57] (1.0x10^{6}M^{-1}).

The data obtained for Crotalus Hb-II $^4K_2$=1.5±0.24x10^{7}M^{-1} showed that the assembly is a stable tetramer, being less dissociated than human hemoglobin (Fig. **5**). If the hemoglobin had the behavior proposed for *Liophis miliaris* hemoglobin, the elution values would reach those indicated by the "αβ dimer" label in Fig. **5**, or at least close to it.

The data allow us to estimate that, under the experimental conditions used for the functional studies (60mM/heme), Hb-II exhibits 97% of tetramers in solution. Dumolin *et al.* [58] analyzed the dissociation constant of human hemoglobin, in the range of concentrations between 0.02 and 4mM. They found that human hemoglobin, at 1mM, possess about 60% tetramers in solution, whereas under similar conditions, Hb-II possesses 78% as tetramers.

Concerning the results for the hemoglobin from *Liophis miliaris*, they suggest that this protein is tetrameric in solution, since the value obtained for the association constant was 2.89±0.35x10^{7}M^{-1}, one order of magnitude higher than the respective value for human Hb (2.0x10^{6}M^{-1}) determined in our laboratory, in agreement with the value reported in literature: 1.2x10^{6}M^{-1} [16]. Although the calculation exactness is subject to uncertainty for not having more data on the lowest concentrations, the behavior is clearly opposed to the assumption [2,4] that *L. miliaris'* Hb is unstable and promptly dissociates.

The mutant hemoglobin Hirose [59], which possesses a substituted Trp37β(C3) → Ser shows a high degree of dissociation in the oxygenated form, with a dissociation constant of 1.0M. Human mutant Hb Rothschild (Trp37β(C3) → Arg) is also in the dimeric form in the oxygenated form [60].

This suggests an important role of residue Trp37β(C3) in the maintenance of the tetrameric assembly form in hemoglobins. Another important substitution in the dimeric interface is found in Kansas Hb (Asn102β(G4) → Thr). This variant hemoglobin shows higher degree of dissociation in the oxygenated form. The dissociation constant determined for Kansas hemoglobin was $2.0x10^{-4}$M, in 0.1M phosphate buffer pH 7.5, at 20°C. Kansas Hb in the concentration of 100μM possesses approximately 50% of dimers [61].

Our determination of $^4K_2$ allowed us to estimate the percentage of tetramers and dimers from *Liophis*' hemolysate both *in vitro* and *in vivo*. Under *in vitro* conditions (60μM: experimental conditions), the percentage of dimers would be 2.5%, and *in vivo*, inside the red blood cells, in a concentration near 5mM [61], the percentage of dimers would be 0.5%.

Studies involving O_2-binding functional properties and tetramer-dimer equilibrium of some variant hemoglobins [10], all of them having substitutions affecting the residue Glu101β(G3): Hb Rush (Gln), Hb British Columbia (Lys), Hb Alberta (Gly), and Hb Potomac (Asp), concluded that these substitutions modify the O_2-binding properties, but not the dimer-tetramer equilibrium. The apparent physiological advantages of Hb dissociation are not supported by results either, which shows that when tetrameric human oxyhemoglobin dissociates into dimers, a 17-fold increase in the oxidation rate from 0.25(%/h) to 4.3(%/h) occurs [62].

Since the tetramer dissociation mechanism is postulated to occur *in vivo* (at high hemoglobin concentrations), we used SAXS to analyze the behavior of this protein in solution. In addition, this method allowed us to choose the pH, saline concentration, and the presence or absence of allosteric effectors.

Furthermore, the radius of gyration and maximum dimension of a macromolecule are important parameters that characterize molecular species with different molecular mass, such as the hemoglobin tetramer (64kDa) and its dimeric form (32kDa).

The results showed that *Liophis'* Hb possesses radius of gyration of 24.0Å and maximum dimension of 64.0Å, which confirms the presence of tetrameric forms in solution, as determined in the AGFC and osmotic stress method.

The data (Fig. **8**) show that Hb-II possesses a radius of gyration of 23.6Å at pH 7.0 and 23.8Å at pH 8.0. The maximum dimension of this protein was 63Å.

These results agree with reported values for human hemoglobin, whose radius of gyration is 22.8Å at pH 7.0 and 23.2Å at pH 8.0. The maximum dimension for human hemoglobin and Hb-II was 63Å. However, these data are very different compared to the dimeric hemoglobin, which has a radius of gyration of 17Å and maximum dimension of 50Å.

Moreover, the analysis of molecular modeling showed that the residues Glu43β(CD2) and Glu101β(G3) are not essential in the maintenance of the tetrameric form, not supporting the theory proposed in the literature for the Hbs of *Liophis miliaris* and other species of snakes [1, 2, 4, 7, 8].

In conclusion, in order to analyze the postulated dimerization process from *Liophis miliaris*' and *Crotalus durissus terrificus'* hemoglobins, we used functional and analytical techniques as well as molecular modeling. Our data show that the hemoglobins of both snakes are stable tetrameric structures both *in vitro* and *in vivo*. The data obtained allow us to conclude that none of *Liophis miliaris* or *Crotalus durissus terrificus* main hemoglobins undergo the process of dimerization proposed by Matsuura *et al.* [2].

ACKNOWLEDGEMENTS

We thank Prof. Dr. Arif Cais for expert assistance with snake handling and Ms. Eliani Ohira for technical assistance. This work was financially supported by grants from FAPESP and CNPq (Brazil). F.R.L. was supported by a scholarship from FAPESP.

CONFLICT OF INTEREST

The authors confirm that this chapter contents have no conflict of interest.

DISCLOSURE

The chapter submitted for eBook series titled "Advances in Protein and Peptide Sciences Volume 1" is an update of our article published in Current Protein and Peptide Science (Invited Review), Volume 10, Number 3 (June), Pages 196-215, 2009 (PMID: 19519451), with additional text and references.

REFERENCES

[1] Focesi Jr.; A. Ogo, S.H.; Matsuura, M.S.A.; Say, J.C. *J. Med. Biol. Res.*, **1987**, 20, 861-864.

[2] Matsuura, M.S.A.; Fushitani, K.; Riggs, A.F. *J. Biol. Chem.*, **1989**, 264, 5515-5521.

[3] Oyama Jr., S.; Nagatomo, C.L.; Bonilla, G.O.; Matsuura, M.S.A.; Focesi Jr., A. *Comp. Biochem. Physiol.*, **1993**, 105B, 271-275.

[4] Bonafé, C.F.; Matsukuma, A.Y.; Matsuura, M.S.A. *J. Biol. Chem.*, **1999**, 274, 1196-1198.

[5] Ogo, S.H.; Abe, A.S.; Focesi Jr., A. *Comp. Biochem. Physiol.,* **1979**, 63A, 285-289.

[6] Focesi Jr., A.; Bonilla, G.O.; Nagatomo, C.L.; Matsuura, M.S.A. *Comp. Biochem. Physiol.*, **1992**, 103, 985-989.

[7] Bonilla, G.O.; Focesi Jr., A.; Bonaventura, C.; Bonaventura, J.; Cashon, R. E. *Comp. Biochem. Physiol.*, **1994**, 109A, 1085-1095.

[8] Matsuura, M.S.A.; Ogo, S.H.; Focesi Jr., A. *Comp. Biochem. Physiol.*, **1987**, 86, 683-687.

[9] Focesi Jr., A.; Ogo, S.H.; Matsuura, M.S.A. *Comp. Biochem. Physiol.*, **1990**, 96B, 119-122.

[10] Shih, D. T-B.; Jones, R.T.; Imai, K.; Tyuma, I. *J. Biol. Chem.*, **1985**, 260, 5919-5924.

[11] Rossi-Fanelli, A.; E. Antonini. *Arch. Biochem. Biophys.*, 1958, 77, 478-492.

[12] Colombo, M.F. And Bonilla-Rodriguez, G.O. (**1996**) *J. Biol. Chem.* 271, 4895-4899.

[13] Colowich, S.P.; N.O. Kaplan. *Methods in enzymology*: *Hemoglobins*, Academic Press: New York, **1981**.

[14] Amiconi, G.; Bertollini, A.; Bellelli, A.; Coletta, M.; Condó, S.G.; Brunori. M. *Eur. J. Biochem.*, **1985**, 150, 387-393.

[15] Ackers, G. K. *The Proteins*, 1rd ed., Academic Press: New York, **1975**.

[16] Nenortas, E.; Beckett, D. *Analytical Biochemistry*., **1994**, 222, 366-373.

[17] Perutz, M.F.; Shih, D.T.; Williamson, D. *J. Mol. Biol.*, **1994**, 239, 555-560.

[18] Chiancone, E. *J. Biol. Chem.*, **1968**, 243, 1212-1219.

[19] Giardina, B.; Amiconi, G. *Methods Enzymol.*, **1981**, 76, 417-427.

[20] Colombo, M.F.; Bonilla-Rodriguez, G.O. *J. Biol. Chem.*, **1996**, 271, 4895-4899.

[21] Colombo, M.F.; Rau, D.C.; Parsegian, V.A. *Science.*, **1992**, 256, 655-659.

[22] Peres, P.; Azevedo Jr., W. F.; Bonilla-Rodriguez, G.O. *European Journal of Biochemistry*, **2004**, 271, 4270-4274.

[23] Benech, R.; Macduff, G.; Benesch, R.E. *Anal. Biochem.*, 1985, 11, 81-87.

[24] Wyman, J. *Adv. Prot. Chem.*, **1964**, 223-286.

[25] Wyman, J.; Gill, S.J. *Binding and Linkage: Functional Chemistry of Biological Macromolecules*. University Science Books, Mill Valley, California, **1990**.
[26] Kratky, O. *Small Angle X-Ray Scattering*. London: Academic Press, **1982**.
[27] Glatter, O. *J. Appl. Cryst.*, **1977**, 10, 415–421.
[28] Glatter, O.; Kratky, O. *Small Angle X-ray Scattering*. New York: Academic Press, **1982**.
[29] Glatter, O. *Acta Phys. Aust.*, **1980**, 52, 243-256.
[30] Olivieri, J. R. *Estudo das variações conformacionais das albuminas humana e bovina em solução por espalhamento de raio-X a baixo ângulo*. PhD Thesis. University of São Paulo. Brazil. **1992**.
[31] Sali, A.; Potterton, F.; Yuan, H.; van Vlijmen; Karplus, M. *Proteins*, **1995**, 23, 318-326.
[32] Knapp, J.E.; Oliveira, M.A.; Xie, Q.; Ernst, S. R.; Riggs, A.F.; Hackert, M. L. *J. Biol. Chem.*, **1999**, 274, 6411-6420.
[33] Brooks, B.R.; Bruccoleri, R.E.; Olafson, B.D.; States, D.J.; Swaminathan, S.; Karplus, M. *J. Comput. Chem.*, **1983**, 4, 187-217.
[34] Braun, W.; Go, N. *J. Mol. Biol.*, **1985**, 186, 611-626.
[35] CCP4. *Acta Cryst.*,**1994**, D50, 760-763.
[36] Laskowski, R.A.; MacArthur, D.K.; Smith, D.T.; Jones, *E.G.*; Hutchinson, A.L.; Morris, A.L.; Naylor, D.; Moss, D.S.E.; Thorton, J. M. *PROCHECK v.3.0 - Programam to Check Stereochemistry Quality of Protein structures Operating Instructions.* [s.l.:s.n.], **1994**.
[37] Baldwin, J.; Chothia, C. *J. Mol. Biol.*, **1979**, 129, 175-220.
[38] Abraham, D. J.; Kellogg, G.E.; Holt, J.M.; Ackers, G.K. *J. Mol. Biol.*, **1997**, 272, 613-632.
[39] Shaanan, B. *J. Mol. Biol.*, **1983**, 171, 31-59.
[40] Mercedes-Silva, M.; Rogers, P.H.; Arnone, A. *J. Biol. Chem.*, **1992**, 267, 17248-17256.
[41] Tame, J.R.; Wilson, J.C.; Weber, R.E. *J. Mol. Biol.*, **1996**, 259, 749-760.
[42] Guex, N.; Peitsch, M.C. *Electrophoresis*, **1997**, 18, 2714-2723.
[43] Bondi, A. Van der Waals volumes and radii. *J. Phys. Chem.*, **1964**, 68, 441-451.
[44] Altschul, S.F.; Gish, W.; Miller, W.; Myers, E.W.; Lipman, D.J. *J. Mol. Biol.*, **1990**, 215, 403-410.
[45] Sanchez, R.; Sali, A. *Curr. Opin. Struct. Biol.*, **1997**, 7, 206-214.
[46] Eguchi, Y.; T. Eguchi. *Journal of Protein Chemistry*, **2002**, 21, 215-221.
[47] Weber, R.E. In: *Animal Nutrition and Transport Processes. 2. Transport, Respiration and Excretion: Comparative and Environmental Aspects*.; Truchot Ed. J-P and Lahlou; 1990,Vol. *6*, pp. 58-75.
[48] Powers, D. Molecular ecology of teleost fish hemoglobin: strategies for adapting to changing environment. *Am. Zool.*, **1980**, 20: 139-162.
[49] Bonilla, G.O.; A. Focesi Jr.; C. Bonaventura, J. Bonaventura; R.E. Cashon. Functional properties of the hemoglobin from the south american *Mastigodryas bifossatus*. Comp. Biochem. Physiol., **1994b**, 109B, 1085-1095.
[50] De Rosa, M.C.; Sanna, M.T.; Messana, I.; Castagnola, M.; Galtieri, A.; Tellone, E.; Scatena, R.; Botta, B.; Botta, M.; Giardina, B. *Biophysical Chemistry*, **1998**, 72, 323-335.
[51] Colombo, M.F.; Seixas, F.A.V. *Biochemistry.*, **1999**, 38, 11741-11748.
[52] Hundahl, C.; Fago, A.; Malte, H.; Weber, R.E. *J. Biol. Chem.*, **2003**, 278, 42769-42773.
[53] Kodkjaer, P.; Wang, T.; Taylor, T.W.; Abe, A.S. *Comp. Biochem. Physiol. A Mol. Integr. Physiol.*, **2000**, 127, 49-54.
[54] Müller, G.; Fago, A.; Weber, R.E. *J. Exp. Biol.*, **2003**, 206, 1389-1395.
[55] Nichols, W. L.; Zimm, B. H.; Ten Eyck, L. F. *J. Mol. Biol.*, **1997**, 270, 598-615.

[56] Royer, W.E.; Pardanani, A.; Gibson, Q.H.; Peterson, E.S.; Friedman, J.M. *Proc. Natl. Acad. Sci.* **1996**, 93, 14526-14531.

[57] Valdes Jr., R.; Vickers, L.P.; Halvorson, H.R.; Ackers, G.K. *Proc. Natl. Acad. Sci.*, **1978**, *75,* 5493-5496.

[58] Dumoulin, A.; Manning, L.R.; Jenkins, W.T.; Winslow, R.M.; Manning, J.M. *J. Biol. Chem.*, **1997**, 272, 31.326-31.332.

[59] Imamura, J.S.; Yanase, T.; Atha, D.; Riggs, A.; Bonaventura, J.; Bonaventura, C. *J. Biol. Chem.*, **1978**, 253, 87-94.

[60] Sharma, V.S.; Newton, G.L.; Ranney, H.M.; Ahmed, F.; Harris, J.W.; Danish, E.H. *J. Mol. Biol.*, **1980**, 144, 267-280.

[61] Atha, D. *J. Biol. Chem.*, **1979**, 259, 12390-12398.

[62] Sode, F. (**1991**) *Comp. Biochem. Physiol.* 100B, 697-703.

Send Orders for Reprints to reprints@benthamscience.net
Frontiers in Protein and Peptide Sciences, Vol. 1, 2014, 31-62

CHAPTER 2

Major Enzymes of Endocannabinoid Metabolism

Balapal S. Basavarajappa[1,2,3,*]

[1]*Division of Analytical Psychopharmacology, New York State Psychiatric Institute, New York, USA,* [2]*Department of Psychiatry, College of Physicians & Surgeons, Columbia University, New York, NY 10032, USA and* [3]*Nathan Kline Institute for Psychiatric Research, Orangeburg, NY 10962, USA*

Abstract: In recent years, there have been an increasing number of investigations of the pathways involved in the metabolism of endocannabinoids (eCBs) following the discovery of cannabinoid receptors (CB) and their endogenous ligands, such as anandamide (AEA) and 2-arachidonoylglycerol (2-AG). The *in vivo* biosynthesis of AEA has been shown to occur through several pathways mediated by *N*-acylphosphatidylethanolamide-phospholipase D (NAPE-PLD), a secretory PLA_2; α-β-hydrolase 4 (ABH4); glycerophosphodiesterase (GDE1); PLC; and lyso-PLD. 2-AG, a second eCB, is generated through the action of selective enzymes, such as phosphatidic acid phsophohydrolase, diacylglycerol lipase (DAGL), phosphoinositide-specific PLC (PI-PLC), and lyso-PLC. In the adult brain these enzymes are localized in the postsynaptic plasma membrane, consistent with their role in generating 2-AG for use in retrograde transmission. A putative membrane transporter is involved in the cellular uptake and release of eCBs. AEA is metabolized by fatty acid amidohydrolase (FAAH), while 2-AG is metabolized by FAAH, monoacylglycerol lipase (MAGL), and serine hydrolase α-β-hydrolase domain 6 and 12 (ABHD6 and 12). The author presents an integrative overview of current research on the enzymes involved in the metabolism of eCBs and discusses possible therapeutic interventions for various diseases, including addiction.

Keywords: FAAH, Endocannabinoids, CNS, NAPE-PLD, MAGL, CB1 receptors.

1. INTRODUCTION

In the central nervous system (CNS), the endocannabinoids (eCBs) arachidonoyl ethanolamine (anandamide, AEA) and 2-arachidonylglycerol (2-AG) are produced and inactivated by neurons and glia [1, 2]. These eCBs have been shown to mimic the major active component of marijuana, Δ^9-tetrahydrocannabinol

***Address correspondence to Balapal S. Basavarajappa:** Nathan Kline Institute for Psychiatric Research, 140 Old Orangeburg Rd, Orangeburg, NY-10962, USA; Tel: 845-398-3234 or 5454; Fax: 845-398-5451; E-mail: Basavaraj@nki.rfmh.org

Ben M. Dunn (Ed)

(Δ^9THC), binds to members of the cannabinoid family of G-protein-coupled receptors (GPCRs). To date, two cannabinoid receptors have been cloned (CB1 and CB2) and additional eCBs that exhibit cannabimimetic activity have also been discovered [3-5].

Unlike classical neurotransmitters and neuropeptides, eCBs are not stored in intracellular compartments; rather they are produced on demand by receptor-stimulated cleavage of lipid precursors by various enzymes [6-11] and released from neurons immediately afterwards [6, 8-12]. Following their release, the signaling function of the eCBs is terminated through the action of a specific reuptake protein or transporter [11, 13, 14] and subsequent hydrolysis of the molecules by either fatty acid amide hydrolase (FAAH) [15] or monoacylglycerol lipase (MAGL) [16] or by other uncharacterized enzymatic activities known to hydrolyze eCBs *in vivo*. The various enzymes involved in eCB metabolism offer intriguing opportunities for targeted drug development. This review will provide an overview of our current understanding of the critical enzymes involved in eCB biosynthesis and degradation and discuss possible therapeutic interventions for various diseases, including addiction.

2. ENZYMES INVOLVED IN ANANDAMIDE BIOSYNTHESIS AND DEGRADATION

AEA, has been identified and quantitated throughout the human brain as well as in peripheral tissues [17, 18]. Signal transduction and ligand binding studies have suggested that it can act at both the CB1 (K_i 61 nM) and CB2 (K_i 1930 nM) receptors, although it may be more efficacious at CB1 [19]. AEA behaves as an affinity-driven CB1 receptor agonist. Thus its efficacy at CB1 receptors, although higher than that of Δ^9-THC, is often found to be lower than for other cannabinoid agonists, *e.g.*, (+)-WIN55212-2 or CP-55940.

AEA is widely distributed in the brain and throughout peripheral tissues [18, 20]. Its levels vary by a factor of 4 to 6 in different regions of the rat brain, with the highest levels in the striatum and brainstem and the lowest levels in the cerebellum and cortex [17, 21]. AEA has been found in both rat and human brains in regions that contain high densities of CB1 receptors (*e.g.* the hippocampus, cerebellum, and striatum) and also in a region that is sparse in CB1 receptors, the thalamus [18]. In the rat, the concentration of AEA in the thalamus was approximately twice that measured in the cerebellum. It is clear from these data that the relative regional abundance of AEA in the brain does not correlate with the distribution of CB1 receptors. AEA levels in the brain are equivalent to those of other neurotransmitters, such as dopamine and serotonin, but are at least 10-fold lower than the levels reported for GABA and glutamate. AEA has also been found in peripheral tissues, such as the spleen, which expresses high levels of CB2 receptors—a finding that lends support to the idea that AEA is an agonist at both CB1 and CB2 receptors [18]. Small amounts of AEA were also found in the human heart and thalamus as well as in rat skin, whereas only trace quantities were detected in human serum, plasma, and cerebrospinal fluid [18]. It should be note that the absolute levels of eCB in various brain regions are not of paramount importance; instead, it is the variation of eCB levels under physiological or pathological conditions that is most important for understanding the role of eCBs in normal neurotransmission and in a variety of disease conditions.

The enzymes responsible for the biosynthesis and degradation of AEA have been characterized in various tissues. The postulated biosynthetic pathway for AEA involves the energy-independent condensation of ethanolamine and free arachidonic acid [22, 23]. The biochemical characteristics and inhibitor kinetics observed for the enzyme catalyzing this condensation reaction were found to be essentially identical to those described for the enzyme fatty acid amide hydrolase (FAAH; see below) [15, 22, 23]. However, several lines of evidence derived from both physical and biological data argue that the production of AEA *in vivo* does not occur through the energy-independent condensation of free arachidonic acid and ethanolamine [6, 24, 25]. Instead, the AEA precursor is *N*-arachidonylphosphatidylethanolamine (N-ArPE), which is believed to originate from the transfer of arachidonic acid from the sn-1 position of 1,2-sn-di-arachidonylphosphatidylcholine to phosphatidylethanolamine, catalyzed by a calcium-dependent N-acyltransacylase (NAT) (Fig. **1**). N-ArPE is then cleaved by an *N*-acylphosphatidylethanolamine (NAPE)-specific phospholipase D (PLD) (NAPE-PLD) [6, 26, 27], which releases AEA and phosphatidic acid. It is not clear whether the *N*-acyltransferase (NAT) or the NAPE-specific PLD controls the rate-limiting step of AEA synthesis [28-30]. The NAT enzyme has been identified biochemically in both brain and testis microsomal preparations [7, 31, 32]. However, the detailed purification, characterization and cloning of a specific NAT remains to be performed. On the other hand, it was recently found that the formation of N-ArPE was catalyzed in a Ca^{2+}-independent manner by an enzyme termed Ca^{2+}-independent *N*-acyltransferase (iNAT), also referred to as PLA/AT-5 [33]. Another pathway that utilizes the conversion of NAPE into 2-lysol-NAPEs *via* the action of secretory PLA_2 has also been reported. 2-Lysol-N-ArPEs are then converted into *N*-acyl-ethanolamides, which include AEA, *via* a selective lyso-phospholipase, D (lyso-PLD) [34]. The NAPE-PLD (see details below) has been purified, cloned, and characterized [35]. It was recently reported that oleoylethanolamide (OEA) generation was followed by the enhanced accumulation of OEA-generating NAPEs through Ca^{2+}-independent NAT in the proximal small intestine [36].

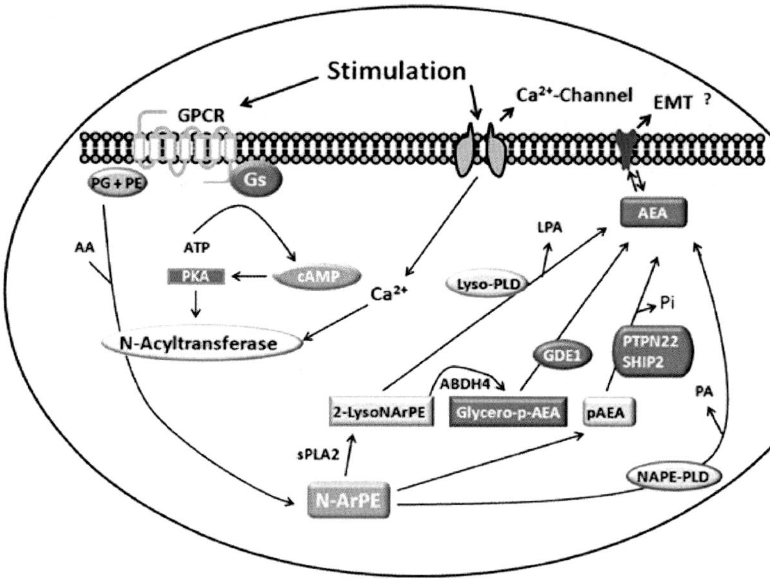

Figure 1: The potential enzymes involved in anandamide biosynthesis. Stimulation of adenylate cyclase and cAMP-dependent protein kinase potentiate the *N*-acyltransferase (Ca^{2+}-dependent transacylase) NAT. A fatty arachidonic acid chain is transferred by NAT from the sn-1 position of phospholipids (phosphoglycerides, PG) to the primary amine of phosphatidylethanolamine (PE) in a Ca^{2+}-dependent manner, forming an *N*-arachidonyl phosphatidylethanolamine (N-ArPE). This N-ArPE intermediate is then hydrolyzed by a phospholipase D (PLD)-like enzyme to yield anandamide (AEA). In addition, possible alternative pathways for AEA formation from N-ArPE are shown. Once synthesized, AEA can be transported to the outside of the cell through a process that has not yet been well characterized. EMT, eCB membrane transporter.

Recent studies have proposed the existence of NAPE-PLD-independent pathways for AEA synthesis in the mouse brain and in RAW264.7 macrophages [37]. AEA is generated from NAPE in a two-step process involving the PLC-catalyzed cleavage of NArPE to yield pAEA, which is subsequently dephosphorylated by a protein tyrosine phosphatase(s), PTPN22 [37] (Fig. **1**). A recent study using NAPE-PLD knockout mice characterized a lysophospholipase D (lyso-PLD) that hydrolyzed lyso-NAPEs to generate N-acylethanolamines in the rat brain. It was found that this pathway could be catalyzed sequentially by α/β-hydrolase 4 (Abh4) and the glycerophosphodiesterase GDE1 [38, 39]. However, the enzyme lyso-PLD remains to be cloned and characterized. Recently, using NAPE-PLD KO mice, it was shown that N-acylethanolamines can be formed from 1-alkenyl-2-acyl-glycero-3-phospho [n-acyl]ethanolamines (pNAPEs) in the brain through a NAPE-PLD-independent pathway as well as *via* direct release by NAPE-PLD [40]. Recent study suggests the participation of GDE1 in generating AEA in neonatal mice. It was shown that postnatal ethanol treatment enhances GDE1 protein levels through transcriptional activation of respective gene in neonatal mice [41].

3. NAPE-PLD

NAPE-PLD has been cloned from mouse, rat, and human, and the three enzymes were found to be highly conserved, with more than 89% identity at the amino acid level [35]. NAPE-PLD belongs to the zinc metallohydrolase family of the β-hydrolase fold and is structurally unrelated to other known PLD enzymes [35]. NAPE-PLD's lack of transphosphatidylation activity indicates that the enzyme is not only structurally but also functionally distinct from PLD members of the HKD/phosphatidyltransferase gene family [42]. NAPE-PLD uses not only *N*-arachidonyl-PE (a precursor of AEA) as a substrate but also other NAPEs with different *N*-acyl groups [35]. NAPE-PLD is present in almost all organs of the mouse, with higher specific activities in the brain, kidney, and testis [35]. In bovine organs, the brain exhibits the highest NAPE-PLD specific activity, followed by the kidney, spleen, lung, heart, and liver. In all animal species, the brain consistently displays high NAPE-PLD activity. Whether there are isozymes of NAPE-PLD is not currently known. An age-dependent change in NAPE-PLD activity was reported recently in the rat brain [43]. The mechanisms regulating the expression of NAPE-PLD mRNA and protein are not known, such that the generation of specific inhibitors of NAPE-PLD and gene-disrupted animals will be of great use for future studies. Indeed, in a recent study, AEA biosynthesis was unaffected in NAPE-PLD knockout mice, suggesting that other enzymes [44] or isozymes of NAPE-PLD also participate in the synthesis of AEA. These NAPE-PLD (-/-) mice possess lower brain levels of saturated NAEs but essentially unchanged levels of polyunsaturated NAEs, including AEA, suggesting the existence of additional enzymatic pathways for the biosynthesis of NAEs *in vivo*. The enzyme α/β-hydrolase 4 (Abh4), which selectively hydrolyzes NAPEs and lyso-NAPEs bearing both saturated and polyunsaturated *N*-acyl chains, was recently described in NAPE-PLD (-/-) mice [38]. This report supports the existence of a NAPE-PLD-independent route for NAE biosynthesis, and suggests that Abh4 may have a role in this metabolic pathway by acting as a lyso-NAPE-selective lipase [38]. In recent study, developmental alcohol was shown to enhance NAPE-PLD protein levels through transcriptional activation of respective gene in neonatal mice [41].

As a putative neuromodulator, AEA that is released into the synaptic cleft is expected to be rapidly inactivated. In general, two mechanisms are known to remove eCBs from the synaptic cleft to ensure rapid signal inactivation: re-uptake (see detailed discussion below) and enzymatic degradation by FAAH (see below).

4. ENZYMES INVOLVED IN 2-ARACHIDONYLGLYCEROL BIOSYNTHESIS AND DEGRADATION

2-AG has been characterized as a unique monoacylglycerol species isolated from the canine gut [45] and rat brain [46] and also as an endogenous cannabinoid receptor ligand. According to its chemical structure, this new putative eCB is an arachidonyl ester rather than an amide [47-49]; it was found to bind to both CB1 (K_i 2.4 μM) and CB2 receptors. The CB1 receptor binding activity of 2-AG was 24 times less potent than that of AEA. 2-AG caused the typical effects of Δ^9-THC, such as antinociception, immobility, immunomodulation, and inhibition of electrically evoked contractions of the mouse vas deferens [45, 50, 51]. However, further studies are necessary to determine the relative importance of 2-AG in the human body and brain. Information regarding tissue levels of 2-AG is still limited. Brain tissue concentrations of 2-AG are approximately 200-fold higher than those of AEA [17]. The spatial distributions of the two eCBs are similar in different regions of the brain, with the highest concentrations found in the brainstem, medulla, limbic forebrain, striatum, and hippocampus and the lowest concentrations in the cortex, diencephalon, mesencephalon, hypothalamus, and cerebellum (for review see [29]). However, no correlation was found between 2-AG concentrations and CB1 receptor distribution. 2-AG was also detected in the peripheral nervous system, including the sciatic nerve, lumbar spinal cord, and lumbar dorsal root ganglion. 2-AG was also detected in the rat and bovine retinas (for review see [29]).

Generation of 2-AG was first described in ionomycin-stimulated N18TG2 cells [52], in electrically stimulated rat hippocampal slices, and in ionomycin-stimulated neurons [53]. Rapid generation of 2-AG was also observed in rat brain homogenate during incubation with Ca^{2+} [54] and in thrombin- or A23187-stimulated human umbilical vein endothelial cells [55]. Intraperitoneal injection of picrotoxin stimulated 2-AG levels in the rat brain [46]. Generation of 2-AG was also observed in cerebellar granular neurons in culture after chronic alcohol exposure [10]. Substantial amounts of this stimulus-induced 2-AG were released from the cells [10, 55, 56].

2-AG biosynthesis occurs *via* two possible routes in neurons, which are illustrated in Fig. **2**. Phospholipase C (PLC)-mediated hydrolysis of membrane phospholipids may produce diacylglycerol (DAG), which may subsequently be converted to 2-AG by diacylglycerol lipase (DAGL) activity [50, 57]. Alternatively, the formation of DAG may involve the hydrolysis of phosphatidic acid through Mg^{2+}- and Ca^{2+}-dependent PA phosphohydrolase activity [58, 59]. There is a large amount of literature available on the biochemical properties and molecular structure of mammalian phospholipase C [60]. On the other hand, the molecular cloning of DAGLα was reported only recently [61]. Both DAGLα and DAGLβ are associated with the cell membrane and are stimulated by Ca^{2+} and blocked by DAGL inhibitors, such as p-hydroxy-mercuri-benzoate, $HgCl_2$, and RHC-80267. DAGLα appears to be targeted to postsynaptic spines; it is highly enriched at the base of the spine neck and in the adjacent somatodendritic membrane but is excluded from the main body of the spines and from the excitatory synapses in cerebellar Purkinje cells. In hippocampal pyramidal cells, DAGLα can be distributed at the spine head, neck, or at both structures [62, 63]. The difference in the precise localization of DAGLα in different neuron types suggests that the specificity and efficiency of eCB-mediated retrograde suppression of neurotransmission depends not only on CB1 expression levels in presynaptic elements but also on the distance between the postsynaptic site of 2-AG production and the presynaptic CB1 receptor. Alternatively, phospholipase A1 (PLA1) may generate a lysophospholipid, which may be hydrolyzed to 2-AG by lyso-PLC activity [50]. The enzyme activities (PLA1 and lyso-PLC) involved in this second pathway are well characterized [64-66]. LysoPI-specific PLC is distinct from various other types of PLCs that act on other inositol phospholipids and is located in the synaptosomes [64-66]. Under certain conditions, 2-AG can also be synthesized through the conversion of 2-arachidonyl lysophosphatidic acid (LPA) (Fig. **2**) *via* a phosphatase [67]. In certain conditions, 2-AG can also be synthesized in intact N18TG2 [58] and cultured rat microglial [59] cells through the hydrolysis of PA rather than through the activation of PLC. Various structurally distinct inhibitors of PLC and DGL activities prevent 2-AG formation in cultures of cortical neurons, suggesting that the PLC/DGL pathway may play a primary role in this process [53].

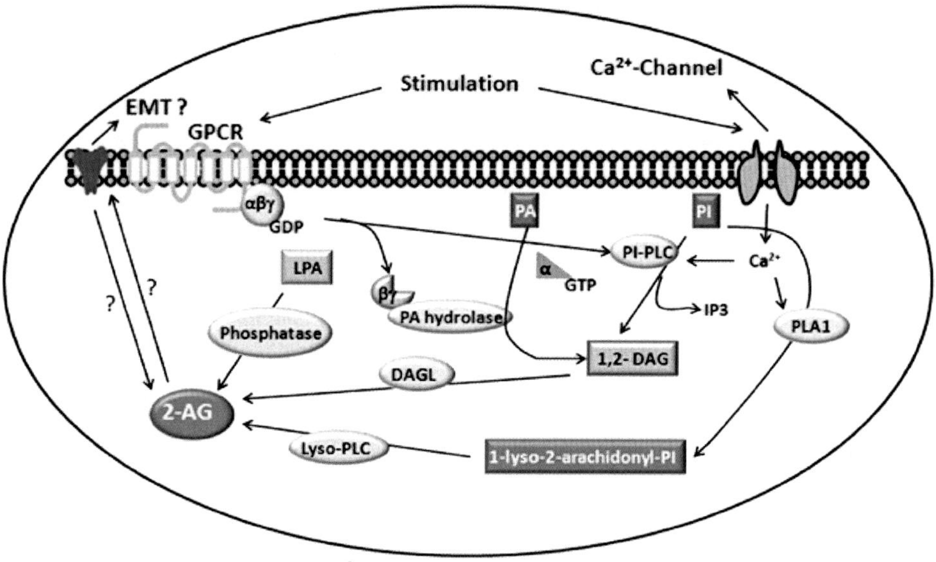

Figure 2: The enzymes involved in the biosynthetic pathways of 2-arachidonylglycerol. Intracellular Ca^{2+} initiates 2-AG biosynthesis by inducing the formation of diacylglycerol (DAG) in the membrane by stimulating the phosphatidyl-inositol-phospholipase C (PI-PLC) pathway. Alternatively, the formation of DAG involves the hydrolysis of phosphatidic acid through Mg^{2+} and Ca^{2+}-dependent PA phsophohydrolase activity. 2-AG is the product of DAG-lipase (DAGL) acting on DAG. The second pathway involves hydrolysis of PI by phospholipase A1 (PLA_1) and hydrolysis of the resultant lyso-PI by a specific lyso-PLC. In certain conditions 2-AG can also be synthesized through the conversion of 2-arachidonyl lysophosphatidic acid (LPA) by a phosphatase to yield 2-AG. Once synthesized, 2-AG can be transported to the outside of the cell through a process that has not yet been characterized. EMT, eCB membrane transporter.

2-AG is inactivated by reuptake *via* a membrane transport molecule, the 'eCB membrane transporter' (EMT) (see below for details) [11-13, 68-71], and *via* subsequent intracellular enzymatic degradation [6, 72, 73] by monoacylglycerol lipase (MAGL) (see below for details) in the same manner as other monoacylglycerols [74]. Similarly, 2-AG is metabolized by porcine MAGL from brain cytosol and particulate fractions [75]. FAAH (see below) has also been shown to metabolize 2-AG (Fig. **3**) [76, 77]. 2-AG is metabolized to 2-arachidonyl LPA through the action of monoacylglycerol kinase (s). 2-Arachidonyl LPA is then converted into 1-steroyl-2-arachidonyl PA [78]. 1-Steroyl-2-arachidonyl PA is further utilized in the "PI cycle" or is used in the *de novo* synthesis of phosphatidylcholine (PC) and phosphatidylethanolamine (PE). Furthermore, 2AG is metabolized by the enzymatic oxygenation of 2-AG *via* cyclooxygenase-2 (COX-2) into prostaglandin H_2 (PGH_2) glycerol esters. The biological activity and role of oxygenated 2-AG have not yet been determined. Inhibition of COX-2 prolongs depolarization-induced suppression of inhibition (DSI), suggesting that COX-2 limits eCB action in retrograde signaling and synaptic plasticity [79-81]. Recently, two new enzymes that are expressed in the brain were found to hydrolyze 2-AG (serine hydrolase α-β-hydrolase domain 6 and 12; ABHD6 and ABHD12) [38]. ABHD6 was found to be located postsynaptically, and its selective inhibition induced CB1-dependent long-term depression (LTD) in adult mouse cortex [82].

5. UPTAKE OF ENDOCANNABINOIDS

Although the enzymes responsible for eCB (AEA and 2-AG) metabolism have been well characterized, the processes involved in uptake have been the subject of much controversy. AEA appears to be inactivated by a two-step process involving the transport of this lipid into cells [69] (Fig. **1**) followed by intracellular hydrolysis by the integral membrane enzyme FAAH. However, with respect to the inactivation of AEA, only FAAH has been molecularly characterized and structurally studied. Indeed, the actual mechanism of AEA uptake or, more generally, the movement of this fatty acid amide not only through facilitated diffusion by a protein transporter but also through passive diffusion across the plasma membrane remains an enigmatic and controversial subject. Some authors have suggested that AEA uptake occurs by a facilitated diffusion process mediated by a protein transporter [11, 13, 68, 69], whereas others have described it as a passive diffusion process driven by FAAH [72, 73]. For a current discussion of these two points of view, see the recent papers [83, 84].

AEA uptake by cells occurs *via* diffusion through the cell membrane, facilitated by a saturable, temperature-dependent and selective transport system [6]. The eCB membrane transporter (EMT) has been identified in most cells analyzed to date [70] (Figs. **3** and **4**), and EMT inhibitors capable of enhancing AEA actions *in vitro* and *in vivo* have been developed [69, 85]. Additionally, structure-activity

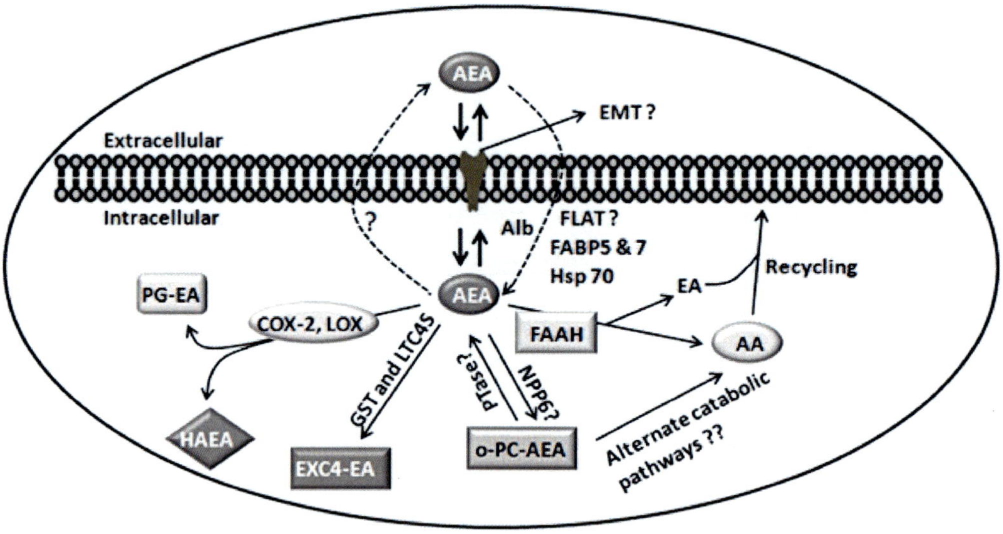

Figure 3: Scheme illustrating the potential proteins and enzymes involved in AEA uptake and degradation. AEA can be internalized by neurons through an as yet unidentified transport mechanism, "the eCB transporter (EMT)". Once inside neurons, AEA is rapidly cleaved by the hydrolytic enzyme fatty acid amide hydrolase (FAAH) to give arachidonic acid (AA) and ethanolamine (EA). Alternatively, different lipoxygenase (LOXs) and cyclooxygenase-2 (COX-2) can metabolize AEA, generating hydroxyl derivatives of AEA (HAEAs) and prostaglandins-ethanolamides (PG-EAs), respectively. On the other hand, in FAAH (-/-) mice, AEA is converted into o-phosphorylcholine (o-PC)-AEA by an unidentified biosynthetic pathway and o-PC-AEA can in turn be catabolized to AA by a choline-specific phosphodiesterase (NPP6). In certain conditions, AEA can be metabolized by GST and LTC4S to EXC4-EA. EMT, eCB membrane transporter.

relationship studies have been carried out on the EMT with a large variety of AEA analogs [14, 70, 86-89]. It was established that at least one to four *cis* double bonds in the fatty acyl chain are necessary for binding to the EMT or for transport into cells [86]. On the other hand, it was also observed that the ethanolamine 'head' of AEA could be substituted with bulkier groups, especially aromatic structures, to yield compounds still capable of binding to the EMT [86, 87]. Finally, it was reported that the EMT is present in endothelial cells [90], is activated by nitric oxide (NO) [71, 91-93], and is inhibited by the plant cannabinoids Δ^9-THC and cannabidiol at micromolar concentrations [94]. The role of intracellular catabolism of AEA in driving, in part, the EMT has also been highlighted [73, 94]. Recently, it was found that a partially cytosolic variant of the intracellular AEA-degrading enzyme, FAAH-1, which lacks amidase activity and is termed FAAH-like AEA transporter (FLAT), bound AEA but not 2-AG or other N-acylethanolamines with a high affinity and acted as an intracellular sink for AEA transport across the membrane in neurons and astrocytes. However, FLAT is certainly not the long-sought EMT, primarily because of its cytosolic location and the fact that it is unable to recognize 2-AG [95]. Nonetheless, FLAT is potentially very important from a therapeutic point of view because its pharmacological inhibition will block the cellular uptake of AEA, but not 2-AG or other lipid mediators.

In addition, several intracellular AEA carrier proteins have been proposed [96-98]. It was found that fatty acid binding proteins (FABPs), specifically FABP5 and FABP7 but not FABP3, increased the rate of AEA uptake and the subsequent hydrolysis in COS-7 and N18GT2 cells [97]. Cytosolic heat shock protein 70 (Hsp70), found in the cytosolic preparations of mouse keratinocytes, binds AEA and albumin as binding proteins [98]. In addition, overexpression of Hsp70 increased AEA uptake into SHSY5Y cells [98]. The binding of AEA to albumin has been well investigated *in vitro* [96]. Further research is required to characterize the relative importance of the individual carriers to AEA uptake under native conditions and in the intact brain.

The uptake of 2-AG by cells was first observed in rat basophilic RBL-2H3 and mouse neuroblastoma N18TG2 cells and was shown to be inhibited by unsaturated monoacylglycerols, such as 2-oleoyl- and 2-linoleoyl-glycerols [99]. In studies carried out in RBL-2H3 and J774 cells [76, 94, 100] it was reported that 2-AG did not effectively inhibit the uptake of [^{14}C] AEA when the two substances were present at the same concentration, thus suggesting that the EMT does not recognize 2-AG as a substrate. However, recently, a study carried out in human astrocytoma cells [68] confirmed that 2-AG is taken up by cells and showed for the first time that this process can be inhibited by the previously developed EMT inhibitor, AM404 [69].

Despite the accumulation of new data from various studies, little is known regarding the identity of the transport molecule for AEA and 2-AG. Cloning and *in vitro* expression of the transporter gene will be required for a clear understanding of the eCB signaling system, including the biosynthetic, uptake, and degradation pathways.

Soon after the discovery of AEA, an enzyme that catalyzes the hydrolysis of AEA to arachidonic acid and ethanolamine was detected in rat brain tissue [22] (Fig. **3**). This enzyme is variously known as anandamide amidase, anandamide hydrolase, anandamide amidohydrolase, or fatty acid amide hydrolase (FAAH) (see below). In addition to its hydrolysis by FAAH, AEA is metabolized by COX-2, lipoxygenase (LOX), and cytochrome P450. COX-2 has been shown to metabolize AEA into PGE_2-ethanolamide (PGE_2-EA) [101]. 12- and 15-LOX, which are non-heme iron-containing enzymes, convert AEA into 12- and 15-hydroxy-AEA (12- and 15-HAEA) *in vitro*, respectively [102, 103]. Cytochrome P450 also metabolizes AEA into several polar lipids [104]. cDNA cloning and functional expression of the enzyme termed "*N*-acylethanolamine-hydrolyzing acid amidase" (NAAA) from human, rat, and mouse have been reported recently. NAAA exhibited no homology to FAAH but belonged to the choloylglycine hydrolase family [105]. This NAE-hydrolyzing enzyme (NAAA) was revealed to be a glycoprotein localizing mainly in lysosomes [105]. Recently, in the absence of FAAH, exogenously injected AEA was shown to be converted into o-phosphorylcholine (o-PC)-AEA in the brain and spinal cord. The choline-specific phosphodiesterase NPP6 was found to convert PC-NAE into NAE [106] (Fig. **3**). Further research is required to elucidate the exact mechanism and enzymes involved in this pathway of AEA metabolism. Recently, catabolism of AEA through the concerted action of 15-lipoxygenase-1 (15-LO-1) and glutathione S-transferases or leukotriene (LT) C4-synthase to yield epoxide eoxin (EX) A4 (EXC4-EA) suggested the existence of an additional mechanism to terminate AEA signaling in species that express 15-LO-1 [107].

6. FAAH

The actual enzymes involved in fatty acid amide (FAA) metabolism remained unknown until the late 1990s, when a rat oleamide hydrolase activity was affinity purified and its cDNA cloned [15]. Oleamide amidase may be involved in AEA hydrolysis because the same enzyme was shown to catalyze the hydrolysis of AEA and oleamide before FAAH was cloned [108]. FAAH is a membrane-bound enzyme that belongs to the family of amidase proteins. Members of this family are characterized by a highly conserved region that is rich in serine, glycine, and alanine residues. This "amidase signature" region (AS) is common to more than 80 amidases and corresponds to amino acids 215–257 in mammalian FAAH. Recombinantly expressed FAAH was shown to hydrolyze oleamide, AEA, and several other endogenous FAAs. 2-Arachidonylglycerol also appears to be a substrate of FAAH *in vitro* [75] and thus FAAH is generally required for inactivation of eCBs in the brain (Figs. **3** and **4**). FAAH was first isolated from rat liver through its ability to catalyze the hydrolysis of the sleep-inducing lipid oleamide [15, 109].

The mouse and human FAAH genes, each composed of 15 exons, are localized on chromosomes 1 and 4, respectively [110]. Recently, the first X-ray crystal structure of FAAH was reported [111]. This 2.8 Å structure of FAAH in a complex with the irreversible inhibitor methoxy arachidonyl fluorophosphonate revealed several unusual features of the enzyme. First, the core catalytic machinery of FAAH is composed of a serine–serine–lysine catalytic triad (S241–S217–K142), in contrast to the serine–histidine–aspartate triad typical of most serine hydrolases. These results are consistent with previous enzymological studies indicating that S241 and K142 play key catalytic roles as the nucleophile and acid/base, respectively [112]. The structure of FAAH also revealed that this enzyme possesses a remarkable collection of channels that appear to grant it simultaneous access to both the membrane and cytoplasmic compartments of the cell, possibly to facilitate substrate binding, product release, and catalytic turnover. These unusual mechanistic and structural features of FAAH have inspired new strategies for the design of specific inhibitors (*e.g.*, URB-597) that display high selectivity for this enzyme relative to the hundreds of serine hydrolases present in the human proteome [113].

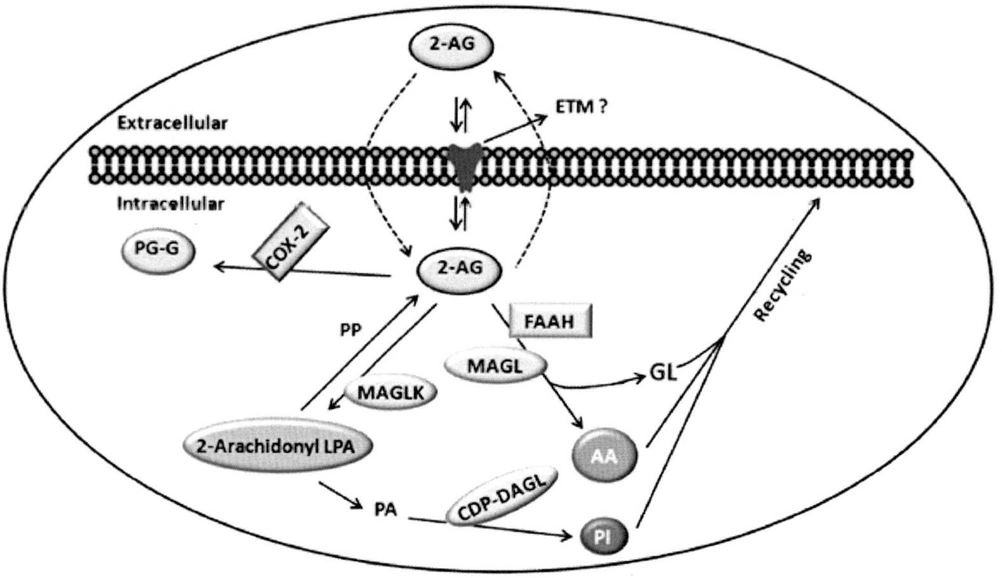

Figure 4: Schematic summary of the proteins and enzymes involved in 2-AG uptake and degradation. 2-Arachidonylglycerol can be internalized by neurons through an as yet unidentified transport mechanism, "the eCB membrane transporter (EMT)". Once inside neurons, 2-AG can be hydrolyzed by the enzymes fatty acid amide hydrolase (FAAH) or monoacylglycerol lipase (MAGL) or serine hydrolase α-β-hydrolase domain 6 and 12 (ABHD6 and ABHD12). Alternatively, 2-AG can be metabolized to 2-arachidonyl LPA (2-AG-LPA) through the action of monoacyl glycerol kinase(s) (MAGLK). 2-AG-LPA is then converted into 1-steroyl-2-arachidonyl PA. 1-Steroyl-2-arachidonyl PA is further utilized in the *de novo* synthesis of PC and PE. Furthermore, 2AG is metabolized by enzymatic oxygenation of 2-AG by COX-2 into PGH_2 glycerol esters (PG-G). EMT, eCB membrane transporter.

Despite biochemical and cell biological studies supporting a role for AEA as an endogenous CB1 agonist, the behavioral effects elicited by AEA are weak and transient compared with those produced by Δ^9-THC [114]. The limited pharmacological activity of AEA may be due to its rapid catabolism *in vivo*, as this lipid is hydrolyzed to arachidonic acid within minutes of exogenous administration [115]. Nonetheless, the relative contribution made by FAAH to the hydrolysis of anandamide *in vivo* remained largely unclear until it was cloned and a mouse model was generated in which this enzyme was genetically deleted (FAAH knockout) [116]. FAAH KO mice were shown to be born at the expected Mendelian frequency and were viable, fertile, and normal in their general cage behavior. Tissues from FAAH KO mice were found to display a 50–100-fold reduction in hydrolysis rates for AEA and related FAAs. In contrast to FAAH wild-type mice, in which administered AEA failed to produce significant behavioral effects, FAAH KO mice exhibited robust responses to AEA, becoming hypomotile, analgesic, cataleptic, and hypothermic. Furthermore, all of the behavioral effects of AEA in FAAH KO mice were blocked by pre-treatment with the CB1 receptor antagonist SR141716A (rimonabant), indicating that AEA acts as a potent and selective CB1 agonist in these animals. Consistent with this notion, AEA in brain homogenates from FAAH KO mice showed an approximately 15-fold higher apparent binding affinity for the CB1 receptor [117].

These FAAH KO mice have been shown to have elevated (10–15-fold) levels of AEA and other *N*-acylethanolamines (NAEs) in several brain regions, including the hippocampus, cortex, and cerebellum [116, 118]. Interestingly, these increased levels of AEA and NAEs in the brain correlated with a CB1-dependent reduction in pain sensation in FAAH KO mice [116]. Taken together, these observations suggest that FAAH is a key enzyme involved in the catabolism of AEA and other NAEs *in vivo* and suggest that pain pathways are under the influence of an FAAH-regulated eCB tone. However, it should be noted that FAAH KO mice exhibited normal motility, body weight, and body temperature [116], suggesting that in these animals, several other neurobehaviors affected by exogenously applied CB1 agonists were not controlled by AEA tone.

6.1. Localization and Distribution of FAAH in the Brain

The majority of FAAH in the mouse brain is associated with neurons; this is particularly interesting and most likely directly relevant to eCB signaling. The expression of FAAH appears not to be confined to neuronal somata but to also be extend into areas containing axonal and dendritic processes as well as into non-neuronal areas. This is most clearly observed in hippocampal formation where, although the highest concentration of FAAH is clearly located in the somata of hippocampal pyramidal cells and granule cells of the dentate gyrus, there is also a widespread presence of FAAH in the adjacent layers, which contain their axons and dendrites [119]. The most striking and intensely stained FAAH-immunoreactive neurons were found in the mesencephalic trigeminal nucleus (Me5) [119]. In addition to neuronal FAAH expression, oligodendrocytes and ventricular ependymal cells of the mouse brain [119] also express FAAH. The functional significance of FAAH expression in these non-neuronal cells is not yet known. More typically, FAAH and CB1 are anatomically associated, with FAAH-immunoreactive neuronal somata surrounded by CB1-immunoreactive fibers, such as in the rat cerebellar cortex, hippocampus, and neocortex [120]. FAAH-expressing neurons were also found in brain regions (thalamus, mid-brain, and hind brain) known to express few or no CB1 receptors. However, there are clearly a number of regions of the brain (*e.g.*, thalamic nuclei, mesencephalic trigeminal nuclei, cerebellar nuclei) where FAAH-expressing neurons are widespread but lack associated CB1 expression. In regions of the brain such as these, the role of FAAH is likely to be unrelated to CB1-dependent eCB signaling mechanisms [119]. This may be due to the presence of potential AEA targets in the brain other than CB1. This could explain the lack of complementarity between CB1 and FAAH distribution in some brain areas. For example, in these brain areas, AEA may instead act as an endogenous ligand of TRPV1 receptors [121].

FAAH is likely to influence retrograde synaptic signaling molecules in several brain regions, including the cerebellar cortex, where eCBs mediate retrograde suppression of both excitatory and inhibitory synapses onto Purkinje cells [122-124]. In the cerebellar cortex, FAAH is located in Purkinje cell somata, while CB1 is located on the presynaptic terminals of granule cells (glutamatergic parallel fibers), basket cells (GABAergic), and, most likely, climbing fibers (glutamatergic). Depolarization of Purkinje cells causes transient eCB-mediated inhibition of excitatory inputs from parallel fibers and climbing fibers, with maximal inhibition occurring after approximately 5 s, followed by a gradual restoration of basal excitatory input within 50 s [123]. Because FAAH is located in the somato-dendritic compartment of Purkinje cells, the duration of the inhibitory action of Purkinje cell-derived eCBs is likely to be largely influenced by the uptake and FAAH-mediated inactivation of eCBs in those cells [119]. FAAH is also present in fibers of the olfactory nerves that project into the olfactory bulb glomeruli, where they form synapses with the dendrites of glomerular mitral cells. This is of particular interest because FAAH is targeted to the axonal compartment of these neurons, whereas in all other regions of the mouse and rat brain it is targeted to the somato-dendritic compartment. It also provides a unique example of a synapse at which FAAH is located both presynaptically (olfactory-receptor neuron terminals) and postsynaptically (mitral cells) [119]. The functional significance of this complex pattern of expression has yet to be investigated. Using FAAH inhibitors, it was recently shown that FAAH inhibition does not affect depolarization-induced suppression of inhibition (DSI) [125, 126].

The localization of FAAH protein in the brain, although strikingly complementary with CB1 receptors in many regions of the forebrain and cerebellar cortex, is not always associated with CB1 receptor-expressing neurons. For example, there is a population of striatal GABAergic medium-spiny projection neurons that target the CB1 receptor to their axonal terminals in the globus pallidus (GP), entopeduncular nucleus (EP), and substantia nigra pars reticulata (SNR). The density of CB1 receptors in these output nuclei is particularly high, while FAAH-expressing neurons are sparse or absent [119].

The existence of a second mammalian AS enzyme with FAAH activity [127] has also been proposed. The FAAH-2 gene exists in primates and in distantly related vertebrates but not in mice and rats. This FAAH-2 enzyme exhibits an overlapping but distinct tissue distribution, substrate selectivity, and inhibitor sensitivity compared to the original FAAH enzyme (FAAH-1; discussed above). Both FAAH-1 and FAAH-2 share 20% amino acid sequence identity. Similar to FAAH-1, FAAH-2 possesses an N-terminal transmembrane domain and an AS sequence containing the serine-serine-lysine catalytic triad, along with other amino acid residues required for enzyme activity [127]. FAAH-2 showed greater activity with monounsaturated (C18: 1 NAE) substrates than with polyunsaturated (C20: 4 NAE) acyl chains. Moreover, FAAH-2 is situated within the membrane, whereas FAAH-1 has a predominantly cytoplasmic orientation on the membrane [127]. These observations indicate that FAAH-2 may be important for the regulation of monounsaturated lipid amides in the CNS and peripheral tissues; however, further investigation is needed.

A naturally occurring single nucleotide polymorphism in the human *FAAH* gene, 385A, was found to be strongly associated with an eCB risk factor in obesity [128], street drug use, and problematic drug/alcohol use. This association seems especially relevant to illegal drug use because neither alcohol nor nicotine abuse alone showed any significant relationship to the 385C→A missense mutation. The association of a naturally occurring human *FAAH* mutation with problematic drug use provides further support that the endogenous cannabinoid system plays an important role in the neural circuits that underlie drug abuse and dependence [129], including alcoholism [47]. Inhibition of FAAH by URB-597 at any dose tested did not increase the reinforcing efficacy of heroin, suggesting that FAAH is not involved in heroin addiction [130]. Mice deficient in FAAH have reduced levels of the peptide product of the cocaine- and amphetamine-related transcript (CART) in nerve fibers and terminals in several brain regions implicated in appetite control [131].

The FAAH inhibitor N-(pyridin-3-yl)-4-(3-(5-(trifluoromethyl)pyridin-2-yloxy)benzyl)-piperdine-1-carboxamide (PF-3845) reduced the intensity of opioid dependence through the activation of CB1 receptors [132]. Inhibition of FAAH reduces neuropathic pain through CB1 and CB2 receptors but not through TRPV1 or opioid receptors [133]. Pharmacological and genetic disruption of FAAH in mice enhances nicotine reward and withdrawal. However, in rats, pharmacological blockade of FAAH significantly inhibits nicotine reward while having no effect in nicotine withdrawal [134]. This species difference in FAAH may be due to slight differences in the structure, despite the high level of sequence homology and overall similarity in temperature-dependence. Additionally, the functional properties of mouse and rat FAAHs are different in their interaction with lipid bilayers and exogenous inhibitors [135]. There is strong evidence that some of the pharmacological action of alcohol is mediated by the eCB signaling system [136-138]. Although FAAH activity is not affected by chronic alcohol treatment in cerebellar granular neurons [11], mice exposed to chronic alcohol showed decreased FAAH activity in the cortex [139]. Overwhelming evidence suggests that the eCB signaling system regulates alcohol drinking behavior in various rodent models (see recent reviews [136-138]). FAAH has been shown to regulate alcohol drinking behavior in mice [47]. Increased consumption and preference for alcohol, decreased sensitivity to alcohol-induced sedation, and faster recovery from alcohol-induced motor incoordination were observed in both male and female FAAH knockout mice as well as in wild-type mice that received URB-597, a specific inhibitor of FAAH activity [140]. These results suggest that impaired FAAH function may present a phenotype of high voluntary alcohol consumption, and identify FAAH as both a regulator of eCB function and a possible therapeutic target for alcohol-related disorders. Decreased FAAH expression and activity was found in the prefrontal cortex (PFC) of alcohol-preferring rats with a compensatory down-regulation of CB1 signaling [141]. Furthermore, intra-PFC injections of the FAAH inhibitor URB-597 increased alcohol self-administration in Wistar rats [141]. Taken together, these data indicate that the enzymes regulating the eCB signaling system may have positively or negatively reinforcing effects on drugs of abuse or alcohol and should be taken into consideration during the development of potential therapeutic drugs targeting the eCB system for the treatment of drugs of abuse, including alcoholism.

6.2. MAG Lipase

MAGL (Fig. **4**) has been shown to regulate eCB levels in the CNS [16]. Recently, MAGL has been cloned from mouse adipose tissue [142] as well as rat brain [16]. The human MAGL cDNA was also cloned and possesses 84% identity with the mouse MAGL [143] and 85% identity with the rat MAGL. The mouse MAGL gene is located on chromosome 6 in a region with known homology to human chromosome 3q21 [142]. The mouse and human MAGL genes consist of seven exons ranging in size from 90 bp to over 2580 bp; the intron sizes are not conserved between the two species [142]. The mouse MAGL contains 302 amino acids with a molecular weight of 33 kD; its mRNA is a single transcript of 4 kb [143] and is expressed in various regions of the mouse brain. The catalytic triad was identified as Ser-122, Asp-239, and His-269 [142], and these residues are conserved between mouse and human MAGL. MAGL is found in all rat organs [142] and has a heterogeneous distribution in the rat brain, with the highest levels of expression being in regions where the CB1 receptor is abundant, such as the hippocampus, cortex, and cerebellum [16]. MAGL is localized to axon terminals of granule cells and CA3 pyramidal cells of the hippocampus [144]. That MAGL is expressed in presynaptic terminals [16, 144], which suggests that it has a role in terminating retrograde signaling at presynaptic neurons [145]. Recently, MAGL KO mice have been generated to show that MAGL acts as a distinct metabolic node coupling eCBs to prostaglandin signaling networks in the nervous system. This suggests that pharmacological inhibition of MAGL may be a new and potentially safer way to suppress the proinflammatory cascades that underlie neurodegenerative disorders [146]. 2-AG elevation in MAGL KO mice caused tonic activation and down-regulation of CB1 receptors [147] and selectively enhanced theta burst stimulation (TBS)-induced long-term potentiation (LTP) in the CA1 region of hippocampal slices but had no significant effect on LTP induced by high-frequency stimulation or long-term depression induced by low-frequency stimulation. In addition, MAGL KO mice also exhibited enhanced learning, as shown by improved performance in novel object recognition tasks and the Morris water maze. These results indicate that genetic deletion of MAGL causes profound changes in eCB signaling, long-term synaptic plasticity, and learning behavior [148]. The MAGL inhibitor JZL184 attenuated spontaneous withdrawal signs in morphine-dependent mice [132]. Ethanol intake for 15–30 days caused a marked impairment of heparin-releasable MAGL activity in rat heart tissue [149]. The fact that MAGL inhibition in mice increases ethanol consumption suggests that elevated levels of 2-AG in key brain regions of the reward system are involved in mediating the ethanol-seeking behavior of mice [150]. Inhibition of MAGL reduces neuropathic pain *via* a CB1- but not CB2-dependent receptor mechanism [133, 151].

7. CONCLUSION

Although redundancy exists in the metabolic pathways and molecular targets of eCBs (for review, see [152]) the enzymes involved in eCB metabolism are emerging as important regulators of eCB function at peripheral and neuronal levels. In the case of AEA and 2-AG, both FAAH, MAGL, NAT, NAPE-PLD, GDE1 or EMT appear to play more imperative roles in controlling their cellular activities. Considering the various biological and physiological functions of eCBs in animal tissues, the enzymes involved in the biosynthesis and metabolism of eCBs have been attractive targets for many diseases. Achieving a better understanding of these enzymes in pathophysiological conditions will contribute strongly to the development of selective inhibitors, which may ultimately prove useful for the development of novel therapeutic drugs.

ACKNOWLEDGEMENTS

The author acknowledges support in part by grants from the National Institutes of Health (NIH AA11031 and AA019443) and the New York State Psychiatric Institute.

CONFLICT OF INTEREST

The authors confirm that this chapter contents have no conflict of interest.

DISCLOSURE

The chapter submitted for eBook series titled "Frontiers in Protein and Peptide Science" Volume 1" is an update of our article "Critical enzymes involved in endocannabinoid metabolism" published in Protein and Peptide Letter (Invited Review), Volume 14, Number 3, Pages 237-246, 2007 (PMID: 17346227).

REFERENCES

[1] Horne, E., and Stella, N. (2008) The ins and outs of endocannabinoid signaling in healthy and diseased brain, *Future Lipidol 3*, 435-452.

[2] Stella, N. (2009) Endocannabinoid signaling in microglial cells, *Neuropharmacology 56 Suppl 1*, 244-253.

[3] Hanus, L., Abu-Lafi, S., Fride, E., Breuer, A., Vogel, Z., Shalev, D. E., Kustanovich, I., and Mechoulam, R. (2001) 2-arachidonyl glyceryl ether, an endogenous agonist of the cannabinoid CB1 receptor., *Proc Natl Acad Sci U S A. 98*, 3662-3665.

[4] Mechoulam, R., and Fride, E. (1995) *In cannabinoid receptors*, Academic Press.

[5] Porter, A. C., Sauer, J. M., Knierman, M. D., Becker, G. W., Berna, M. J., Bao, J., Nomikos, G. G., Carter, P., Bymaster, F. P., Leese, A. B., and Felder, C. C. (2002) Characterization of a novel endocannabinoid, virodhamine, with antagonist activity at the CB1 receptor., *J Pharmacol Exp Ther. 301*, 1020-1024.

[6] Di Marzo, V., Fontana, A., Cadas, H., Schinelli, S., Cimino, G., Schwartz, J. C., and Piomelli, D. (1994) Formation and inactivation of endogenous cannabinoid anandamide in central neurons., *Nature 372*, 686-691.

[7] Cadas, H., di Tomaso, E., and Piomelli, D. (1997) Occurrence and biosynthesis of endogenous cannabinoid precursor, N-arachidonoyl phosphatidylethanolamine, in rat brain., *J. Neurosci. 17*, 1226-1242.

[8] Mechoulam, R., Fride, E., and Di Marzo, V. (1998) Endocannabinoids., *Eur. J. Pharmacol. 359*, 1-18.

[9] Basavarajappa, B. S., and Hungund, B. L. (1999) Chronic Ethanol Increases the Cannabinoid Receptor Agonist, Anandamide and its Precursor N-Arachidonyl phosphatidyl ethanolamine in SK-N-SH Cells., *J. Neurochem 72*, 522-528.

[10] Basavarajappa, B. S., Saito, M., Cooper, T. B., and Hungund, B. L. (2000) Stimulation of cannabinoid receptor agonist 2-arachidonylglycerol by chronic ethanol and its modulation

by specific neuromodulators in cerebellar granule neurons., *Biochemica Biophysica Acta 1535*, 78-86.

[11] Basavarajappa, B. S., Saito, M., Cooper, T. B., and Hungund, B. L. (2003) Chronic ethanol inhibits the anandamide transport and increases extracellular anandamide levels in cerebellar granule neurons, *Eur. J. Pharmacol. 466*, 73-83.

[12] Giuffrida, A., Parsons, L. H., Kerr, T. M., Rodriguez de Fonseca, F., Navarro, M., and Piomelli, D. (1999) Dopamine activation of endogenous cannabinoid signaling in dorsal striatum., *Nat. Neurosci. 2*, 358-363.

[13] Hillard, C. J., Edgemond, W. S., Jarrahian, A., and Campbell, W. B. (1997) Accumulation of N-arachidonoylethanolamine (anandamide) into cerebellar granule cells occurs *via* facilitated diffusion., *J. Neurochem. 69*, 631-638.

[14] Piomelli, D., Beltramo, M., Glasnapp, S., Lin, S. Y., Goutopoulos, A., Xie, X. Q., and Makriyannis, A. (1999) Structural determinants for recognition and translocation by the anandamide transporter., *Proc. Natl. Acad. Sci. U. S. A. 96*, 5802-5807.

[15] Cravatt, B. F., Giang, D. K., Mayfield, S. P., Boger, D. L., Lerner, R. A., and Gilula, N. B. (1996) Molecular characterization of an enzyme that degrades neuromodulatory fatty-acid amides., *Nature 384*, 83-87.

[16] Dinh, T. P., Carpenter, D., Leslie, F. M., Freund, T. F., Katona, I., Sensi, S. L., Kathuria, S., and Piomelli, D. (2002) Brain monoglyceride lipase participating in endocannabinoid inactivation., *Proc Natl Acad Sci U S A. 99*, 10819-10824.

[17] Bisogno, T., Berrendero, F., Ambrosino, G., Cebeira, M., Ramos, J. A., Fernandez-Ruiz, J. J., and Di Marzo, V. (1999) Brain regional distribution of endocannabinoids: implications for their biosynthesis and biological function., *Biochem Biophys Res Commun. 256*, 377-380.

[18] Felder, C., Nielsen, A., Briley, E., Palkovits, M., Priller, J., Axelrod, J., Nguyen, D., Richardson, J., RM., R., Koppel, G., Paul, S., and Becker, G. (1996) Isolation and measurement of the endogenous cannabinoid receptor agonist, anandamide, in brain and peripheral tissues of human and rat., *FEBS Lett. 393*, 231-235.

[19] Felder, C. C., Joyce, K. E., Briley, E. M., Mansouri, J., Mackie, K., Blond, O., Lai, Y., Ma, A. L., and Mitchell, R. L. (1995) Comparison of the pharmacology and signal transduction of the human cannabinoid CB1 and CB2 receptors., *Mol. Pharmacol. 48*, 443-450.

[20] Felder, C. C., Briley, E. M., Axelrod, J., Simpson, J. T., Mackie, K., and Devane, W. A. (1993) Anandamide, an endogenous cannabimimetic eicosanoid, binds to the cloned human cannabinoid receptor and stimulates receptor-mediated signal transduction., *Proc. Natl. Acad. Sci. U.S.A., 90*, 7656-7660.

[21] Yang, H. Y., Karoum, F., Felder, C., Badger, H., Wang, T. C., and Markey, S. P. (1999) GC/MS analysis of anandamide and quantification of N-arachidonoylphosphatidylethanolamides in various brain regions, spinal cord, testis, and spleen of the rat, *J. Neurochem. 72*, 1959-1968.

[22] Deutsch, D. G., and Chin, S. A. (1993) Enzymatic synthesis and degradation of anandamide, a cannabinoid receptor agonist., *Biochem. Pharmacol. 46*, 791-796.

[23] Ueda, N., Kurahashi, Y., Yamamoto, S., and Tokunaga, T. (1995) Partial purification and characterization of the porcine brain enzyme hydrolyzing and synthesizing anandamide., *J Biol Chem. 270*, 23823-23827.

[24] Kurahashi, Y., Ueda, N., Suzuki, H., Suzuki, M., and Yamamoto, S. (1997) Reversible hydrolysis and synthesis of anandamide demonstrated by recombinant rat fatty-acid amide hydrolase., *Biochem Biophys Res Commun. 237*, 512-515.

[25] Katayama, K., Ueda, N., Katoh, I., and Yamamoto, S. (1999) Equilibrium in the hydrolysis and synthesis of cannabimimetic anandamide demonstrated by a purified enzyme., *Biochim Biophys Acta. 1440*, 205-214.

[26] Natarajan, V., Reddy, P. V., Schmid, P. C., and Schmid, H. H. (1981) On the biosynthesis and metabolism of N-acylethanolamine phospholipids in infarcted dog heart., *Biochim Biophys Acta. 664*, 445-448.

[27] Schmid, P. C., Reddy, P. V., Natarajan, V., and Schmid, H. H. (1983) Metabolism of N-acylethanolamine phospholipids by a mammalian phosphodiesterase of the phospholipase D type., *J Biol Chem. 258*, 9302-9306.

[28] Di Marzo, V. (1998) 'Endocannabinoids' and other fatty acid derivatives with cannabimimetic properties: biochemistry and possible, *Biochim. Biophys. Acta 1392*, 153-175.

[29] Sugiura, T., Kobayashi, Y., Oka, S., and Waku, K. (2002) Biosynthesis and degradation of anandamide and 2-arachidonoylglycerol and their possible physiological significance., *Prostaglandins Leukot Essent Fatty Acids. 66*, 173-192.

[30] Hansen, H. H., Hansen, S. H., Schousboe, A., and Hansen, H. S. (2000) Determination of the phospholipid precursor of anandamide and other N-acylethanolamine phospholipids before and after sodium azide-induced toxicity in cultured neocortical neurons., *J Neurochem. 75*, 861-871.

[31] Sugiura, T., Kondo, S., Sukagawa, A., Tonegawa, T., Nakane, S., Yamashita, A., and Waku, K. (1996) Enzymatic synthesis of anandamide, an endogenous cannabinoid receptor ligand, through N-acylphosphatidylethanolamine pathway in testis: involvement of Ca(2+)-dependent transacylase and phosphodiesterase activities., *Biochem. Biophys. Res. Commun. 218*, 113-117.

[32] Sugiura, T., Kondo, S., Sukagawa, A., Tonegawa, T., Nakane, S., Yamashita, A., and Waku, K. (1996) N-arachidonoylethanolamine (anandamide), an endogenous cannabinoid receptor ligand, and related lipid molecules in the nervous tissues., *Lipid Mediat Cell Signal 14*, 51-56.

[33] Uyama, T., Ikematsu, N., Inoue, M., Shinohara, N., Jin, X. H., Tsuboi, K., Tonai, T., Tokumura, A., and Ueda, N. (2012) Generation of N-Acylphosphatidylethanolamine by Members of the Phospholipase A/Acyltransferase (PLA/AT) Family, *J. Biol. Chem.*

[34] Sun, Y. X., Tsuboi, K., Okamoto, Y., Tonai, T., Murakami, M., Kudo, I., and Ueda, N. (2004) Biosynthesis of anandamide and N-palmitoylethanolamine by sequential actions of phospholipase A2 and lysophospholipase D, *Biochem. J. 380*, 749-756.

[35] Okamoto, Y., Morishita, J., Tsuboi, K., Tonai, T., and Ueda, N. (2004) Molecular characterization of a phospholipase D generating anandamide and its congeners., *J Biol Chem. 279*, 5298-5305.

[36] Fu, J., Astarita, G., Gaetani, S., Kim, J., Cravatt, B. F., Mackie, K., and Piomelli, D. (2007) Food intake regulates oleoylethanolamide formation and degradation in the proximal small intestine, *J. Biol. Chem. 282*, 1518-1528.

[37] Liu, J., Wang, L., Harvey-White, J., Osei-Hyiaman, D., Razdan, R., Gong, Q., Chan, A. C., Zhou, Z., Huang, B. X., Kim, H. Y., and Kunos, G. (2006) A biosynthetic pathway for anandamide, *Proc. Natl. Acad. Sci. U. S. A. 103*, 13345-13350.

[38] Simon, G. M., and Cravatt, B. F. (2006) Endocannabinoid biosynthesis proceeding through glycerophospho-N-acyl ethanolamine and a role for alpha/beta-hydrolase 4 in this pathway, *J. Biol. Chem. 281*, 26465-26472.

[39] Simon, G. M., and Cravatt, B. F. (2008) Anandamide biosynthesis catalyzed by the phosphodiesterase GDE1 and detection of glycerophospho-N-acyl ethanolamine precursors in mouse brain, *J. Biol. Chem. 283*, 9341-9349.

[40] Tsuboi, K., Okamoto, Y., Ikematsu, N., Inoue, M., Shimizu, Y., Uyama, T., Wang, J., Deutsch, D. G., Burns, M. P., Ulloa, N. M., Tokumura, A., and Ueda, N. (2011) Enzymatic formation of N-acylethanolamines from N-acylethanolamine plasmalogen through N-acylphosphatidylethanolamine-hydrolyzing phospholipase D-dependent and -independent pathways, *Biochim. Biophys. Acta 1811*, 565-577.

[41] Subbanna, S., Shivakumar, M., Psychoyos, D., Xie, S., and Basavarajappa, B. S. (2013) Anandamide-CB1 Receptor Signaling Contributes to Postnatal Ethanol-Induced Neonatal Neurodegeneration, Adult Synaptic and Memory Deficits, *Journal of neuoscience 33*, 6350-6366.

[42] Liscovitch, M., Czarny, M., Fiucci, G., and Tang, X. (2000) Phospholipase D: molecular and cell biology of a novel gene family, *Biochem. J. 345 Pt 3*, 401-415.

[43] Moesgaard, B., Petersen, G., Jaroszewski, J. W., and Hansen, H. S. (2000) Age dependent accumulation of N-acyl-ethanolamine phospholipids in ischemic rat brain. A (31)P NMR and enzyme activity study, *J. Lipid Res. 41*, 985-990.

[44] Leung, D., Saghatelian, A., Simon, G. M., and Cravatt, B. F. (2006) Inactivation of N-acyl phosphatidylethanolamine phospholipase D reveals multiple mechanisms for the biosynthesis of endocannabinoids, *Biochemistry (Mosc). 45*, 4720-4726.

[45] Mechoulam, R., Ben-Shabat, S., Hanus, L., Ligumsky, M., Kaminski, N. E., Schatz, A. R., Gopher, A., Almog, S., Martin, B. R., Compton, D. R., Pertwee, R. G., Griffin, G., Bayewitch, M., Barg, J., and Vogel, Z. (1995) Identification of an endogenous 2-monoglyceride, present in canine gut, that binds to cannabinoid receptors., *Biochem. Pharmacol. 50*, 83-90.

[46] Sugiura, T., Yoshinaga, N., Kondo, S., Waku, K., and Ishima, Y. (2000) Generation of 2-arachidonoylglycerol, an endogenous cannabinoid receptor ligand, in picrotoxinin-administered rat brain., *Biochem. Biophys. Res. Commun. 271*, 654-658.

[47] Basavarajappa, B. S., Yalamanchili, R., Cravatt, B. F., Cooper, T. B., and Hungund, B. L. (2006) Increased ethanol consumption and preference and decreased ethanol sensitivity in female FAAH knockout mice, *Neuropharmacology 50*, 834-844.

[48] Basavarajappa, B. S., Yalamanchili, R., Cooper, T. B., and Hungund, B. L. (2007) Endocannabinoid System., in *Handbook of Neurochemistry and Molecular Neurobiology* (Hamon, M., and Sylvester, V. E., Eds.), Springer, NY.

[49] Howlett, A. C., Barth, F., Bonner, T. I., Cabral, G., Casellas, P., Devane, W. A., Felder, C. C., Herkenham, M., Mackie, K., Martin, B. R., Mechoulam, R., and Pertwee, R. G. (2002) International Union of Pharmacology. XXVII. Classification of cannabinoid receptors., *Pharmacol. Rev. 54*, 161-202.

[50] Sugiura, T., Kondo, S., Sukagawa, A., Nakane, S., Shinoda, A., Itoh, K., Yamashita, A., and Waku, K. (1995) 2-Arachidonoylglycerol: a possible endogenous cannabinoid receptor ligand in brain., *Biochem. Biophys. Res. Commun. 215*, 89-97.

[51] Sugiura, T., Kodaka, T., Kondo, S., Tonegawa, T., Nakane, S., Kishimoto, S., Yamashita, A., and K., W. (1996) 2-Arachidonoylglycerol, a putative endogenous cannabinoid receptor

ligand, induces rapid, transient elevation of intracellular free Ca2+ in neuroblastoma x glioma hybrid NG108-15 cells., *Biochem. Biophys. Res. Commun. 229*, 58-64.

[52] Bisogno, T., Sepe, N., Melck, D., Maurelli, S., De Petrocellis, L., and Di Marzo, V. (1997) Biosynthesis, release and degradation of the novel endogenous cannabimimetic metabolite 2-arachidonoylglycerol in mouse neuroblastoma cells., *Biochem. J. 322*, 671-677.

[53] Stella, N., Schweitzer, P., and Piomelli, D. (1997) A second endogenous cannabinoid that modulates long-term potentiation., *Nature 388*, 773-778.

[54] Kondo, S., Kondo, H., Nakane, S., Kodaka, T., Tokumura, A., K., W., and Sugiura, T. (1998) 2-Arachidonoylglycerol, an endogenous cannabinoid receptor agonist: identification as one of the major species of monoacylglycerols in various rat tissues, and evidence for its generation through CA2+-dependent and -independent mechanisms., *FEBS. Lett. 429*, 152-156.

[55] Sugiura, T., Kodaka, T., Nakane, S., Kishimoto, S., Kondo, S., and Waku, K. (1998) Detection of an endogenous cannabimimetic molecule, 2-arachidonoylglycerol, and cannabinoid CB1 receptor mRNA in human vascular cells: is 2-arachidonoylglycerol a possible vasomodulator?, *Biochem. Biophys. Res. Commun. 243*, 838-843.

[56] Bisogno, T., Sepe, N., De Petrocellis, L., and Di Marzo, V. (1997) Biosynthesis of 2-arachidonoyl-glycerol, a novel cannabimimetic eicosanoid, in mouse neuroblastoma cells., *Adv. Exp. Med. Biol. 433*, 201-204.

[57] Prescott, S. M., and Majerus, P. W. (1983) Characterization of 1,2-diacylglycerol hydrolysis in human platelets. Demonstration of an arachidonoyl-monoacylglycerol intermediate, *J. Biol. Chem. 258*, 764-769.

[58] Bisogno, T., Melck, D., De Petrocellis, L., and Di Marzo, V. (1999) Phosphatidic acid as the biosynthetic precursor of the endocannabinoid 2-arachidonoylglycerol in intact mouse neuroblastoma cells stimulated with ionomycin., *Journal of Neurochemistry 72*, 2113-2119.

[59] Carrier, E. J., Kearn, C. S., Barkmeier, A. J., Breese, N. M., Yang, W., Nithipatikom, K., Pfister, S. L., Campbell, W. B., and Hillard, C. J. (2004) Cultured rat microglial cells synthesize the endocannabinoid 2-arachidonylglycerol, which increases proliferation *via* a CB2 receptor-dependent mechanism, *Mol. Pharmacol. 65*, 999-1007.

[60] Cocco, L., Faenza, I., Fiume, R., Maria Billi, A., Gilmour, R. S., and Manzoli, F. A. (2006) Phosphoinositide-specific phospholipase C (PI-PLC) beta1 and nuclear lipid-dependent signaling, *Biochim. Biophys. Acta 1761*, 509-521.

[61] Bisogno, T., Howell, F., Williams, G., Minassi, A., Cascio, M. G., Ligresti, A., Matias, I., Schiano-Moriello, A., Paul, P., Williams, E. J., Gangadharan, U., Hobbs, C., Di Marzo, V., and Doherty, P. (2003) Cloning of the first sn1-DAG lipases points to the spatial and temporal regulation of endocannabinoid signaling in the brain, *J. Cell Biol. 163*, 463-468.

[62] Katona, I., Urban, G. M., Wallace, M., Ledent, C., Jung, K. M., Piomelli, D., Mackie, K., and Freund, T. F. (2006) Molecular composition of the endocannabinoid system at glutamatergic synapses, *J. Neurosci. 26*, 5628-5637.

[63] Yoshida, T., Fukaya, M., Uchigashima, M., Miura, E., Kamiya, H., Kano, M., and Watanabe, M. (2006) Localization of diacylglycerol lipase-alpha around postsynaptic spine suggests close proximity between production site of an endocannabinoid, 2-arachidonoyl-glycerol, and presynaptic cannabinoid CB1 receptor, *J. Neurosci. 26*, 4740-4751.

[64] Tsutsumi, T., Kobayashi, T., Ueda, H., Yamauchi, E., Watanabe, S., and Okuyama, H. (1994) Lysophosphoinositide-specific phospholipase C in rat brain synaptic plasma membranes, *Neurochem. Res. 19*, 399-406.

[65] Ueda, H., Kobayashi, T., Kishimoto, M., Tsutsumi, T., and Okuyama, H. (1993) A possible pathway of phosphoinositide metabolism through EDTA-insensitive phospholipase A1 followed by lysophosphoinositide-specific phospholipase C in rat brain, *J. Neurochem. 61*, 1874-1881.

[66] Ueda, H., Kobayashi, T., Kishimoto, M., Tsutsumi, T., Watanabe, S., and Okuyama, H. (1993) The presence of Ca(2+)-independent phospholipase A1 highly specific for phosphatidylinositol in bovine brain, *Biochem. Biophys. Res. Commun. 195*, 1272-1279.

[67] Nakane, S., Oka, S., Arai, S., Waku, K., Ishima, Y., Tokumura, A., and Sugiura, T. (2002) 2-Arachidonoyl-sn-glycero-3-phosphate, an arachidonic acid-containing lysophosphatidic acid: occurrence and rapid enzymatic conversion to 2-arachidonoyl-sn-glycerol, a cannabinoid receptor ligand, in rat brain, *Arch. Biochem. Biophys. 402*, 51-58.

[68] Beltramo, M., and Piomelli, D. (2000) Carrier-mediated transport and enzymatic hydrolysis of the endogenous cannabinoid 2-arachidonylglycerol, *Neuroreport 11*, 1231-1235.

[69] Beltramo, M., Stella, N., Calignano, A., Lin, S. Y., Makriyannis, A., and Piomelli, D. (1997) Functional role of high-affinity anandamide transport, as revealed by selective inhibition., *Science 277*, 1094-1097.

[70] Hillard, C. J., and Jarrahian, A. (2000) The movement of N-arachidonoylethanolamine (anandamide) across cellular membranes., *Chem. Phys. Lipids 108*, 123-134.

[71] Maccarrone, M., van der Stelt, M., Rossi, A., Veldink, G. A., Vliegenthart, J. F., and Agro, A. F. (1998) Anandamide hydrolysis by human cells in culture and brain., *J. Biol. Chem. 273*, 32332-32339.

[72] Day, T. A., Rakhshan, F., Deutsch, D. G., and Barker, E. L. (2001) Role of fatty acid amide hydrolase in the transport of the endogenous cannabinoid anandamide., *Mol Pharmacol. 59*, 1369-1375.

[73] Deutsch, D. G., Glaser, S. T., Howell, J. M., Kunz, J. S., Puffenbarger, R. A., Hillard, C. J., and Abumrad, N. (2001) The cellular uptake of anandamide is coupled to its breakdown by fatty-acid amide hydrolase., *J. Biol. Chem. 276*, 6967-6973.

[74] Konrad, R. J., Major, C. D., and Wolf, B. A. (1994) Diacylglycerol hydrolysis to arachidonic acid is necessary for insulin secretion from isolated pancreatic islets: sequential actions of diacylglycerol and monoacylglycerol lipases, *Biochemistry (Mosc). 33*, 13284-13294.

[75] Goparaju, S. K., Ueda, N., Yamaguchi, H., and Yamamoto, S. (1998) Anandamide amidohydrolase reacting with 2-arachidonoylglycerol, another cannabinoid receptor ligand., *FEBS Lett. 422*, 69-73.

[76] Di Marzo, V., Bisogno, T., Sugiura, T., Melck, D., and De Petrocellis, L. (1998) The novel endogenous cannabinoid 2-arachidonoylglycerol is inactivated by neuronal- and basophil-like cells: connections with anandamide., *Biochem. J. 331*, 15-19.

[77] Goparaju, S. K., Kurahashi, Y., Suzuki, H., Ueda, N., and Yamamoto, S. (1999) Anandamide amidohydrolase of porcine brain: cDNA cloning, functional expression and site-directed mutagenesis(1), *Biochim. Biophys. Acta 1441*, 77-84.

[78] Simpson, C. M., Itabe, H., Reynolds, C. N., King, W. C., and Glomset, J. A. (1991) Swiss 3T3 cells preferentially incorporate sn-2-arachidonoyl monoacylglycerol into sn-1-stearoyl-2-arachidonoyl phosphatidylinositol, *J. Biol. Chem. 266*, 15902-15909.

[79] Kim, J., and Alger, B. E. (2004) Inhibition of cyclooxygenase-2 potentiates retrograde endocannabinoid effects in hippocampus, *Nat. Neurosci. 7*, 697-698.

[80] Sang, N., Zhang, J., and Chen, C. (2006) PGE2 glycerol ester, a COX-2 oxidative metabolite of 2-arachidonoyl glycerol, modulates inhibitory synaptic transmission in mouse hippocampal neurons, *J Physiol 572*, 735-745.

[81] Slanina, K. A., Roberto, M., and Schweitzer, P. (2005) Endocannabinoids restrict hippocampal long-term potentiation *via* CB1, *Neuropharmacology 49*, 660-668.

[82] Marrs, W. R., Blankman, J. L., Horne, E. A., Thomazeau, A., Lin, Y. H., Coy, J., Bodor, A. L., Muccioli, G. G., Hu, S. S., Woodruff, G., Fung, S., Lafourcade, M., Alexander, J. P., Long, J. Z., Li, W., Xu, C., Moller, T., Mackie, K., Manzoni, O. J., Cravatt, B. F., and Stella, N. (2010) The serine hydrolase ABHD6 controls the accumulation and efficacy of 2-AG at cannabinoid receptors, *Nat. Neurosci. 13*, 951-957.

[83] Glaser, S. T., Abumrad, N. A., Fatade, F., Kaczocha, M., Studholme, K. M., and Deutsch, D. G. (2003) Evidence against the presence of an anandamide transporter, *Proc. Natl. Acad. Sci. U. S. A. 100*, 4269-4274.

[84] Hillard, C. J., and Jarrahian, A. (2003) Cellular accumulation of anandamide: consensus and controversy, *Br. J. Pharmacol. 140*, 802-808.

[85] Di Marzo, V., Bisogno, T., Melck, D., Ross, R., Brockie, H., Stevenson, L., Pertwee, R., and De Petrocellis, L. (1998) Interactions between synthetic vanilloids and the endogenous cannabinoid system, *FEBS Lett. 436*, 449-454.

[86] Melck, D., Bisogno, T., De Petrocellis, L., Chuang, H., Julius, D., Bifulco, M., and Di Marzo, V. (1999) Unsaturated long-chain N-acyl-vanillyl-amides (N-AVAMs): vanilloid receptor ligands that inhibit anandamide-facilitated transport and bind to CB1 cannabinoid receptors, *Biochem. Biophys. Res. Commun. 262*, 275-284.

[87] Jarrahian, A., Manna, S., Edgemond, W., Campbell, W., and Hillard, C. (2000) Structure—Activity Relationships Among N-Arachidonylethanolamine (Anandamide) Head Group Analogues for the Anandamide Transporter, *J. Neurochem. 74*, 2597-2606.

[88] Palmer, S. L., Khanolkar, A. D., and Makriyannis, A. (2000) Natural and synthetic endocannabinoids and their structure -activity relationships., *Curr. Pharm. Des. 6*, 1381-1397.

[89] Di Marzo, V., Bisogno, T., and De Petrocellis, L. (2000) Endocannabinoids: New targets for drug development, *Curr. Pharm. Des. 6*, 1361-1380.

[90] Maccarrone, M., Attina, M., Bari, M., Cartoni, A., Ledent, C., and Finazzi-Agro, A. (2001) Anandamide degradation and N-acylethanolamines level in wild-type and CB1 cannabinoid receptor knockout mice of different ages., *J. Neurochem. 78*, 339-348.

[91] Maccarrone, M., Bari, M., Menichelli, A., Del Principe, D., and Agro, A. F. (1999) Anandamide activates human platelets through a pathway independent of the arachidonate cascade, *FEBS Lett. 447*, 277-282.

[92] Maccarrone, M., Bari, M., Lorenzon, T., Bisogno, T., Di Marzo, V., and Finazzi-Agro, A. (2000) Anandamide uptake by human endothelial cells and its regulation by nitric oxide, *J. Biol. Chem. 275*, 13484-13492.

[93] Maccarrone, M., Fiorucci, L., Erba, F., Bari, M., Finazzi-Agro, A., and Ascoli, F. (2000) Human mast cells take up and hydrolyze anandamide under the control of 5-lipoxygenase and do not express cannabinoid receptors, *FEBS Lett. 468*, 176-180.

[94] Rakhshan, F., Day, T. A., Blakely, R. D., and Barker, E. L. (2000) Carrier-mediated uptake of the endogenous cannabinoid anandamide in RBL-2H3 cells., *J. Pharmacol. Exp. Ther. 292*, 960-967.

[95] Fu, J., Bottegoni, G., Sasso, O., Bertorelli, R., Rocchia, W., Masetti, M., Guijarro, A., Lodola, A., Armirotti, A., Garau, G., Bandiera, T., Reggiani, A., Mor, M., Cavalli, A., and

Piomelli, D. (2012) A catalytically silent FAAH-1 variant drives anandamide transport in neurons, *Nat. Neurosci. 15*, 64-69.

[96] Bojesen, I. N., and Hansen, H. S. (2003) Binding of anandamide to bovine serum albumin, *J. Lipid Res. 44*, 1790-1794.

[97] Kaczocha, M., Glaser, S. T., and Deutsch, D. G. (2009) Identification of intracellular carriers for the endocannabinoid anandamide, *Proc. Natl. Acad. Sci. U. S. A. 106*, 6375-6380.

[98] Oddi, S., Fezza, F., Pasquariello, N., D'Agostino, A., Catanzaro, G., De Simone, C., Rapino, C., Finazzi-Agro, A., and Maccarrone, M. (2009) Molecular identification of albumin and Hsp70 as cytosolic anandamide-binding proteins, *Chem. Biol. 16*, 624-632.

[99] Ben-Shabat, S., Fride, E., Sheskin, T., Tamiri, T., Rhee, M. H., Vogel, Z., Bisogno, T., De Petrocellis, L., Di Marzo, V., and Mechoulam, R. (1998) An entourage effect: inactive endogenous fatty acid glycerol esters enhance 2-arachidonoyl-glycerol cannabinoid activity, *Eur. J. Pharmacol. 353*, 23-31.

[100] Di Marzo, V., De Petrocellis, L., Sepe, N., and Buono, A. (1996) Biosynthesis of anandamide and related acylethanolamides in mouse J774 macrophages and N18 neuroblastoma cells., *Biochem. J. 316*, 977-984.

[101] Ross, R. A., Craib, S. J., Stevenson, L. A., Pertwee, R. G., Henderson, A., Toole, J., and Ellington, H. C. (2002) Pharmacological characterization of the anandamide cyclooxygenase metabolite: prostaglandin E2 ethanolamide, *J. Pharmacol. Exp. Ther. 301*, 900-907.

[102] Kozak, K. R., Crews, B. C., Morrow, J. D., Wang, L. H., Ma, Y. H., Weinander, R., Jakobsson, P. J., and Marnett, L. J. (2002) Metabolism of the endocannabinoids, 2-arachidonylglycerol and anandamide, into prostaglandin, thromboxane, and prostacyclin glycerol esters and ethanolamides, *J. Biol. Chem. 277*, 44877-44885.

[103] Ueda, N., Yamamoto, K., Yamamoto, S., Tokunaga, T., Shirakawa, E., Shinkai, H., Ogawa, M., Sato, T., Kudo, I., Inoue, K., and *et al.* (1995) Lipoxygenase-catalyzed oxygenation of arachidonylethanolamide, a cannabinoid receptor agonist, *Biochim. Biophys. Acta 1254*, 127-134.

[104] Burstein, S. H., Rossetti, R. G., Yagen, B., and Zurier, R. B. (2000) Oxidative metabolism of anandamide, *Prostaglandins Other Lipid Mediat. 61*, 29-41.

[105] Tsuboi, K., Sun, Y. X., Okamoto, Y., Araki, N., Tonai, T., and Ueda, N. (2005) Molecular characterization of N-acylethanolamine-hydrolyzing acid amidase, a novel member of the choloylglycine hydrolase family with structural and functional similarity to acid ceramidase, *J. Biol. Chem. 280*, 11082-11092.

[106] Mulder, A. M., and Cravatt, B. F. (2006) Endocannabinoid Metabolism in the Absence of Fatty Acid Amide Hydrolase (FAAH): Discovery of Phosphorylcholine Derivatives of N-Acyl Ethanolamines, *Biochemistry (Mosc). 45*, 11267-11277.

[107] Forsell, P. K., Brunnstrom, A., Johannesson, M., and Claesson, H. E. (2012) Metabolism of anandamide into eoxamides by 15-lipoxygenase-1 and glutathione transferases, *Lipids 47*, 781-791.

[108] Maurelli, S., Bisogno, T., De Petrocellis, L., Di Luccia, A., Marino, G., and Di Marzo, V. (1995) Two novel classes of neuroactive fatty acid amides are substrates for mouse neuroblastoma 'anandamide amidohydrolase', *FEBS Lett. 377*, 82-86.

[109] Cravatt, B. F., Prospero-Garcia, O., Siuzdak, G., Gilula, N. B., Henriksen, S. J., Boger, D. L., and Lerner, R. A. (1995) Chemical characterization of a family of brain lipids that induce sleep, *Science 268*, 1506-1509.

[110] Wan, M., Cravatt, B. F., Ring, H. Z., Zhang, X., and Francke, U. (1998) Conserved chromosomal location and genomic structure of human and mouse fatty-acid amide hydrolase genes and evaluation of clasper as a candidate neurological mutation, *Genomics 54*, 408-414.

[111] Bracey, M. H., Hanson, M. A., Masuda, K. R., Stevens, R. C., and Cravatt, B. F. (2002) Structural adaptations in a membrane enzyme that terminates endocannabinoid signaling, *Science 298*, 1793-1796.

[112] Patricelli, M. P., Lovato, M. A., and Cravatt, B. F. (1999) Chemical and mutagenic investigations of fatty acid amide hydrolase: evidence for a family of serine hydrolases with distinct catalytic properties, *Biochemistry (Mosc). 38*, 9804-9812.

[113] McKinney, M. K., and Cravatt, B. F. (2005) Structure and function of fatty acid amide hydrolase, *Annu. Rev. Biochem. 74*, 411-432.

[114] Adams, I. B., Compton, D. R., and Martin, B. R. (1998) Assessment of anandamide interaction with the cannabinoid brain receptor: SR 141716A antagonism studies in mice and autoradiographic analysis of receptor binding in rat brain, *J. Pharmacol. Exp. Ther. 284*, 1209-1217.

[115] Willoughby, K. A., Moore, S. F., Martin, B. R., and Ellis, E. F. (1997) The biodisposition and metabolism of anandamide in mice, *J. Pharmacol. Exp. Ther. 282*, 243-247.

[116] Cravatt, B. F., Demarest, K., Patricelli, M. P., Bracey, M. H., Giang, D. K., Martin, B. R., and Lichtman, A. H. (2001) Supersensitivity to anandamide and enhanced endogenous cannabinoid signaling in mice lacking fatty acid amide hydrolase., *Proc. Natl. Acad. Sci. USA. 98*, 9371-9376.

[117] Lichtman, A. H., Hawkins, E. G., Griffin, G., and Cravatt, B. F. (2002) Pharmacological activity of fatty acid amides is regulated, but not mediated, by fatty acid amide hydrolase *in vivo*, *J. Pharmacol. Exp. Ther. 302*, 73-79.

[118] Clement, A. B., Hawkins, E. G., Lichtman, A. H., and Cravatt, B. F. (2003) Increased seizure susceptibility and proconvulsant activity of anandamide in mice lacking fatty acid amide hydrolase, *J. Neurosci. 23*, 3916-3923.

[119] Egertova, M., Cravatt, B. F., and Elphick, M. R. (2003) Comparative analysis of fatty acid amide hydrolase and cb(1) cannabinoid receptor expression in the mouse brain: evidence of a widespread role for fatty acid amide hydrolase in regulation of endocannabinoid signaling., *Neuroscience. 119*, 481-496.

[120] Egertova, M., Giang, D. K., Cravatt, B. F., and Elphick, M. R. (1998) A new perspective on cannabinoid signalling: complementary localization of fatty acid amide hydrolase and the CB1 receptor in rat brain., *Proc. R. Soc. Lond. B. Biol. Sci. 265*, 2081-2085.

[121] Maione, S., Bisogno, T., de Novellis, V., Palazzo, E., Cristino, L., Valenti, M., Petrosino, S., Guglielmotti, V., Rossi, F., and Di Marzo, V. (2006) Elevation of endocannabinoid levels in the ventrolateral periaqueductal grey through inhibition of fatty acid amide hydrolase affects descending nociceptive pathways *via* both cannabinoid receptor type 1 and transient receptor potential vanilloid type-1 receptors, *J. Pharmacol. Exp. Ther. 316*, 969-982.

[122] Kreitzer, A. C., and Regehr, W. G. (2001) Retrograde inhibition of presynaptic calcium influx by endogenous cannabinoids at excitatory synapses onto Purkinje cells., *Neuron. 29*, 717-727.

[123] Kreitzer, A. C., and Regehr, W. G. (2001) Cerebellar depolarization-induced suppression of inhibition is mediated by endogenous cannabinoids., *J. Neurosci. 21*, RC174.

[124] Maejima, T., Ohno-Shosaku, T., and Kano, M. (2001) Endogenous cannabinoid as a retrograde messenger from depolarized postsynaptic neurons to presynaptic terminals, *Neurosci. Res. 40*, 205-210.

[125] Makara, J. K., Mor, M., Fegley, D., Szabo, S. I., Kathuria, S., Astarita, G., Duranti, A., Tontini, A., Tarzia, G., Rivara, S., Freund, T. F., and Piomelli, D. (2005) Selective inhibition of 2-AG hydrolysis enhances endocannabinoid signaling in hippocampus, *Nat. Neurosci. 8*, 1139-1141.

[126] Szabo, B., Urbanski, M. J., Bisogno, T., Di Marzo, V., Mendiguren, A., Baer, W. U., and Freiman, I. (2006) Depolarization-induced retrograde synaptic inhibition in the mouse cerebellar cortex is mediated by 2-arachidonoylglycerol, *J Physiol 577*, 263-280.

[127] Wei, B. Q., Mikkelsen, T. S., McKinney, M. K., Lander, E. S., and Cravatt, B. F. (2006) A second fatty acid amide hydrolase with variable distribution among placental mammals, p M606646200.

[128] Sipe, J. C., Waalen, J., Gerber, A., and Beutler, E. (2005) Overweight and obesity associated with a missense polymorphism in fatty acid amide hydrolase (FAAH), *Int J Obes (Lond) 29*, 755-759.

[129] Sipe, J. C., Chiang, K., Gerber, A. L., Beutler, E., and Cravatt, B. F. (2002) A missense mutation in human fatty acid amide hydrolase associated with problem drug use, *Proc. Natl. Acad. Sci. U. S. A. 99*, 8394-8399.

[130] Solinas, M., Panlilio, L. V., Tanda, G., Makriyannis, A., Matthews, S. A., and Goldberg, S. R. (2005) Cannabinoid agonists but not inhibitors of endogenous cannabinoid transport or metabolism enhance the reinforcing efficacy of heroin in rats, *Neuropsychopharmacology 30*, 2046-2057.

[131] Osei-Hyiaman, D., Depetrillo, M., Harvey-White, J., Bannon, A. W., Cravatt, B. F., Kuhar, M. J., Mackie, K., Palkovits, M., and Kunos, G. (2005) Cocaine- and amphetamine-related transcript is involved in the orexigenic effect of endogenous anandamide, *Neuroendocrinology 81*, 273-282.

[132] Ramesh, D., Ross, G. R., Schlosburg, J. E., Owens, R. A., Abdullah, R. A., Kinsey, S. G., Long, J. Z., Nomura, D. K., Sim-Selley, L. J., Cravatt, B. F., Akbarali, H. I., and Lichtman, A. H. (2011) Blockade of endocannabinoid hydrolytic enzymes attenuates precipitated opioid withdrawal symptoms in mice, *J. Pharmacol. Exp. Ther. 339*, 173-185.

[133] Kinsey, S. G., Long, J. Z., O'Neal, S. T., Abdullah, R. A., Poklis, J. L., Boger, D. L., Cravatt, B. F., and Lichtman, A. H. (2009) Blockade of endocannabinoid-degrading enzymes attenuates neuropathic pain, *J. Pharmacol. Exp. Ther. 330*, 902-910.

[134] Muldoon, P. P., Lichtman, A. H., Parsons, L. H., and Damaj, M. I. (2012) The role of fatty acid amide hydrolase inhibition in nicotine reward and dependence, *Life Sci.*

[135] Di Venere, A., Dainese, E., Fezza, F., Angelucci, B. C., Rosato, N., Cravatt, B. F., Finazzi-Agro, A., Mei, G., and Maccarrone, M. (2012) Rat and human fatty acid amide hydrolases: Overt similarities and hidden differences, *Biochim. Biophys. Acta.*

[136] Basavarajappa, B. S. (2005) Endocannabinoid system in the development of tolerance to alcohol., *Klinik and Forschung (Journal of Clinical Research). 11*, 16-19.

[137] Basavarajappa, B. S. (2007) Endocannabinoid Signaling and Alcohol Addiction., in *New Research on Alcoholism* (Baye, D. R., Ed.), pp 1-55, Nova Science Publishers, Inc, New York.

[138] Colombo, G., Serra, S., Vacca, G., Carai, M. A., and Gessa, G. L. (2005) Endocannabinoid system and alcohol addiction: pharmacological studies, *Pharmacol. Biochem. Behav. 81*, 369-380.

[139] Vinod, K. Y., Yalamanchili, R., Xie, S., Cooper, T. B., and Hungund, B. L. (2006) Effect of chronic ethanol exposure and its withdrawal on the endocannabinoid system, *Neurochem. Int. 49*, 619-625.

[140] Blednov, Y. A., Cravatt, B. F., Boehm, S. L., II, Walker, D., and Harris, R. A. (2006) Role of Endocannabinoids in Alcohol Consumption and Intoxication: Studies of Mice Lacking Fatty Acid Amide Hydrolase, *Neuropsychopharmacology*.

[141] Hansson, A. C., Bermudez-Silva, F. J., Malinen, H., Hyytia, P., Sanchez-Vera, I., Rimondini, R., Rodriguez de Fonseca, F., Kunos, G., Sommer, W. H., and Heilig, M. (2006) Genetic Impairment of Frontocortical Endocannabinoid Degradation and High Alcohol Preference, *Neuropsychopharmacology*.

[142] Karlsson, M., Contreras, J. A., Hellman, U., Tornqvist, H., and Holm, C. (1997) cDNA cloning, tissue distribution, and identification of the catalytic triad of monoglyceride lipase. Evolutionary relationship to esterases, lysophospholipases, and haloperoxidases, *J. Biol. Chem. 272*, 27218-27223.

[143] Karlsson, M., Reue, K., Xia, Y. R., Lusis, A. J., Langin, D., Tornqvist, H., and Holm, C. (2001) Exon-intron organization and chromosomal localization of the mouse monoglyceride lipase gene, *Gene 272*, 11-18.

[144] Gulyas, A. I., Cravatt, B. F., Bracey, M. H., Dinh, T. P., Piomelli, D., Boscia, F., and Freund, T. F. (2004) Segregation of two endocannabinoid-hydrolyzing enzymes into pre- and postsynaptic compartments in the rat hippocampus, cerebellum and amygdala, *Eur. J. Neurosci. 20*, 441-458.

[145] Straiker, A., and Mackie, K. (2005) Depolarization-induced suppression of excitation in murine autaptic hippocampal neurones, *J Physiol 569*, 501-517.

[146] Nomura, D. K., Morrison, B. E., Blankman, J. L., Long, J. Z., Kinsey, S. G., Marcondes, M. C., Ward, A. M., Hahn, Y. K., Lichtman, A. H., Conti, B., and Cravatt, B. F. (2011) Endocannabinoid hydrolysis generates brain prostaglandins that promote neuroinflammation, *Science 334*, 809-813.

[147] Chanda, P. K., Gao, Y., Mark, L., Btesh, J., Strassle, B. W., Lu, P., Piesla, M. J., Zhang, M. Y., Bingham, B., Uveges, A., Kowal, D., Garbe, D., Kouranova, E. V., Ring, R. H., Bates, B., Pangalos, M. N., Kennedy, J. D., Whiteside, G. T., and Samad, T. A. (2010) Monoacylglycerol lipase activity is a critical modulator of the tone and integrity of the endocannabinoid system, *Mol. Pharmacol. 78*, 996-1003.

[148] Pan, B., Wang, W., Zhong, P., Blankman, J. L., Cravatt, B. F., and Liu, Q. S. (2011) Alterations of endocannabinoid signaling, synaptic plasticity, learning, and memory in monoacylglycerol lipase knock-out mice, *J. Neurosci. 31*, 13420-13430.

[149] Sevilla, E., Valette, A., Gastaldi, M., Boyer, J., and Verine, A. (1991) Effect of ethanol intake on lipoprotein lipase activity in rat heart, *Biochem. Pharmacol. 41*, 2005-2009.

[150] Gutierrez-Lopez, M. D., Llopis, N., Feng, S., Barrett, D. A., O'Shea, E., and Colado, M. I. (2010) Involvement of 2-arachidonoyl glycerol in the increased consumption of and preference for ethanol of mice treated with neurotoxic doses of methamphetamine, *Br. J. Pharmacol. 160*, 772-783.

[151] Bisogno, T., Ortar, G., Petrosino, S., Morera, E., Palazzo, E., Nalli, M., Maione, S., and Di Marzo, V. (2009) Development of a potent inhibitor of 2-arachidonoylglycerol hydrolysis with antinociceptive activity *in vivo*, *Biochim. Biophys. Acta 1791*, 53-60.

[152] Piscitelli, F., and Di Marzo, V. (2012) "Redundancy" of endocannabinoid inactivation: new challenges and opportunities for pain control, *ACS chemical neuroscience 3*, 356-363.

CHAPTER 3

Molecular Modeling of Two CYP2C19 SNPs and its Implications for Personalized Drug Design

Jing-Fang Wang[1,2], Li-Na Ma[1], Dong-Qing Wei[1,*], Chao Chen[3], Yixue Li[2] and Kuo-Chen Chou[4]

[1]College of Life Science and Biotechnology, Shanghai Jiaotong University, Shanghai 200240, China, [2]Bioinformatics Center, Key Lab of Systems Biology, Shanghai Institutes for Biological Sciences, Graduate School of the Chinese Academy of Sciences, Chinese Academy of Sciences, Shanghai 200031, China, [3]National Micro-measurement Center, Xibei University, Xian 710069, China and [4]Gordon Life Science Institute, 13784 Torrey Del Mar Drive, San Diego, CA 92130, USA

Abstract: CYP2C19 is an important member of the cytochrome P-450 enzyme superfamily and plays a significant role in the drug metabolism. In order to gain insights for developing personalized drugs, the structure-activity relationships of two SNPs, W120R and I331V, with the ligands of CEC, Fluvoxamine, Lescol and Ticlopidine were investigated through the structure-activity relationship approach. By means of a series of docking studies, the binding pockets of the two SNPs for the four compounds are explicitly defined that will be very useful for conducting mutagenesis studies, providing insights into personalization of drug treatments and stimulating novel strategies for finding desired personalized drugs.

Keywords: CYP2C19, cytochrome P-450, structure-activity relationship, CEC, fluvoxamine, lescol, ticlopidine, personalized drug design.

1. INTRODUCTION

New and important drug interactions are being identified on a daily basis. About 75% of drug transformation reactions involve catalysis by the Cytochrome P450 enzymes (CYPs), which are located in the intestines, liver, lung, kidney and brain [1,2]. A significant number of CYPs have to be identified yet. As known, drug

*Address correspondence to Dong-Qing Wei: College of Life Science and Biotechnology, Shanghai Jiaotong University, Shanghai 200240, China; E-mail: dqwei@sjtu.edu.cn

metabolism occurs in two phases: Phase I involves oxidation, reduction, and hydrolysis; Phase II involves synthesis and conjugation. The CYPs are considered to be involved in Phase I oxidative reactions with the abilities of utilizing haem iron to oxidize molecules [3]. Therefore, mammalian CYPs are significant for the detoxification of foreign substances and the control of the level of endogenous compounds, such as vitamin D metabolism, cholesterol synthesis, hormone synthesis and breakdown, *et al.* [4]. CYP 450 interactions generally result from one of two processes, inhibition and induction. Induction means that a substance stimulates the synthesis of the enzyme and metabolic capacity is increased. Inhibition means competitive binding at an enzyme's binding sites [5]. A drug with a high affinity for an enzyme will slow the metabolism of any low affinity drug. Though there is a high concentration of CYPs in the liver, these enzymes can be detected to exist throughout our body, where they often play specialised roles.

People can carry a series of alleles of CYP genes, which have a little variation in their genetic sequence attributable to nucleotide changes or polymorphisms (SNP) [6]. Certain alleles are found to lead to reduced enzymatic function, thereby interfering with production of the drug metabolites [7]. Moreover, the level of expression of a P450 is determined by non-coding SNPs [8,9].

These polymorphic variations are capable of causing individual and population differences in the tolerance to toxins and drugs. Because CYPs are specific in terms of which drugs they clear and which they activate, SNP variations in different CYP genes may have different effects from one substrate (or reaction) to another [10,11]. Therefore, there is an exigent problem that we should design personal drug with the knowledge of structural information for different persons who have the same illness [12].

CYP2C19 is an important member of CYP superfamily [13]. It can affect a series clinically important drugs, for example, proton pump inhibitors (omeprazole [14], lanzoprazole, and rabeprazole), phenytoin, tricyclic antidepressants (imipramine, amitryptiline), propranolol and benzodiazepines (diazepam) [15]. Marked interethnic differences in the polymorphism frequency [4, 13-16] have led to 28 variant alleles, from CYP2C19*1 to CYP2C19*28, being documented. The highly polymorphisms of CYP2C19 can lead to different function in metabolism of related drugs [17]. For example, patients with a loss-of-function CYP2C19 allele did poor active metabolite of clopidogrel, thus lead to diminishing platelet inhibition, and an increased risk of major adverse cardiovascular events, including stent thrombosis [18].

In our previous study, the 3D structure of CYP2C19 has been built using structural bioinformatics methods [19], and its interactions with ligand CEC, Fluvoxamine, Lescol and Ticlopidine have been simulated [20]. These works provide advantages for our further investigations. In addition, the similar methods have been used to study the subtypes 1, 2, 3, and 5 of GABA-A receptor [21], BACE2 [22], H5N1-NA [23,24], GFAT [25], cathepsin-E [26], alpha 7 nicotinic acetylcholine receptor [27], New Delhi metallo-beta-lactamse [28], xylose reductases [29] as well as human potassium channel and sodium channel [30,31]. In current study, two SNPs of CYP2C19, W120R and I331V, have been chosen as the investigation targets. Docking studies were carried out to seek the most favorable binding sites of the two SNPs interacting with ligands mentioned above. Subsequently, the molecular dynamics simulations (MD) were used to verify the docking results. Finally, some significant results were given for the further personal drug design.

2. COMPUTATIONAL METHODS

We firstly modeled the 3D structures of the two SNPs based on the computational structure of CYP2C19. The backbones were kept the same that of the template with the amino acid residues Trp120 and ILe331 mutated to Arg and Val, respectively. And the computational 3D structures of the two SNPs obtained according to the aforementioned methods are shown in Fig. **1**. Subsequently, the new structures were optimized by energy minimization. Finally, the computational structures obtained were used for further docking studies.

The molecular docking with the Metropolis algorithm was used to find the most favorable binding interaction, which was performed by our in-house program SIMM (Shanghai Molecular Modeling). In our docking studies, the ligands, CEC, Fluvoxamine, Lescol and Ticlopidine, were flexible. The program generates a diversified set of conformations by making random changes to the ligand coordinates [32,33]. When a new conformation of the ligand was generated, the search for favorable binding configurations was conducted within a specified 3D docking box, using either simulated annealing [34] or Tabu search [35]. Both methods seek to optimize the purely spatial contacts as well as electrostatic interactions. The interaction energy was calculated using the electrostatic and van der Waals potential fields. In all our computations, the CHARMM22 force field parameters were used [36]. The computational structures of two SNPs were taken as the receptor, which four ligands, CEC, Fluvoxamine, Lescol and Ticlopidine, were docked to.

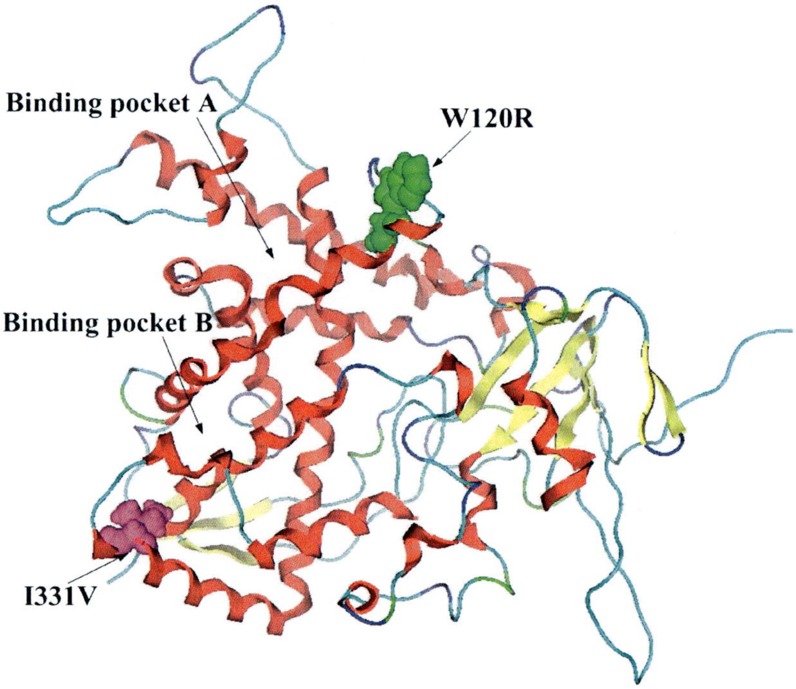

Figure 1: The computational 3D structures of SNP W120R and I331V, where the backbone is rendered in slab ribbon, and the mutation sites are colored in green and pink for W120R and I331V, respectively.

CYP2C19 has the multifaceted and flexible structural feature. To reflect this dynamic feature, one of the feasible approaches is using the molecular dynamics (MD) tools [37,38]. The MD simulations can solve the classical equations of motions for a system formed by target protein (SNPs) and small ligands (CEC, Fluvoxamine, Lescol and Ticlopidine) under specified ensembles. We have used this method to successfully investigate various problems such as lipid membranes [39], T1 lipase, M2 proton channel [40], CYP2E1 [41, 42], HHH Syndrome [43] and human FXa [44]. In current case, the energy favorable structures derived by the docking operation mentioned above were further studied with the MD simulations triggered by hydrogen bond braking and making events of the ligand and receptor interactions. Thus, the final results obtained would provide further conformational searching information in space. The MD simulations were performed at a constant temperature 300K and normal pressure. With the side chains being allowed to move freely, all the backbone atoms were fixed to maintain the correct protein 3D structure. In order to preferably represent the motions of heat in and out of the system, some fictitious degrees of freedom were added to the system. This operation could generate a series of conformations in the important phase space area, which can provide configuration and momentum information for each relevant atom, and the system thermodynamic properties can be calculated.

The aforementioned approach by combining the molecular docking operation with the MD simulations would provide very useful clues for the possible binding sites of the two SNPs with the ligands CEC, Fluvoxamine, Lescol and Ticlopidine molecules and their interaction modes. The similar molecular docking approach have been utilized for investigating the binding interaction of CYP2D6 with ligands [45], as well as alpha 7 nAChR dimer with GTS-21 [46] and DMXBA [47], and for anti-SARS drug screening [48-50], and M2 proton channel [51], human Rab5a [52] and drug resistance in HIV-1 protease [53].

3. RESULTS AND DISCUSSIONS

In the computational 3D structure of CYP2C19, two binding pockets, named A and B, are considered to be more energy favorable than others, as shown in Fig. **1**. Binding pocket B is also the active site of wildtype CYP2C19 for drug metabolisms, as discussed in our previous work [11]. In order to find the most energy favorable binding pocket, all the four ligands were docked to the two pockets A and B, respectively. Listed in Tables **1** and **2** are the binding energies obtained by docking CEC, Fluvoxamine, Lescol and Ticlopidine to SNPs W120R and I331V, respectively. For SNPI331V, as shown from Table **2**, the binding pocket B is better. But the same situation does not appear for SNP W120R. As shown from Table **1**, for Fluvoxamine and Lescol, binding pocket A is preferred, which means that they may not be metabolized by SNP W120R. While for CEC and Ticlopidine, binding pocket B is preferred.

Table 1: List of interaction energies (kcal/mol) obtained by docking CEC, Fluvoxamine, Lescol, and Ticlopidine to SNP W120R, respectively.

Ligands	**E (Electronic)**		**E (van der Waals)**		**E (Binding)**	
	A	B	A	B	A	B
CEC	-6.0021	-3.1040	-10.1027	-15.1207	-15.0080	-18.2248
Fluvoxamine	-3.7223	-4.7834	-15.7543	-14.3985	-19.4766	-19.1819
Lescol	-5.8674	-4.2639	-14.4406	-15.7880	-20.3081	-20.0519
Ticlopidine	-0.2002	-1.2017	-14.2631	-16.1514	-14.4633	-17.3532

Table 2: List of interaction energies (kcal/mol) obtained by docking CEC, Fluvoxamine, Lescol, and Ticlopidine to SNP I331V, respectively.

Ligands	**E (Electronic)**		**E (van der Waals)**		**E (Binding)**	
	A	B	A	B	A	B
CEC	-1.7356	-4.8003	-13.2505	-14.3010	-14.9860	-19.1012
Fluvoxamine	-3.8445	-4.7782	-12.0220	-17.7722	-15.8665	-22.5504
Lescol	-5.2894	-9.7284	-16.0395	-15.2188	-21.3289	-24.9472
Ticlopidine	-0.7782	-1.2675	-14.3404	-15.4698	-15.1186	-16.7373

According to [54], the binding pocket of a receptor for its ligand is defined by those residues that have at least one heavy atom (*i.e.*, other than hydrogen) with a distance 5 from a heavy atom of the ligand. The amino acid residues in forming the binding pockets of SNP W120R and I331V are given in Tables **3** and **4**, respectively.

For SNP W120R, 9 residues in the binding pockets of Fluvoxamine and Lescol are shared, that is, Val113, Arg120, Ile123, Ser127, Phe263, Phe267, Leu268, Met271, and Thr290; while residues in the binding pockets of CEC and Ticlopidine are almost the same except Asn324 for Ticlopidine. It can be seen from the Table **3** that there are five hydrogen bonds formed between Lescol and SNP W120, indicating that Lescol has stronger hydrogen bonding interaction with SNP W120 than the other three ligands. The amino acids involved in forming the hydrogen bonds with Lescol include Arg120, Ile123, and Phe267. Of these three, Arg120 is considered to have more contributions to the binding interaction because more hydrogen bonds are formed by it. Besides, Arg120 is also involved in forming hydrogen bonds with Fluvoxamine.

Table 3: Residues in forming the binding pockets of SNP W120R for the four ligands.

Ligands	Binding Pocket Residues[a]					
CEC	Gln136	Glu318	**Val321**	Gly322	Gln444	**Asn445**
	Ile478	Pro479	Val480			
Fluvoxamine	Val113	**Arg120**	**Ile123**	Phe126	Ser127	Phe263
	Cys266	**Phe267**	Leu268	Met271	Thr290	
Lescol	Val113	Arg119	**Arg120**	**Ile123**	Arg124	Ser127
	Phe263	**Phe267**	Leu268	Met271	Thr290	Leu294
Ticlopidine	Gln136	Glu318	Val321	Gly322	Asn324	**Gln444**
	Asn445	Ile478	**Pro479**	Val480		

[a] Residues with bold-face type mean that they have hydrogen bonding interaction with the ligands.

Table 4: Residues in forming the binding pockets of SNP I331V for the four ligands.

Ligands	Binding Pocket Residues[a]					
CEC	Glu132	**Gln136**	Glu137	Gln314	Glu318	Val321
	Phe441	Gln444	Asn445	Ile478	**Pro479**	**Val480**
Fluvoxamine	Glu142	Ile327	**Glu328**	Ile331	Gly332	Asn334
	Arg335	Ser336	Pro337	Gln340	Thr450	Phe451
	Gln454	**Asn455**	Ile488	Pro489	Val490	
Lescol	Arg139	**Glu142**	Gln146	Glu147	Ile331	**Asn334**
	Arg335	Ser336	Pro337	Phe451	Gln454	**Asn455**
Ticlopidine	Glu142	Gln324	Ile327	**Glu328**	**Ile331**	Gly332
	Asn334	Arg335	Ser336	Pro337	Gln340	**Gln454**
	Asn455	Pro489				

[a] Residues with bold-face type mean that they have hydrogen bonding interaction with the ligands.

For SNP I331V, we can find that none is occur in the four pockets, while Ile331, Asn334, Arg335, Ser336, Pro337, Gln454, and Asn455 in three of the four. Also, seen from Table **4** that there are 13 hydrogen bonds formed between Fluvoxamine and SNP I331, indicating that Fluvoxamine has stronger hydrogen binding interaction with SNP I331 than the other three ligands. The amino acids involved in forming the hydrogen bonds with Fluvoxamine include Glu328, Gln454, and

Asn455. Of these three, Glu328 seems to have a little more contributions to the binding interaction for forming more hydrogen bonds. In addition, Asn455 is also involved in forming hydrogen bonds with Lescol, and Gln454 involved in forming hydrogen bonds with Ticlopidine.

Shown in the Figs. **2** and **3** are the lipophilic and hydrophilic surfaces of the two SNPs with the four ligands: (A) CEC, (B) Fluvoxamine, (C) Lescol and (D) Ticlopidine, respectively. Furthermore, it can be seen from the Figs. **2** and **3** that the active cavities of SNP W120R are mainly hydrophilic, while the active cavities of SNP I331V are mainly hydrophobic.

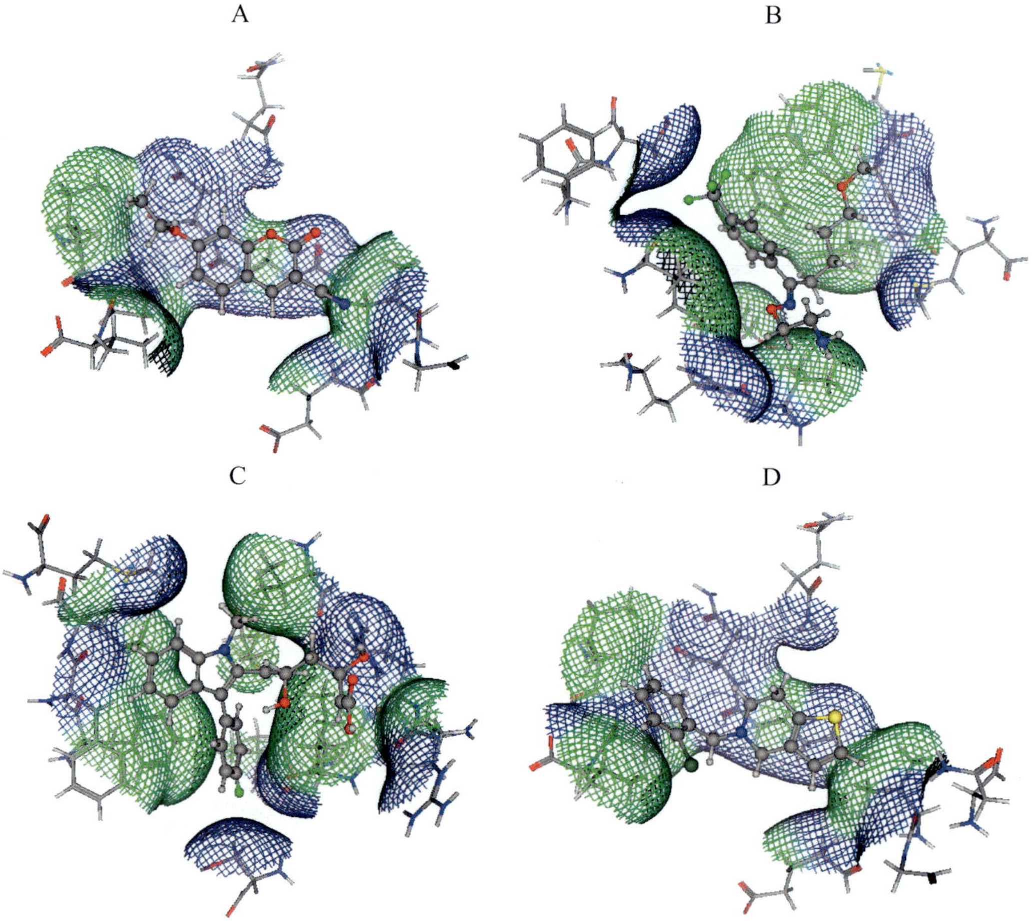

Figure 2: Illustrations showing the lipophilic and hydrophilic surfaces of the binding pockets (or cavities) of SNP W120 for (A) CEC, (B) Fluvoxamine, (C) Lescol, and (D) Ticlopidine, where the lipophilic and hydrophilic surfaces are colored in green and blue, respectively. (For interpretation of the references to color in this figure legend, the reader is referred to the web version of this paper).

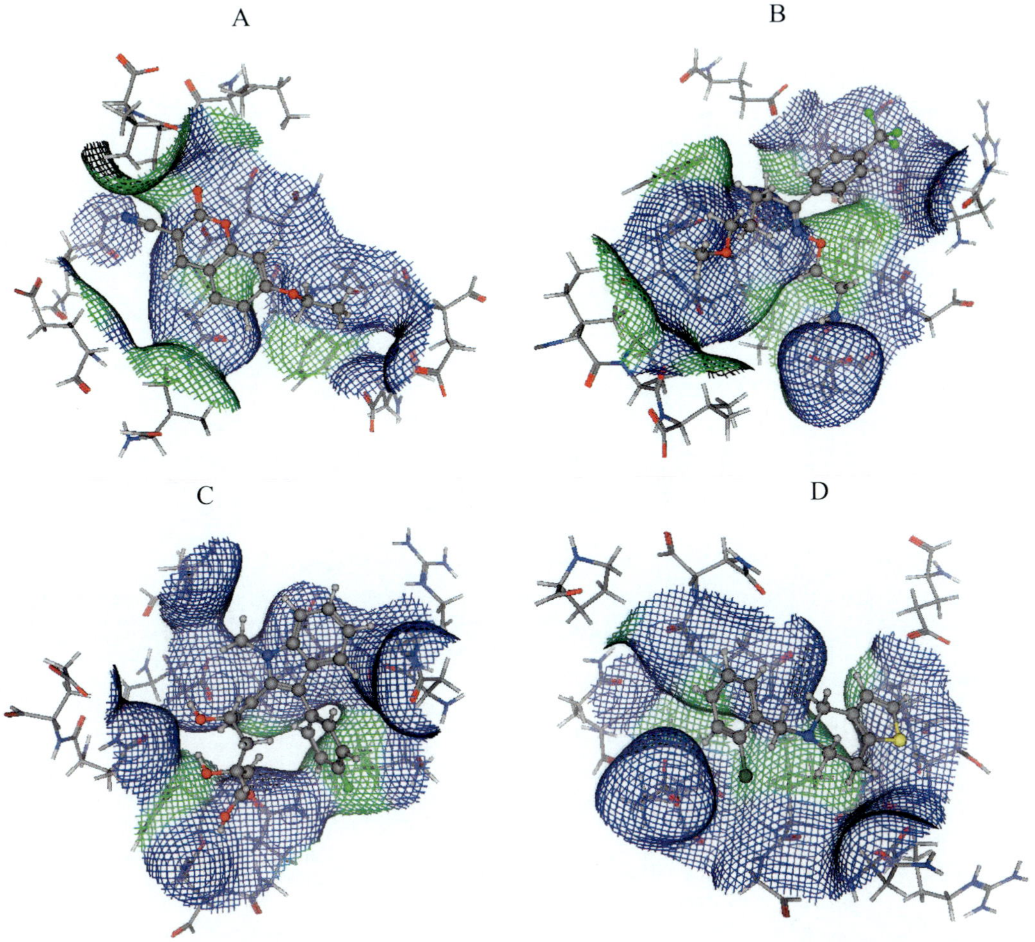

Figure 3: Illustrations showing the lipophilic and hydrophilic surfaces of the binding pockets (or cavities) of SNP I331 for (A) CEC, (B) Fluvoxamine, (C) Lescol, and (D) Ticlopidine, where the lipophilic and hydrophilic surfaces are colored in green and blue, respectively. (For interpretation of the references to color in this figure legend, the reader is referred to the web version of this paper).

4. CONCLUSIONS

The lipophilicity is a significant factor for designing orally active drugs due to the complementary lipophilic and hydrophilic interactions between the proteins and ligands, and the harmonizing lipophilicity and water solubility. The latter is necessary for its absorption from the intestinal tract. Though it is difficult to understand the hydrophobic effects on molecular levels, remarkable changes of the docking free energies between proteins and ligands might be a key relevant factor. The different lipophilicity in the binding pockets or active cavities of SNPs W120R and I331V might be the main reason of the receptor's binding pockets found through this study for CEC, Fluvoxamine, Lescol and Ticlopidine, respectively. In addition, for SNP W120R, Fluvoxamine and Lescol may not be metabolized for their larruping binding sites. All these findings will be very useful for conducting mutagenesis investigations for finding desired drugs or proper treatments according to the characteristics of an individual patient to improve efficacy and reduce the number and adverse drug reactions.

ACKNOWLEDGEMENTS

This work was supported by the grants from Chinese National Science Foundation under the Contract No.10376024 and the Tianjin Commission of Education under the contract No. 2003001, and the Tianjin Commission of Sciences and Technology under the contract No.033801911. Additional supports were from the special fund for intensive computation (043185111-4), Virtual Laboratory for Computational Chemistry of CNIC, and the Supercomputing Center of CNIC, Chinese Academy of Sciences.

CONFLICT OF INTEREST

The authors confirm that this chapter contents have no conflict of interest.

DISCLOSURE

This is an updated report of the earlier review article by Wang JF, Wei DQ, Chen C, Li Y, Chou KC. "Molecular modeling of two CYP2C19 SNPs and its implications for personalized drug design", Protein Peptide Letter 2008, 15(1), 27-32.

REFERENCES

[1] Williams, J. A.; Hyland, R.; Jones, B. C.; Smith, D. A.; Hurst, S.; Goosen, T. C.; Peterkin, V.; Koup, J. R.; Ball, S. E. *Drug Metabolism and Disposition* **2004**, *32*, 1201.

[2] Wienkers, L. C.; Heath, T. G. *Nature reviews Drug discovery* **2005**, *4*, 825.

[3] Tanmay, S.; Shaun, F.; Sanish, D.; Vishal, M.; Nilima, A.; Nithya, J. *BMC Clin. Pharm* **2006**, *6*, 1.

[4] Gonzalez, F. J.; Skodat, R. C.; Kimura, S.; Umeno, M.; Zanger, U. M.; Nebert, D. W.; Gelboin, H. V.; Hardwick, J. P.; Meyer, U. A. **1988**, *331*, 442.

[5] Zhou, J. Q.; Tang, Z. Q.; Zhang, J. N.; Tang, J. C. *Acta Pharmacologica Sinica* **2006**, *27*, 635.

[6] Eitrich, T.; Kless, A.; Druska, C.; Meyer, W.; Grotendorst, J. *Journal of chemical information and modeling* **2007**, *47*, 92.

[7] Ingelman-Sundberg, M.; Sim, S. C.; Gomez, A.; Rodriguez-Antona, C. *Pharmacology & therapeutics* **2007**, *116*, 496.

[8] Zhang, T.; Zhou, Q.; Pang, Y.; Wang, Y.; Jin, C.; Huo, J.; Liu, L. A.; Wei, D. *Interdisciplinary Sciences: Computational Life Sciences* **2012**, *4*, 83.

[9] Guengerich, F. P.; Rendic, S. *Current drug metabolism* **2010**, *11*, 1.

[10] Mega, J. L.; Close, S. L.; Wiviott, S. D.; Shen, L.; Hockett, R. D.; Brandt, J. T.; Walker, J. R.; Antman, E. M.; Macias, W.; Braunwald, E. *New England Journal of Medicine* **2009**, *360*, 354.

[11] Kjellander, B.; Masimirembwa, C. M.; Zamora, I. *Journal of chemical information and modeling* **2007**, *47*, 1234.

[12] Wang, J. F.; Zhang, C. C.; Chou, K. C.; Wei, D. Q. *Current medicinal chemistry* **2009**, *16*, 232.

[13] Sim, S. C.; Ingelman-Sundberg, M. *Hum Genomics* **2010**, *4*, 278.

[14] Desta, Z.; Zhao, X.; Shin, J. G.; Flockhart, D. A. *Clinical pharmacokinetics* **2002**, *41*, 913.

[15] Kumana, C.; Lauder, I.; Chan, M.; Ko, W.; Lin, H. *European journal of clinical pharmacology* **1987**, *32*, 211.

[16] Daly, A. K. *Current topics in medicinal chemistry* **2004**, *4*, 1733.

[17] Kalow, W. *Trends in pharmacological sciences* **1991**, *12*, 102.

[18] Wedlund, P. J. *Pharmacology* **2000**, *61*, 174.

[19] Chou, K. C. *Current medicinal chemistry* **2004**, *11*, 2105.

[20] Wang, J. F.; Wei, D. Q.; Li, L.; Zheng, S. Y.; Li, Y. X.; Chou, K. C. *Biochemical and biophysical research communications* **2007**, *355*, 513.

[21] Chou, K. C. *Biochemical and biophysical research communications* **2004**, *316*, 636.

[22] Chou, K. C. *Journal of Proteome Research* **2004**, *3*, 1069.

[23] Wei, D. Q.; Du, Q. S.; Sun, H.; Chou, K. C. *Biochemical and biophysical research communications* **2006**, *344*, 1048.

[24] Gong, K.; Li, L.; Wang, J. F.; Cheng, F.; Wei, D. Q.; Chou, K. C. *Medicinal Chemistry* **2009**, *5*, 242.

[25] Chou, K. C. *Journal of Proteome Research* **2004**, *3*, 1284.

[26] Chou, K. C. *Biochemical and biophysical research communications* **2005**, *331*, 56.

[27] Chou, K. C. *Biochemical and biophysical research communications* **2004**, *319*, 433.

[28] Wang, J. F.; Chou, K. C. *PLoS One* **2011**, *6*, e18414.

[29] Wang, J. F.; Wei, D. Q.; Lin, Y.; Wang, Y. H.; Du, H. L.; Li, Y. X.; Chou, K. C. *Biochemical and biophysical research communications* **2007**, *359*, 323.

[30] Chou, K. C. *Journal of Proteome Research* **2004**, *3*, 856.

[31] Wang, J. F.; Chou, K. C. *Current drug metabolism* **2010**, *11*, 342.

[32] Kuntz, I. D.; Meng, E. C.; Shoichet, B. K. *Accounts of Chemical research* **1994**, *27*, 117.

[33] Morris, G. M.; Goodsell, D. S.; Halliday, R. S.; Huey, R.; Hart, W. E.; Belew, R. K.; Olson, A. J. *Journal of computational chemistry* **1998**, *19*, 1639.

[34] Chou, K. C.; Carlacci, L. *Protein engineering* **1991**, *4*, 661.

[35] Glover, F. *Interfaces* **1990**, *20*, 74.

[36] MacKerell, A. D.; Bashford, D.; Bellott, M.; Dunbrack, R. L.; Evanseck, J. D.; Field, M. J.; Fischer, S.; Gao, J.; Guo, H.; Ha, S.; Joseph-McCarthy, D.; Kuchnir, L.; Kuczera, K.; Lau, F. T. K.; Mattos, C.; Michnick, S.; Ngo, T.; Nguyen, D. T.; Prodhom, B.; Reiher, W. E.; Roux, B.; Schlenkrich, M.; Smith, J. C.; Stote, R.; Straub, J.; Watanabe, M.; Wiorkiewicz-Kuczera, J.; Yin, D.; Karplus, M. *Journal of Physical Chemistry B* **1998**, *102*, 3586.
[37] Wang, Y.; Wei, D. Q.; Wang, J. F. *Journal of chemical information and modeling* **2010**, *50*, 875.
[38] Wang, J. F.; Gong, K.; Wei, D. Q.; Li, Y. X.; Chou, K. C. *Protein Engineering Design and Selection* **2009**, *22*, 349.
[39] Lian, P.; Wei, D. Q.; Wang, J. F.; Chou, K. C. *PLoS One* **2011**, *6*, e18587.
[40] Wang, J. F.; Wei, D. Q.; Chou, K. C. *Biochemical and biophysical research communications* **2009**, *388*, 413.
[41] Li, J.; Wei, D. Q.; Wang, J. F.; Li, Y. X. *Journal of chemical information and modeling* **2011**, *51*, 3217.
[42] Li, J.; Wei, D. Q.; Wang, J. F.; Yu, Z. T.; Chou, K. C. *Medicinal Chemistry* **2012**, *8*, 208.
[43] Wang, J. F.; Chou, K. C. *PLoS One* **2012**, *7*, e31048.
[44] Wang, J. F.; Hao, P.; Li, Y. X.; Dai, J. L.; Li, X. *Journal of molecular modeling* **2012**, *18*, 2717.
[45] Wang, J.; Zhang, C.; Wei, D.; Li, Y. *Chinese Science Bulletin* **2010**, *55*, 1877.
[46] Wei, D. Q.; Sirois, S.; Du, Q. S.; Arias, H. R.; Chou, K. C. *Biochemical and biophysical research communications* **2005**, *338*, 1059.
[47] Gu, R. X.; Gu, H.; Xie, Z. Y.; Wang, J. F.; Arias, H. R.; Wei, D. Q.; Chou, K. C. *Medicinal Chemistry* **2009**, *5*, 250.
[48] Chou, K. C.; Wei, D. Q.; Du, Q. S.; Sirois, S.; Zhong, W. Z. *Current medicinal chemistry* **2006**, *13*, 3263.
[49] Wei, D. Q.; Zhang, R.; Du, Q. S.; Gao, W. N.; Li, Y.; Gao, H.; Wang, S. Q.; Zhang, X.; Li, A. X.; Sirois, S. *Amino Acids* **2006**, *31*, 73.
[50] Chen, Q.; Zhang, T.; Wang, J. F.; Wei, D. Q. *Current drug metabolism* **2011**, *12*, 436.
[51] Wang, J. F.; Chou, K. C. *Protein Engineering Design and Selection* **2010**, *23*, 663.
[52] Wang, J. F.; Chou, K. C. *Biochemical and biophysical research communications* **2009**, *390*, 608.
[53] Gu, H.; Chen, H.; Wei, D.; Wang, J. *Chinese Science Bulletin* **2010**, *55*, 2677.
[54] Chou, K. C.; Watenpaugh, K. D.; Heinrikson, R. L. *Biochemical and biophysical research communications* **1999**, *259*, 420.

Frontiers in Protein and Peptide Sciences, Vol. 1, **2014, 75-90**

CHAPTER 4

Anti-Apoptotic and Oncogenic Function of HSP27

Yusuf Tutar[1,*], Aykut Özgür[2], Esen Tutar[3], Kübra Açıkalın Coşkun[2] and Lütfi Tutar[4]

[1]Cumhuriyet University, Faculty of Pharmacy, Department of Basic Sciences, Division of Biochemistry, Sivas, Turkey; [2]Gaziosmanpasa University, Faculty of Natural Sciences and Engineering, Department of Bioengineering, Tokat, Turkey; [3]Kahramanmaraş Sütçü İmam University, ÜSKİM Research Center, Central Laboratory, Kahramanmaraş, Turkey and [4]Kahramanmaraş Sütçü İmam University, Faculty of Science and Letters, Department of Biology, Kahramanmaraş, Turkey

Abstract: Heat shock Proteins (HSPs) play essential role in cellular homeostasis and therefore, maintain key proteins at their native state against cellular stress to promote cell survival. ATP independent HSP27 makes a defense against stress factors to facilitate ATP dependent Hsp machinery to perform its function under stress conditions. The machinery involves HSP70, HSP40, HSP60, HSP90, and HSP100 proteins. The expression of inducible HSP27 is at basal levels in normal cells; however, HSP27 is abnormally expressed in breast cancer, ovarian cancer, osteo-sarcoma, endometrial cancer, and leukemia cells. Elevated HSP27 in oncogenic cells resists anti-cancer therapy since it protects the cell against spontaneous apoptosis. Furthermore, HSP27 inhibits apoptotic signals at regulatory points which may explain the role of HSP27 in tumor progression and resistance to therapy. Phosphorylation and oligomerization of HSP27 regulate its biochemical function. Recent literature reports on HSP27 structure, function, and its role in oncogenesis and apoptosis are discussed in this review.

Keywords: Apoptosis, Cancer, Heat Shock protein 27, Protein folding.

1. HEAT SHOCK PROTEINS AND CELLULAR RESPONSE TO STRESS

Heat shock proteins (HSPs) were observed in *Drosophila* salivary gland chromosome puffs in response to transient stress for the first time. These sets of

***Address correspondence to Yusuf Tutar:** Department of Basic Sciences, Division of Biochemisty, Faculty of Pharmacy, Cumhuriyet University, Sivas, Turkey, 58140; Tel: +90 346-219-1010 ext. 3907; Fax: +90 346-219-1634; E-mail: ytutar@cumhuriyet.edu.tr; ytutar@yahoo.com

Ben M. Dunn (Ed)

highly conserved proteins are observed in all organisms from bacteria to animals and play a defense mechanism role for protection of cells against several harmful conditions. Stress influences hydration and redox state of the cell and increases misfolded protein levels. Structural alterations by this perturbation may lessen or diminish protein function since proteins perform their cellular function by keeping their three dimensional structure in native-properly folded state [1]. Therefore, cells employ HSP defense system to maintain intact protein structure.

The proteins have been classified into five families according to their molecular weight as HSP100, HSP90, HSP70, HSP60, and small HSPs. Small HSPs including HSP27, act in an ATP independent manner while high molecular weight HSPS are ATP dependent chaperones. HSPs can be categorized into two main groups; folder and holder in terms of substrate protein processing. Folder types HSPs function as molecular chaperones and bind to denatured-partially folded substrate proteins. This magic touch prevents stress induced aggregation and maintains substrate protein solubility by preventing exposure of substrate protein hydrophobic regions. Holder types HSPs are called small HSPs (sHSP). Several lines of evidence suggest that sHSPs activity does not require ATP energy. This is particularly important when the level of ATP is extremely low in cell under stress conditions [2]. sHSPs have a conserved C-terminal domain called α-crystalline (Fig. **1**). The α-crystalline Walker B type ATP binding motif and its enhanced chaperone activity in the presence of ATP contradict with sHsp ATP independent function [3-**7**].

sHSPs may interact without sequence and molecular weight specificity to a wide range of substrate proteins [8-11]. This interaction may "hold" the substrate proteins and prevent substrate protein aggregation. Once the stress is over, substrates are released from sHSPs and substrate proteins are processed according to their folded state [1, 12-17].

Soluble partly folded-molten globular proteins are refolded by ATP dependent chaperone network; HSP70-HSP40. However, large insoluble aggregates are processed by HSP100 proteins. HSP100 is a hexameric ATP-dependent chaperone and either breaks large aggregates into smaller ones or extracts a single peptide into its central pore. Then, Hsp70-Hsp40 complex further helps substrate protein folding process [18-22]. HSP27 keeps the hydrophobic regions of substrate proteins away from each other by patching these regions; however, when HSP27 cannot prevent protein aggregation itself, foldase-HSPs come into play to inhibit the aggregation.

1.1. Structure of HSP27

HSP27 belongs to sHsp family and shares a conserved C-terminal domain; the α-crystallin which is identical to the vertebrate eye lens α-crystallin (Fig. **1**). The α-crystallin domain consists of ß-sheets and these secondary structures play key roles to form stable dimers [23,24].

Following α-crystallin motif at the C-terminal, sHSPs have a polar sequence with high flexibility. This feature may regulate solubility of sHSPs and may provide stability to sHsp and sHsp-substrate protein complex [25].

HSP27 oligomers consist of stable dimers formed by two monomeric neighbor α-crystallin-domains. The dimers aggregate to form unstable oligomers (Fig. **2**) [24].

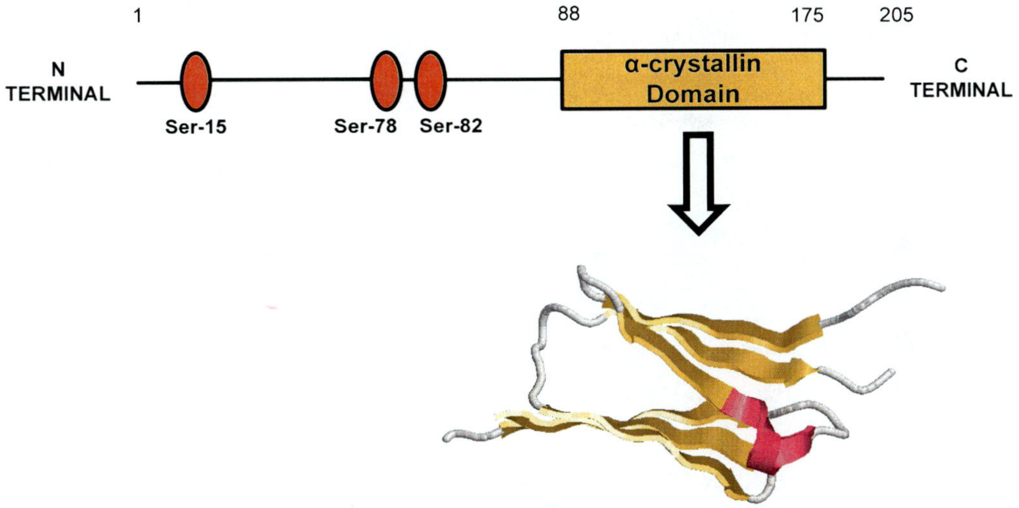

Figure 1: Schematic of the primary sequence (Top) and crystal structure of α-crystalline part (Bottom) of HspB1 (human analog of HSP27; PDB 3q9p). Figure also shows phosphorylation sites at 15, 72, and 82. The C-terminal part of the protein is conserved and consists of mostly ß-sheet structures. ß-sheet structures facilitate protein-protein interaction through increased surface area.

The sHSP family with a molecular range of 12-43 kDa has the smallest subunits; however, sHSPs form oligomeric complexes of 9-40 subunits *in vivo* [26,27].

Oligomerization of HSP27 is a dynamic process and is regulated by cellular conditions. There is a balance between dimers-tetramers and oligomers consisting 16 to 32 units as other sHSPs (Fig. **2**).

Heat shock up-regulates or induces HSP27 expression as a consequence of transcriptional activation. HSP27 transcriptional activation is mediated by heat shock factor (HSF-1). Trimerized HSF-1 binds heat shock elements (HSE) located in Hsp promoter region. This coordinately regulated synthesis accumulates HSPs during and after heat stress transiently. The crystal structure of HSP27 is difficult to obtain due to the hydrophobic nature of the protein. However, several studies revealed HSP27 structure-function relationships. Phosphorylations of HSP27 and cell stress status shift the balance between the dimer and the oligomer forms (Figs. **1**, **2**). Human HSP27 is phosphorylated at Ser-15, Ser-78, and Ser-82 residues. Stress triggers expression and phosphorylation of HSP27. Simulation of p38 MAP kinase by different agents leads activation of MAPKAP kinase 2/3 and this kinase phosphorylates HSP27 (Fig. **2A**, **2B**) [28].

Balance in favor of deoligomerization increases chaperone activity and monomeric HSP27 binds unfolded substrate proteins to keep them competent for refolding by HSP70 [29, 30].

Phosphorylation of HSP27 favors small subunit forms which enable its entrance to the nucleus. Function of nuclear localization under stress may simulate some key transcription factors to regulate stress response [31-33].

HSP27 also interferes with caspase activation at upstream of mitochondria at higher intracellular level and this stabilizes actin microfilaments under normal conditions. HSP27 achieves this by binding to F-actin and prevents disruption of the cytoskeleton. HSP27 cytoprotective effect to bind and stabilize actin polymerization is shown in Fig. **2** [34].

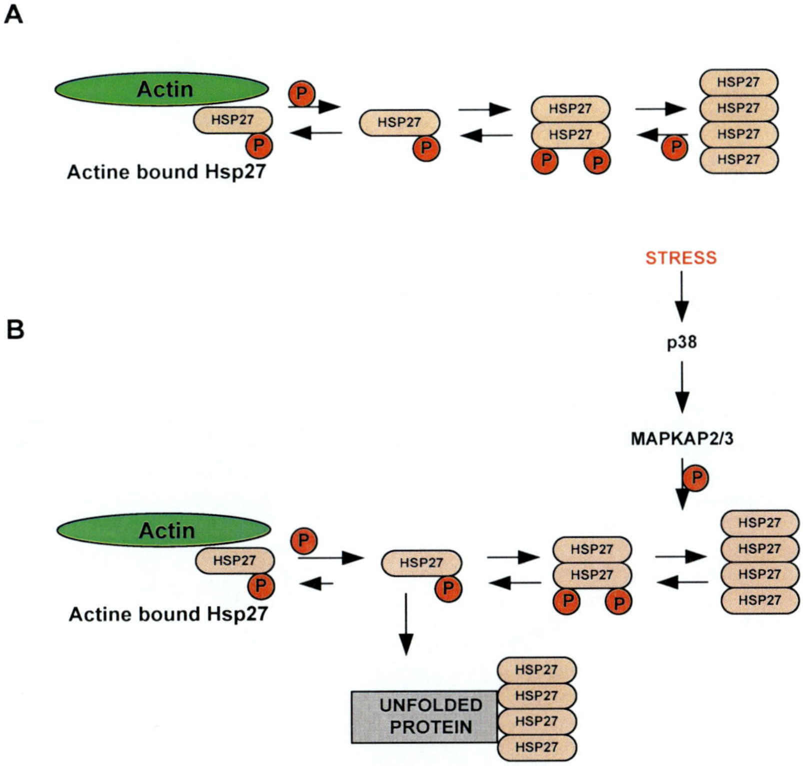

Figure 2: Balance between monomeric and oligomeric form of HSP27. HSP27 behavior at normal (A) and under stress conditions (B) (Redrawn from Ref. [29]).

1.2. Function

The main function of HSP27 is to provide thermotolerance and cell survival under stress. For these purposes, HSP27 inhibits protein aggregation in an ATP independent way and helps cell to overcome stress by preventing partially folded substrate protein interactions. This pre-defense mechanism by HSP27 facilitates irreversible denaturation of substrate proteins in the crowded cellular milieu with no energy cost. Then, Hsp70-Hsp40 complex comes into play to refold these substrate proteins into their functional-native structures (Fig. **3**).

Substrate protein binding to each unit-dimer in HSP27 oligomer is a cooperative process (Fig. **2**) [27, 35]. Substrate binding to HSP27 forms soluble complexes and HSP27 prevents toxic aggregation. However, HSP27 might form large stable insoluble aggregates with protein substrates when HSP27 is overloaded with the substrates. Then, HSP27 cooperates with HSP100 for subsequent resolubilization of the aggregate and refolding of the substrate (Fig. **3**) [1,16,36,37].

Recently identified function of HSP27 has its role in ubiquitin-proteasome system. If substrate proteins are irreversibly denatured then HSP27 activates the proteasome and increases the degradation of ubiquitinated proteins in response to stress signals. HSP27 binds ubiquitinated proteins and 26S proteasome to eliminate the denatured protein.

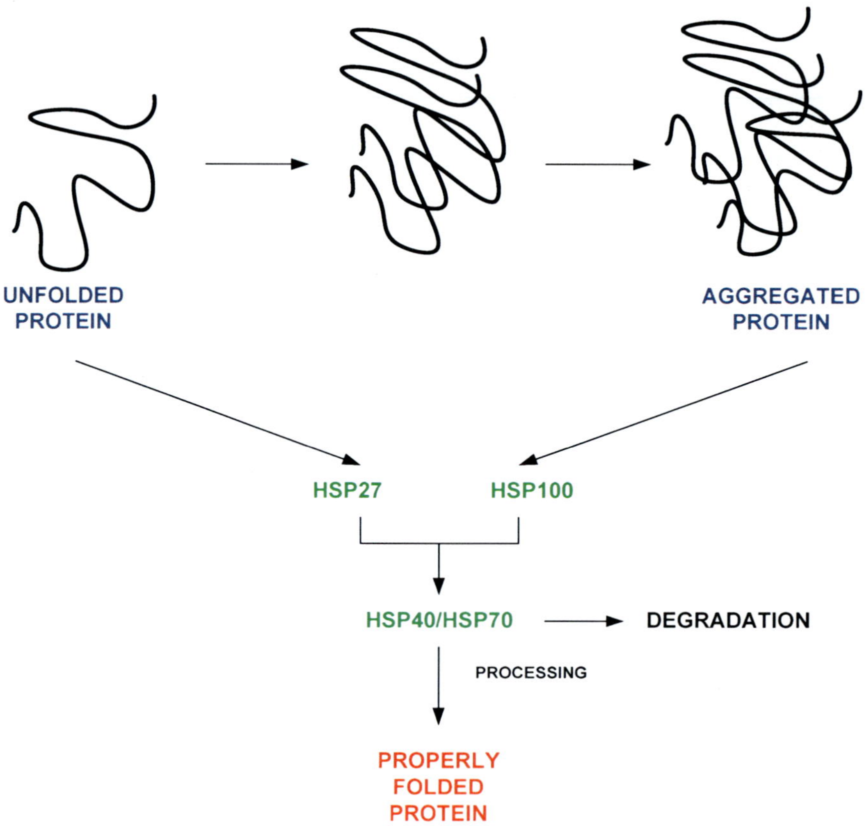

Figure 3: HSP27 in HSP folding-holding mechanism and its coordination with other HSPs.

HSP27 further enhances the activation of the NF-κB pathway. This biochemical process control cell growth, inflammatory, and stress responses [38].

HSP27 is constitutively expressed in many tissues and cell types especially in muscle cells. It is expressed in the cytosol, endoplasmic reticulum, and nucleus. It plays a role in differentiation, cellular proliferation, and development [39]. HSP27 is overexpressed in malignant cells and involved in oncogenesis and in chemotherapy resistance due to its apoptosis inhibition function [39, 40]. Hsps allow cells to adapt changes in their environment and help cells to survive at extreme conditions. In response to stress, induced HSPs take places at regulatory points in the control of apoptosis. Anti-apoptotic (HSP27, HSP70) and pro-apoptotic (HSP60, HSP10) HSPs interact with several cellular proteins in the apoptotic signaling pathway and their expression level determine cell survival in response to death signal. HSP27 helps cell survival and may participate in carcinogenesis. Therefore, in this review the role of HSP27 function in apoptosis and cancer will be discussed.

2. APOPTOSIS

Apoptosis is programmed cell death during development and adult homeostasis. Cytotoxic drugs use apoptosis biochemical pathway to eliminate tumor cells. There are two pathways in apoptotic mechanism: intrinsic-mitochondrial and extrinsic-death receptor pathway. There is also a sub route in the intrinsic pathway: caspase independent pathway (Fig. **4**).

Stress induces multiple pro-apoptotic signal cascades. Mitochondria coordinate these biochemical signals and culminates into apoptosis. The mitochondria transfer the signal by permeabilization of its outer membrane and this releases molecules from inter-membrane space to the cytosol. Released cytochrome *c* interacts with Apafl (apoptotic protease activation factor 1) in the cytosol and then, Apafl activates caspase 9 and caspase 3 respectively. This caspase activation cascade leads to apoptotic cell death. Beside the intrinsic pathway, AIF may use a caspase independent pathway by translocating itself to the nucleus and triggering apoptosis (Fig. **4**).

The extrinsic pathway involves plasma membrane surface death receptors such as FAS and TNFR1. The receptors interact with TRADD (TNF receptor death domain protein) or FADD (Fas associated death domain protein). The proteins interact with Caspase 8 and Caspase 3, and then, the cascade ends up in apoptosis.

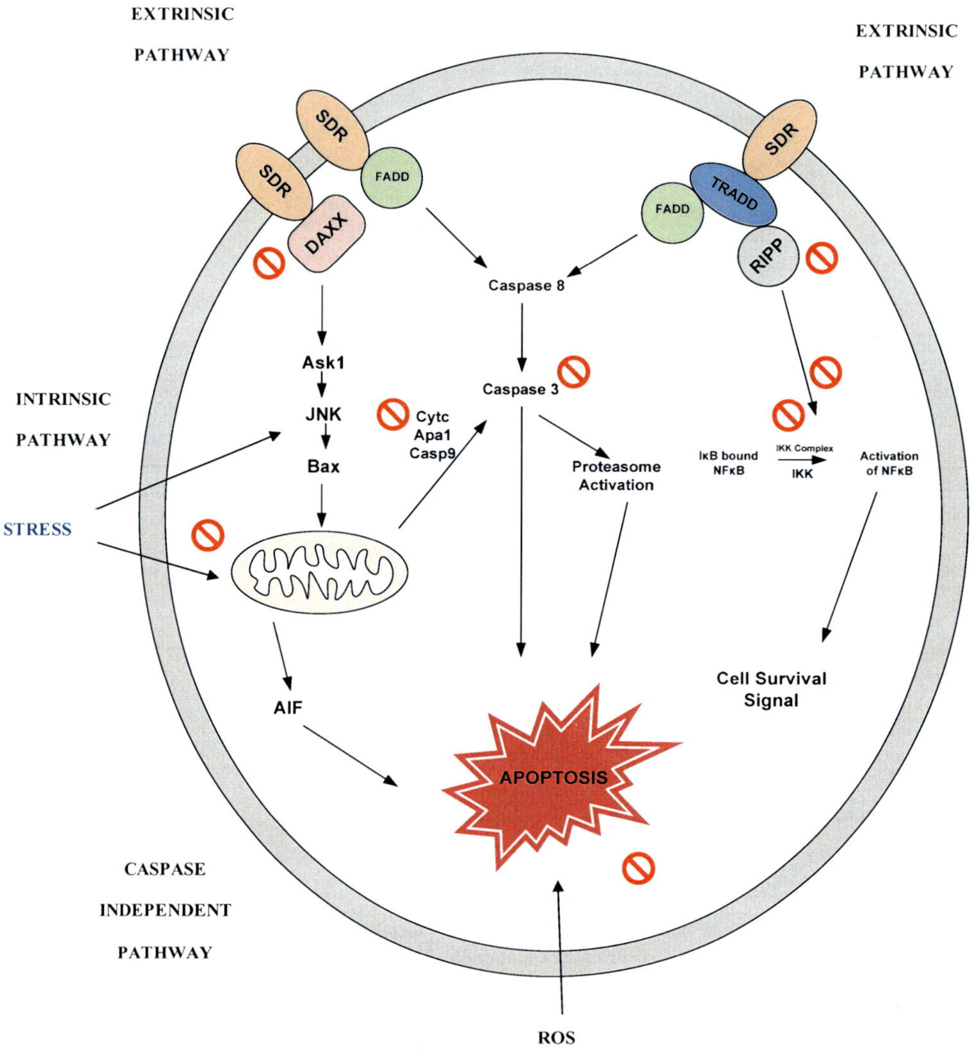

Figure 4: Role of HSP27 in the apoptosis regulation mechanism. Please see the text for details.

Fas induced apoptotic pathway may alternatively recruit the adaptor protein DAXX instead of FADD. DAXX activate ASK1 (Apoptosis Signal Regulated

Kinase 1) and ASK1 activates JNK and apoptosis, respectively. This part of Fas cascade links extrinsic death signals to the intrinsic pathway by the Bcl-2 family (Bax, Bid, Bad, and Bak) and anti-apoptotic (Bcl-XL) members [41-44].

Death induced signals are coupled with survival signals as part of metabolic regulation. Survival and growth factors signals activate a serine/threonine kinase; Akt phosphorylation stabilizes Akt and Receptor Interacting Protein (RIP) also helps this activation. This inhibits downstream apoptotic events and Nuclear Factor-κB (NF-κB) translocates into the nucleus to activate transcription of cell survival genes [44] and NF-κB mediates inhibition of apoptosis.

An exception to anti-apoptotic role of HSP27 is TNF-induced apoptosis where it inhibits IκB degradation and prevents NF-κB mediated cell survival and displays a pro-apoptotic role [45].

Similarly stress may trigger SEK kinase dependent phosphorylation status of JNK and this may lead to apoptosis [46].

3. ROLE OF HSP27 IN APOPTOSIS

HSP27 interacts with key components of the apoptotic pathway as shown in Fig. **4**. Reactive oxygen species production leads to release of pro-apoptotic factors and changes in the structure of mitochondria [47,48].

HSP27 can maintain redox homeostasis and mitochondrial integrity in the cell during stress response. Increase in HSP27 expression copes with cytotoxic stress. HSP27 prevents the assembly of the apoptosome (Cytc, Apaf-1, and Caspase 9) by sequestering cytochrome c from Apaf-1. Therefore, Hsp27 negatively regulates procaspase-9 activation [48, 49] and inhibits intrinsic pathway (Fig. **4**).

Stress dependent actin disruption is inhibited by HSP27 prevention of pro-apoptotic factors translocation onto the mitochondrial membrane and helps actin network integrity [50]. HSP27 also inhibits procaspase-3 activation and DAXX mediated apoptosis. Preventing DAXX translocation to the membrane inhibits interaction between Fas and ASK-1 [51, 52].

HSP27 displays anti-apoptotic properties other than its pro-apoptotic role in NF-κB mediated cell survival. In this pathway HSP27 inhibits IκB degradation and prevents cell survival signal [53].

Oxidative stress leads to reactive oxygen species (ROS) and ROS become cytotoxic at higher concentrations. Apoptosis is induced at moderate stress and necrosis occurs in drastic stress. In apoptosis, ROS deplete glutathione and this triggers the mitochondrial apoptotic pathway. Cell use a variety of defense mechanisms such as vitamins C, E, and the glutathione to protect themselves from ROS. HSP27 up-regulates glutathione level and helps cell survival [54].

4. HSP27 IN CANCER

HSP27 has well described anti-apoptotic activity and is overexpressed in variety of cancers [55]. Therefore, HSP27 was suggested to play an essential role in tumor progression and/or cellular transformation may modulate the basal level of this protein. As described in the previous section HSP27 controls several key points in the apoptotic pathway. Overexpression of HSP27 blocks apoptosis caused by different stimuli and this survival mechanism help tumor development through suppressing the default senescence program of the cell [39, 56, 57].

Two reports indicate HSP27 role in cell migration and in metastases; inhibition of HSP27 phosphorylation suppresses tumor cell migration-invasion and HSP27 regulates transcription of metalloproteases that are essential for metastases [55]. Thus, HSP27 is a versatile cytoprotective protein and may have key roles in malignant cells. There are lines of evidence to support HSP27 role in cancer. Furthermore, increased expression of HSP27 and HSP70 resist chemotherapy [58].

Several clinical studies indicated a positive correlation between HSP27 and the metastatic capacity of tumors [59].

An increase in the level of HSP27 was observed in breast cancer, leukemia, ovarian cancer, endometrial cancer, colorectal cancer, and osteosarcoma [60, 61]. Although molecular basis of this overexpression is not understood in detail, breast, endometrial, and gastric cancer have been associated with metastasis, poor prognosis, and resistance to the therapy. And only poor prognosis has been associated in case of osteosarcoma, prostate, liver, and ovarian cancer. Good prognosis is associated with esophageal cancer and malignant fibrous histiocytomas for HSP27. But HSP27 is not a universal marker for prognosis of head and neck, bladder, renal cancer, and leukemia [58-60, 62]

However, researchers try to elucidate biomarkers currently and HSP27 was identified as biomarker in hepatocellular carcinoma and glioblastoma [63,64].

5. HSP27 AS A PUTATIVE DRUG TARGET

HSP27 expression correlates with metastasis; hence, inhibition of this are potential targets for anti-cancer chemotherapy. However, these efforts did not hit an inhibitor but promising research is underway. Further, overexpression of HSP27 correlates with chemo-resistance in the cancer cells. Anti-apoptotic function of HSP27 and role in drug resistance properties of the protein give a new direction to pharmacological research.

Down regulation of HSP27 by siRNA in gemcitabine-resistant pancreatic cancer cells make the cells sensitive to the drug [65].

Quercetin, tanshinone, and triptolide were determined to inhibit HSP27 function directly or indirectly. Several chemical agents have been studied either alone or with combinational therapy by down regulating HSP27 expression for the treatment of cancer. This enables drug designers to inhibit anti-apoptotic property of HSP27 and to inhibit HSP27 role in the progression of cancer [65].

CONCLUDING REMARKS

This review shows that HSP27 is involved in essential biochemical processes to protect cells under unfavorable conditions. Hsp27 involvement in apoptosis, proteolysis, proliferation and cell motility process along with other macromolecular interactions need profound understanding.

HSP27 cytoprotective function and its anti-apoptotic property help induction and progression of tumor cells. Thus, HSP27 inhibition and down-regulation strategies can be used to develop potential anticancer therapy-drugs.

Studies with human tumor cell lines that are resistant to chemotherapy (head and neck squamous carcinoma, prostate, glioblastoma, and colorectal cancer) have shown induced apoptosis with a decrease in the HSP27 expression levels [66].

Similar apoptosis trends were observed in HSP27 expression and in down regulation using small interference RNA [67]. Development of drugs targeting HSP27 structure and its interaction is on the way and ongoing studies will help to broaden our knowledge about HSP27 and will contribute our understanding of the apoptosis and the cancer.

ACKNOWLEDGEMENTS

Dr. Tutar acknowledges Turkish National Academy of Science-Young Scientist Program (TUBA-GEBIP/28).

CONFLICT OF INTEREST

The authors confirm that this chapter contents have no conflict of interest.

DISCLOSURE

Declared None.

REFERENCES

[1] Tutar, L. and Tutar, Y. Heat shock proteins; an overview. *Curr. Pharm. Biotechnol.*, **2010**, 11(2), 216-222.

[2] Laskowska, E., Matuszewska, E., Kuczyńska-Wiśnik, D. Small heat shock proteins and protein-misfolding diseases. *Curr. Pharm. Biotechnol.*, **2010**, 11(2), 146-157.

[3] Muchowski, P.J. and Clark, J.I. ATP-enhanced molecular chaperone functions of the small heat shock protein human alphaB crystallin. *Proc. Natl. Acad. Sci. U S A.*, **1998**, 95(3), 1004-1009.

[4] Wang, K. and Spector, A. alpha-crystallin prevents irreversible protein denaturation and acts cooperatively with other heat-shock proteins to renature the stabilized partially denatured protein in an ATP-dependent manner. *Eur. J. Biochem.*, **2000**, 267(15), 4705-4712.

[5] Biswas, A. and Das, K.P. Role of ATP on the interaction of alpha-crystallin with its substrates and its implications for the molecular chaperone function. *J. Biol. Chem.*, **2004**, 279(41), 42648-42657.

[6] Barton, K.A., Hsu, C.D. and Petrash, J.M. Interactions between small heat shock protein alpha-crystallin and galectin-related interfiber protein (GRIFIN) in the ocular lens. *Biochemistry*, **2009**, 48(18), 3956-3966.

[7] Ghosh, J.G., Houck, S.A., Doneanu, C.E. and Clark, J.I. The beta4-beta8 groove is an ATP-interactive site in the alpha crystallin core domain of the small heat shock protein, human alphaB crystallin. *J. Mol. Biol.*, **2006**, 364(3), 364-375.

[8] Lindner, R.A., Kapur, A. and Carver, J.A. The interaction of the molecular chaperone, alpha-crystallin, with molten globule states of bovine alpha-lactalbumin. *J. Biol. Chem.*, **1997**, 272(44), 27722-27729.

[9] Carver, J.A., Lindner, R.A., Lyon, C., Canet, D., Hernandez, H., Dobson, C.M. and Redfield, C. The interaction of the molecular chaperone alpha-crystallin with unfolding alpha-lactalbumin: a structural and kinetic spectroscopic study. *J. Mol. Biol.*, **2002**, 318(3), 815-827.

[10] Kumar, M.S., Reddy, P.Y., Sreedhar, B. and Reddy, G.B. Alphab-crystallin-assisted reactivation of glucose-6-phosphate dehydrogenase upon refolding. *Biochem. J.*, **2005**, 391, 335-341.

[11] McHaourab, H.S., Godar, J.A. and Stewart, P.L. Structure and mechanism of protein stability sensors: chaperone activity of small heat shock proteins. *Biochemistry*, **2009**, 48(18), 3828-3837.

[12] Ehrnsperger, M., Gräber, S., Gaestel, M. and Buchner, J. Binding of non-native protein to Hsp25 during heat shock creates a reservoir of folding intermediates for reactivation. *EMBO J.*, **1997**, 16(2), 221-229.

[13] Veinger, L., Diamant, S., Buchner, J. and Goloubinoff, P. The small heat-shock protein IbpB from Escherichia coli stabilizes stress-denatured proteins for subsequent refolding by a multichaperone network. *J. Biol. Chem.*, **1998**, 273(18), 11032-11037.

[14] Lee, G.J. and Vierling, E. A small heat shock protein cooperates with heat shock protein 70 systems to reactivate a heat-denatured protein. *Plant Physiol.*, **2000**, 122(1), 189-198.

[15] Matuszewska, M., Kuczyńska-Wiśnik, D., Laskowska, E. and Liberek, K. The small heat shock protein IbpA of Escherichia coli cooperates with IbpB in stabilization of thermally aggregated proteins in a disaggregation competent state. *J. Biol. Chem.*, **2005**, 280(13), 12292-12298.

[16] Mogk, A., Schlieker, C., Friedrich, K.L., Schönfeld, H.J., Vierling, E. and Bukau, B. Refolding of substrates bound to small Hsps relies on a disaggregation reaction mediated most efficiently by ClpB/DnaK. *J. Biol. Chem.*, **2003**, 278(33), 31033-31042.

[17] Haslbeck, M., Miess, A., Stromer, T., Walter, S. and Buchner, J. Disassembling protein aggregates in the yeast cytosol. The cooperation of Hsp26 with Ssa1 and Hsp104. *J. Biol. Chem.*, **2005**, 280(25), 23861-23868.

[18] Glover, J.R. and Lum, R. Remodeling of protein aggregates by Hsp104. *Protein Pept. Lett.*, **2009**, 16(6), 587-597.

[19] Romanova, N.V. and Chernoff, Y.O. Hsp104 and prion propagation. *Protein Pept. Lett.*, **2009**, 16(6), 598-605.

[20] Glover, J.R. and Lindquist, S. Hsp104, Hsp70, and Hsp40: a novel chaperone system that rescues previously aggregated proteins. *Cell*, **1998**, 94(1), 73-82.

[21] Weibezahn, J., Schlieker, C., Tessarz, P., Mogk, A. and Bukau, B. Novel insights into the mechanism of chaperone-assisted protein disaggregation. *Biol. Chem.*, **2005**, 386(8), 739-744.

[22] Bösl, B., Grimminger, V. and Walter, S. The molecular chaperone Hsp104--a molecular machine for protein disaggregation. *J. Struct. Biol.*, **2006**, 156(1), 139-148.

[23] Kim, K.K., Kim, R. and Kim, S.H. Crystal structure of a small heat-shock protein. *Nature*, **1998**, 394, 595-599.

[24] Van Montfort, R., Slingsby, C. and Vierling, E. Structure and function of the small heat shock protein/alpha-crystallin family of molecular chaperones. *Adv. Protein Chem.*, **2001**, 59, 105-156.

[25] Lelj-Garolla, B. and Mauk, A.G. Self-association of a small heat shock protein. *J. Mol. Biol.*, **2005**, 345(3), 631-642.

[26] Narberhaus, F. Alpha-crystallin-type heat shock proteins: socializing minichaperones in the context of a multichaperone network. *Microbiol. Mol. Biol. Rev.*, **2002**, 66(1), 64-93.

[27] Kocabiyik, S. Essential structural and functional features of small heat shock proteins in molecular chaperoning process. *Protein Pept. Lett.*, **2009**, 16(6), 613-622.

[28] Rogalla, T., Ehrnsperger, M., Preville, X., Kotlyarov, A., Lutsch, G., Ducasse, C., Paul, C., Wieske, M., Arrigo, A.P., Buchner, J. and Gaestel, M. Regulation of Hsp27 oligomerization, chaperone function, and protective activity against oxidative stress/tumor necrosis factor alpha by phosphorylation. *J. Biol. Chem.*, **1999**, 274(27), 18947-18956.

[29] Doshi, B.M., Hightower, L.E. and Lee, J. HSPB1, actin filament dynamics, and aging cells. *Ann. N. Y. Acad. Sci.*, **2010**, 1197, 76-84.

[30] Ehrnsperger, M., Lilie, H., Gaestel, M. and Buchner, J. The dynamics of Hsp25 quaternary structure. Structure and function of different oligomeric species. *J. Biol. Chem.*, **1999**, 274(21), 14867-14874.

[31] de Thonel, A., Vandekerckhove, J., Lanneau, D., Selvakumar, S., Courtois, G., Hazoume, A., Brunet, M., Maurel, S., Hammann, A., Ribeil, J.A., Zermati, Y., Gabet, A.S., Boyes, J., Solary, E., Hermine, O. and Garrido, C. HSP27 controls GATA-1 protein level during erythroid cell differentiation. *Blood*, **2010**, 116(1), 85-96.

[32] Brunet Simioni, M., De Thonel, A., Hammann, A., Joly, A.L., Bossis, G., Fourmaux, E., Bouchot, A., Landry, J., Piechaczyk, M. and Garrido, C. *Oncogene*, **2009**, 28(37), 3332-3344.

[33] Mjahed, H., Girodon, F., Fontenay, M. and Garrido, C. Heat shock proteins in hematopoietic malignancies. *Exp. Cell Res.*, **2012**, 318(15), 1946-1958.

[34] Lavoie, J.N., Lambert, H., Hickey, E., Weber, L.A. and Landry, J. Modulation of cellular thermoresistance and actin filament stability accompanies phosphorylation-induced changes in the oligomeric structure of heat shock protein 27. *Mol. Cell Biol.*, **1995**, 15(1), 505-516.

[35] Bhattacharyya, J., Padmanabha Udupa, *E.G.*, Wang, J. and Sharma, K.K. *Biochemistry*, **2006**, 45(9), 3069-3076.

[36] Jiao, W., Li, P., Zhang, J., Zhang, H. and Chang, Z. Small heat-shock proteins function in the insoluble protein complex. *Biochem. Biophys. Res. Commun.*, **2005**, 335(1), 227-231.

[37] Jego, G., Hazoumé, A., Seigneuric, R. and Garrido, C. Targeting heat shock proteins in cancer. Cancer Lett., 2013, 332(2), 275-285.

[38] Parcellier, A., Schmitt, E., Gurbuxani, S., Seigneurin-Berny, D., Pance, A., Chantôme, A., Plenchette, S., Khochbin, S., Solary, E. and Garrido, C. HSP27 is a ubiquitin-binding protein involved in I-kappaBalpha proteasomal degradation. *Mol. Cell Biol.*, **2003**, 23(16), 5790-5802.

[39] Garrido, C., Schmitt, E., Candé, C., Vahsen, N., Parcellier, A. and Kroemer, G. HSP27 and HSP70: potentially oncogenic apoptosis inhibitors. *Cell Cycle*, **2003**, 2(6), 579-584.

[40] Parcellier, A., Gurbuxani, S., Schmitt, E., Solary, E. and Garrido, C. Heat shock proteins, cellular chaperones that modulate mitochondrial cell death pathways. *Biochem Biophys Res Commun.*, **2003**, 304(3), 505-512.

[41] Chang, H.Y., Nishitoh, H., Yang, X., Ichijo, H. and Baltimore, D. Activation of apoptosis signal-regulating kinase 1 (ASK1) by the adapter protein Daxx. *Science*, **1998**, 281(5384), 1860-1863.

[42] Yang, X., Khosravi-Far, R., Chang, H.Y. and Baltimore, D. Daxx, a novel Fas-binding protein that activates JNK and apoptosis. *Cell*, **1997**, 89(7), 1067-1076.

[43] Gross, A., McDonnell, J.M. and Korsmeyer, S.J. BCL-2 family members and the mitochondria in apoptosis. *Genes Dev.*, **1999**, 13(15), 1899-1911.

[44] Arya, R., Mallik, M. and Lakhotia, S.C. Heat shock genes - integrating cell survival and death. *J. Biosci.*, **2007**, 32(3), 595-610.

[45] Kammanadiminti, S.J. and Chadee, K. Suppression of NF-kappaB activation by Entamoeba histolytica in intestinal epithelial cells is mediated by heat shock protein 27. *J. Biol. Chem.*, **2006**, 281(36), 26112-26120.

[46] Mosser, D.D., Caron, A.W., Bourget, L., Meriin, A.B., Sherman, M.Y., Morimoto, R.I. and Massie, B. The chaperone function of hsp70 is required for protection against stress-induced apoptosis. *Mol. Cell Biol.*, **2000**, 20(19), 7146-7159.

[47] Concannon, C.G., Gorman, A.M. and Samali, A. On the role of Hsp27 in regulating apoptosis. *Apoptosis*, **2003**, 8(1), 61-70.

[48] Concannon, C.G., Orrenius, S. and Samali, A. Hsp27 inhibits cytochrome c-mediated caspase activation by sequestering both pro-caspase-3 and cytochrome c. *Gene Expr.*, **2001**, 9(4-5), 195-201.

[49] Bruey, J.M., Ducasse, C., Bonniaud, P., Ravagnan, L., Susin, S.A., Diaz-Latoud, C., Gurbuxani, S., Arrigo, A.P., Kroemer, G., Solary, E. and Garrido, C. Hsp27 negatively regulates cell death by interacting with cytochrome c. *Nat. Cell Biol.*, **2000**, 2(9), 645-652.

[50] Paul, C., Manero, F., Gonin, S., Kretz-Remy, C., Virot, S. and Arrigo, A.P. Hsp27 as a negative regulator of cytochrome C release. *Mol. Cell Biol.*, **2002**, 22(3), 816-834.

[51] Pandey, P., Farber, R., Nakazawa, A., Kumar, S., Bharti, A., Nalin, C., Weichselbaum, R., Kufe, D. and Kharbanda, S. Hsp27 functions as a negative regulator of cytochrome c-dependent activation of procaspase-3. *Oncogene.*, 2000, 19(16), 1975-1981.

[52] Charette, S.J., Lavoie, J.N., Lambert, H. and Landry, J. Inhibition of Daxx-mediated apoptosis by heat shock protein 27. *Mol. Cell Biol.*, **2000**, 20(20), 7602-7612.

[53] Hall, A.G. Review: The role of glutathione in the regulation of apoptosis. *Eur. J. Clin. Invest.*, **1999**, 29(3), 238-245.

[54] Hall, A.G. Review: The role of glutathione in the regulation of apoptosis. *Eur. J. Clin. Invest.*, **1999**, 29(3), 238-245.

[55] Sherman, M. and Multhoff, G. Heat shock proteins in cancer. *Ann. N. Y. Acad. Sci.*, **2007**, 1113, 192-201.

[56] Jäättelä, M. Escaping cell death: survival proteins in cancer. *Exp. Cell Res.*, **1999**, 248(1), 30-43.

[57] Calderwood, S.K., Khaleque, M.A., Sawyer, D.B. and Ciocca, D.R. Heat shock proteins in cancer: chaperones of tumorigenesis. *Trends Biochem. Sci.*, **2006**, 31(3), 164-172.

[58] Vargas-Roig, L.M., Gago, F.E., Tello, O., Aznar, J.C. and Ciocca, D.R. Heat shock protein expression and drug resistance in breast cancer patients treated with induction chemotherapy. *Int. J. Cancer*, **1998**, 79(5), 468-475.

[59] Ciocca, D.R. and Calderwood, S.K. Heat shock proteins in cancer: diagnostic, prognostic, predictive, and treatment implications. *Cell Stress Chaperones*, **2005**, 10(2), 86-103.

[60] Garrido, C., Ottavi, P., Fromentin, A., Hammann, A., Arrigo, A.P., Chauffert, B. and Mehlen, P. HSP27 as a mediator of confluence-dependent resistance to cell death induced by anticancer drugs. *Cancer Res.*, **1997**, 57(13), 2661-2667.

[61] Khalil, A.A., Kabapy, N.F., Deraz, S.F. and Smith, C. Heat shock proteins in oncology: diagnostic biomarkers or therapeutic targets? *Biochim. Biophys. Acta.*, **2011**, 1816(2), 89-104.

[62] Parcellier, A., Schmitt, E., Brunet, M., Hammann, A., Solary, E. and Garrido, C. Small heat shock proteins HSP27 and alphaB-crystallin: cytoprotective and oncogenic functions. *Antioxid. Redox Signal*, **2005**, 7(3-4), 404-413.

[63] Lee, I.N., Chen, C.H., Sheu, J.C., Lee, H.S., Huang, G.T., Yu, C.Y., Lu, F.J. and Chow, L.P. Identification of human hepatocellular carcinoma-related biomarkers by two-dimensional difference gel electrophoresis and mass spectrometry. *J. Proteome Res.*, **2005**, 4(6), 2062-2069.

[64] Khalil, A.A. Biomarker discovery: a proteomic approach for brain cancer profiling. *Cancer Sci.*, **2007**, 98(2), 201-213.

[65] Kuramitsu, Y., Wang, Y., Taba, K., Suenaga, S., Ryozawa, S., Kaino, S., Sakaida, I. and Nakamura, K. Heat-shock protein 27 plays the key role in gemcitabine-resistance of pancreatic cancer cells. *Anticancer Res.*, **2012**, 32(6), 2295-2299.

[66] Acunzo, J., Katsogiannou, M. and Rocchi, P. Small heat shock proteins HSP27 (HspB1), αB-crystallin (HspB5) and HSP22 (HspB8) as regulators of cell death. *Int. J. Biochem. Cell. Biol.*, **2012**, 44(10), 1622-1631.

[67] Sanchez-Niño, M.D., Sanz, A.B., Sanchez-Lopez, E., Ruiz-Ortega, M., Benito-Martin, A., Saleem, M.A., Mathieson, P.W., Mezzano, S., Egido, J. and Ortiz, A. HSP27/HSPB1 as an adaptive podocyte antiapoptotic protein activated by high glucose and angiotensin II. *Lab. Invest.*, **2012**, 92(1), 32-45.

Send Orders for Reprints to reprints@benthamscience.net

Frontiers in Protein and Peptide Sciences, Vol. 1, **2014, 91-138** **91**

CHAPTER 5

Understanding the Folding Rates and Folding Nuclei of Globular Proteins

Alexei V. Finkelstein[1,2,*], Dmitry N. Ivankov[2,3], Sergiy O. Garbuzynskiy[1] and Oxana V. Galzitskaya[1]

[1]Institute of Protein Research, Russian Academy of Sciences, Pushchino, Moscow Region, 142290, Russian Federation; [2]Pushchino Branch of the Moscow State Lomonosov University; 142290 Pushchino, Moscow Region, Russian Federation and [3]Department of Genome Oriented Bioinformatics, Technische Universitaet Muenchen, Wissenschaftszentrum Weihenstephan, 85354 Freising, Germany

Abstract: The first part of this paper contains an overview of protein structures, their spontaneous formation ("folding"), and the thermodynamic and kinetic aspects of this phenomenon, as revealed by *in vitro* and *in vivo* experiments. It is stressed that universal features of the *in vitro* folding are observed near the point of thermodynamic equilibrium between the native and denatured states of the protein. Here the "two-state" ("denatured state" ↔ "native state") transition proceeds without accumulation of metastable intermediates; this facilitates investigation of the "transition state". This state, which is the most unstable in the folding pathway, and its structured core (a "nucleus") are distinguished by their essential influence on the folding/unfolding kinetics. In the second part of the paper, a theory of protein folding rates and related phenomena is presented. First, it is shown that the protein size determines the range of a protein's folding rates in the vicinity of the point of thermodynamic equilibrium between the native and denatured states of the protein. Then, we present methods for calculating folding and unfolding rates of globular proteins from their sizes, stabilities and either 3D structures or amino acid sequences. Finally, we show that the same theory outlines the location of the protein folding nucleus (*i.e.*, the structured part of the transition state) in reasonable agreement with experimental data.

Keywords: Native protein structure, denatured protein, spontaneous protein folding, folding intermediates, transition state, folding nucleus, folding rate, Levinthal paradox, unfolding rate, mid-transition.

***Address correspondence to Alexei V. Finkelstein:** Institute of Protein Research, Russian Academy of Sciences, Pushchino, Moscow Region, 142290, Russian Federation; Tel/Fax: +7-495-514-0218;
E-mail: afinkel@vega.protres.ru

Ben M. Dunn (Ed)

1. INTRODUCTION

The aim of this paper is to outline the modern understanding of the physical principles of protein structure self-organization.

A protein is a heteropolymer built of amino acid residues. It has a chemically regular backbone chain and a unique (for each protein) sequence of 20 kinds of side groups; sometimes, it also includes "cofactors", which are usually small molecules, and chemical modification of some amino acids [1].

Before considering protein physics, it is useful to remind a reader that proteins exist under various environmental conditions which leave an obvious mark on their structures. Roughly, according to the "environmental conditions" and general structure, proteins can be divided into three large groups [2].

Fibrous proteins form vast aggregates; their structure is usually maintained mainly by interactions between different polypeptide chains.

Membrane proteins exist in water-lacking membranes. Their intramembrane portions have highly regular three-dimensional (3D) structure (like fibrous proteins) but are restricted in size by the membrane thickness (30–40Å).

Water-soluble globular proteins that live in aqueous environments are the most numerous and well studied.

In this review we will deal only with globular proteins. Moreover, we will concentrate mostly on relatively small, "single-domain" proteins that form one compact protein globule. A "single-domain" structure is typical of small, water-soluble globular proteins. Large proteins usually consist of two, three or even more domains [2,3].

The chain of an "operating" protein is usually folded into a "*native*" 3D structure, which is strictly specified except for small fluctuations and (sometimes) small rearrangements, and which exists under normal biological conditions but decays under the action of various denaturants, such as temperature, acid, some chemicals like urea, *etc.* Some (~10%) protein chains, however, have no fixed structure by themselves (these are so-called "disordered" or "natively unfolded" proteins), but they usually obtain it by interacting with other molecules [4,5].

Protein physics is grounded on three fundamental experimental facts: (i) many proteins have well defined (except for small fluctuations) three-dimensional structures [6-8]; (ii) many protein chains are capable of self-organization, *i.e.*, they form their native structures spontaneously in an appropriate environment [9,10], and (iii) the native state of many proteins is separated from the unfolded state of the chain by an "all-or-none" phase transition [11]. The last point ensures the robustness of protein action: as a bulb, the protein either has a correct structure and works correctly, or does not work at all.

2. PHASE TRANSITIONS IN PROTEIN MOLECULES

2.1. Reversible Denaturation of Protein Structures

Under biological conditions the native structure of a protein is rather "solid". However, depending on ambient conditions, the most stable state of a protein molecule may be not solid but molten or even extremely swollen, "unfolded": then the protein "denatures" and loses its native, "working" 3D structure.

Denaturation of a protein is usually reversible: many protein chains are capable of self-organization (usually called "renaturation" or "folding"), *i.e.*, they form their native structures *spontaneously* in an appropriate environment [9,10]. This spontaneous self-organization not only happens to natural proteins produced by a living organism [9], but also to proteins whose amino acid sequences (copied from those of natural proteins) have been synthesized chemically in a test tube [10].

Folding as a physical phenomenon has been known since Anfinsen's famous experiments in the 1960's [9]. Since then, spontaneous folding has been observed for hundreds of proteins [12]. It is now known that a protein can renature (when the ambient conditions return to "physiological" ones) if it is not too large and has not been subjected to substantial chemical modifications after the initial *in vivo* folding (and if the protein solution is sufficiently diluted to avoid aggregation). In this case, a "mild" (without chemical decay) destruction of the protein's native structure (by temperature, denaturant, *etc.*) is reversible, and the native structure is spontaneously restored after environmental conditions have become normal.

Usually protein denaturation and renaturation are studied *in vitro* (in a test tube); then denaturation is caused by an abnormal, "non-physiological" temperature or by an excess of a denaturant (like H^+, OH^- or urea). However, decay of the "solid" protein structure (and its subsequent refolding) can also occur in a living cell, *e.g.*, during trans-membrane transport of proteins.

The spontaneous folding, *i.e.*, the reversibility of protein denaturation, shows that the entire information necessary to build up the protein 3D structure is contained in its amino acid sequence, and that the protein structure itself (to be more exact, the structure of a protein that is not too modified and not too large) is thermodynamically stable. This allows one to use thermodynamics to study and describe de- and renaturation transitions.

Denaturation and folding of water-soluble globular proteins is the most studied, and we will speak about them only.

It is well established that denaturation of small native proteins is a cooperative transition with a simultaneous, abrupt change of various characteristics of the molecule. Narrow transition regions suggest that the transition embraces many amino acid residues.

Moreover, denaturation of a single-domain protein occurs as an "all-or-none" transition [11,13]. The latter means that the transition embraces the domain as a whole, and that only the initial (native) and the final (denatured) states amount to visible quantities, while "semi-denatured" states are unstable and practically absent. (Though, of course, they do exist to a very small extent, since a native molecule cannot come to its denatured state without passing the intermediate forms, and their presence has a crucial effect on kinetics of the transition, which we will discuss below).

The "all-or-none" transition is a microscopic analog of the first-order phase transitions in macroscopic systems (*e.g.*, crystal melting). However, unlike the true phase transitions, the "all-or-none" transitions in proteins have a non-zero temperature width, since this transition embraces a microscopic system. It should be specified that the "all-or-none" denaturation actually concerns small proteins and separate domains of large proteins, while denaturation of a large protein is a sum of denaturations of its domains [14,15].

To prove that melting is an "all-or-none" transition, one has to compare (1) "effective latent heat" of transition calculated from its width (*i.e.*, the amount of heat consumed by one independent melting unit) with (2) "calorimetric heat" of this transition, *i.e.*, the amount of heat consumed by one melting protein molecule [11]. Denaturation of proteins is usually accompanied by a large heat effect, ~1 kcal/mol of amino acid residues; however, the native state of a protein chain is more stable than its unfolded state by no more than a few kcal/mol even under physiological conditions, where the native state is the most stable.

2.2. What do Denatured Proteins Look Like?

Numerous *thermodynamic* experiments have shown that there are no abrupt cooperative transitions within the denatured state of a protein molecule. Therefore, it was initially assumed that the denatured protein is always a very loose random coil (as it is in a "very good" solvent like concentrated solution of urea [12,16-18]).

However, *structural* studies of denatured proteins reported on some large-scale rearrangements within the denatured state, that is, on some stable "intermediates" between the completely unfolded coil and the native state of proteins (Fig. **1**).

Apart from biochemical activity, only two properties always abruptly change during denaturation. These are: (1) the ordering of the environment of aromatic side chains observed by near-UV CD and by NMR, and (2) the rigidity of the globular structure followed (using NMR) by exchange of hydrogens (H) of the protein's polar groups for deuteriums (D) of water, and by acceleration of the protein chain proteolysis [19-21].

A protein chain looks like a random coil in a "very good" solvent, such as a concentrated solution of urea or GuHCl; then, its hydrodynamic volume is proportional to the chain length to the power 3/2 or a little higher. However, in poor solvents (*e.g.*, water) the volume of the denatured globule is often only a little larger than the volume of the native protein. Thus, the above studies have revealed an intermediate of protein unfolding and folding, which is now known as the "*molten globule*" [21] (Fig. **2**).

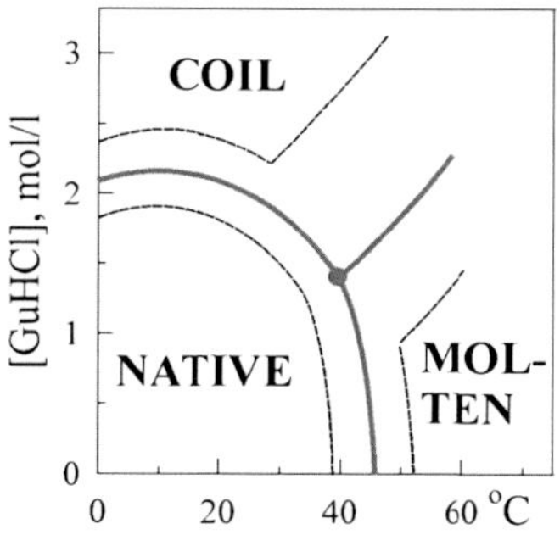

Figure 1: Phase diagram of conformational states (at pH 1.7) of a single-domain protein lysozyme at various temperatures in solution of guanidine hydrochloride (GuHCl) denaturant of various concentrations: the solid NATIVE state, the completely unfolded COIL, and a more compact temperature-denatured state (MOLTEN). The solid line corresponds to the mid-transition, the dashed lines outline the transition zones (from the proportion ≈9:1 in favor of one state to ≈1:9 in favor of another). One can see that the COIL–MOLTEN transition is much wider than the others. Adapted from [18]. Such phase diagram is common for other denaturants and proteins.

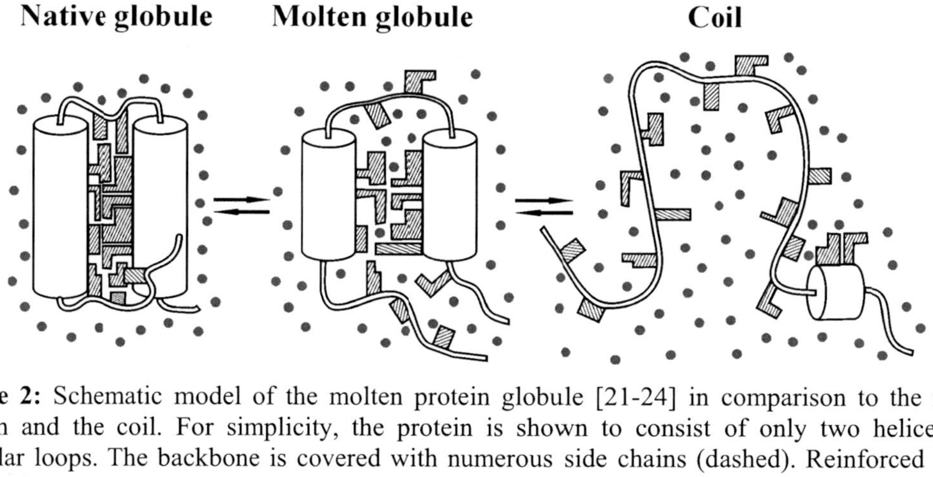

Figure 2: Schematic model of the molten protein globule [21-24] in comparison to the native protein and the coil. For simplicity, the protein is shown to consist of only two helices and irregular loops. The backbone is covered with numerous side chains (dashed). Reinforced by H-bonds, the secondary structures are stable until the globule is "dissolved" by a solvent. Usually, water molecules are unable to do this without a strong denaturant. In the molten globule, side chains lose their close packing, but acquire freedom of movement, *i.e.*, they lose energy but gain entropy. Water molecules (•) move into pores of the molten globule (that appear when the close packing is lost), but, until the denaturant is not too strong, cause no further decay of the globule; a stronger denaturant converts the globule into the coil.

For very many (though not for all) proteins, the "molten globule" arises from a moderate denaturing impact upon the native protein, and decays (turns into a random coil) only under the impact of a concentrated denaturant. The molten globule-like state often occurs after temperature denaturation ("melting"), and this melting has been always observed to be the "all-or-none" transition. The molten globule usually does not undergo cooperative melting with increasing temperature (see Fig. **1**), but its unfolding caused by a strong denaturant looks like a cooperative S-shaped transition.

However, some proteins (especially small ones) unfold directly into a coil without the mediating molten globule state; and many other proteins are converted into the molten globule by some denaturing agents (*e.g.*, by temperature or by acid), while other agents (*e.g.*, urea) directly convert them into a coil. In all these cases, though, the decay of the native state is "all-or-none" transition between the native state and the molten globule (or coil) state, *i.e.*, *via* the first order phase transition between these states [21].

It has been shown that the "all-or-none" folding/unfolding phase transition requires a selected amino acid sequence that only provides a large energy gap between the native and all the other, structurally dissimilar conformations of the protein chain [2,25-27]. As a result of "all-or-none" transition, the protein tolerates, without a change, modification of ambient conditions up to a certain limit, and then melts altogether, like a solid body. This resistance and hardness of the protein, in turn, provides reliability of its biological function, and therefore must be maintained by biological evolution [2].

2.3. Protein Folding *In Vivo*

In a living cell, a protein is synthesized by a ribosome that makes a protein chain (whose sequence is encoded by mRNA) residue by residue, from its N- to C-end, and not quite uniformly: there are temporary rests of the synthesis at the "rare" codons (they correspond to tRNAs which are rare in the cell, and these codons are rare in the cell's mRNAs, too). It is assumed that the pauses may correspond to the boundaries of structural domains that can help a quiet maturation of the domain structures. The biosynthesis takes about a minute and yielding of a "ready" folded protein lasts just as long: the experiment does not see any difference [1,28].

Some enzymes, like prolyl-peptide- or disulfide-isomerases accelerate *in vivo* folding. They catalyze slow, if unaided, *trans*↔*cis* conversions of prolines and the formation (and decay) of chemical S-S bonds between cysteine amino acid residues.

Protein chains fold under the protection of special proteins, called chaperons. These are the cell's trouble-shooters that fight aggregation, since, in a cell, folding takes place in a highly crowded molecular environment. There is no reason to assume, though, that anything other than the amino acid sequence determines protein conformation in the cell [29,30].

It looks as though the biosynthetic machinery (ribosomes + chaperones + ...), besides synthesizing the protein chain, serves only as a kind of incubator, which does not determine the protein structure (at least if the protein is not very large and does not consist of many domains) but rather provides "hothouse" conditions for its maturation, – just like a usual incubator helps a nestling to develop but does not determine what will be developed, a chicken or a duckling.

Unfortunately, it is difficult to follow the *in vivo* folding of a nascent protein chain against the background of the huge ribosome. It is known, though, that the first synthesized domains of multi-domain proteins are able to fold before the biosynthesis of the whole chain is completed [1,30]. Besides, some recent NMR experiments [31] showed that a truncated (*i.e.*, not completely synthesized) chain of a single-domain protein formed no definite structure either in attachment to the ribosome *in vivo*, or when detached from it *in vitro* (although the formation of a molten globule is not ruled out in both cases) The same was true for each domain of a two-domain protein [32]. These experiments showed that there is no principal difference between *in vivo* and *in vitro* folding processes.

However, there is virtually no data on kinetics of the *in vivo* folding events for single-domain proteins. Therefore, most experiments on protein folding are done *in vitro*.

2.4. Protein Folding *In Vitro*

In about 1960, a remarkable discovery was made: it was shown that a globular protein is capable of spontaneous folding *in vitro* [9]. If a protein chain has not been heavily chemically modified after the initial (*in vivo*) folding, and is then gently (without chain damaging) unfolded by temperature, denaturant, *etc.*, the protein spontaneously "renatures", *i.e.*, restores its activity and structure after solvent "normalization". True, effective renaturation requires a careful selection of experimental conditions; otherwise, aggregation (including famous amyloid formation [33,34]) can prevent the protein from folding.

Furthermore, it was demonstrated [10] that the protein chain which had been synthesized chemically, without any cell or ribosome, and placed in the proper ambient conditions, folds into a biologically active protein.

The phenomenon of spontaneous folding of protein native structures allows us to detach, at least to a first approximation, the study of protein folding physics from the study of protein biosynthesis.

Protein folding *in vitro* is the simplest (and therefore, the most interesting for a physicist) case of pure *self*-organization: here nothing "biological" (except for the sequence!) helps the protein chain to fold.

2.4.1. The Levinthal Paradox

The ability of proteins (and RNA) to fold spontaneously immediately raised a fundamental problem that has come to be known as the Levinthal paradox [35,36]. It reads as follows:

On the one hand, the same native state is achieved by various folding processes: *in vivo* on the ribosome, *in vivo* after translocation through the membrane, *in vitro* after denaturation with various agents... The existence of the spontaneous and correct folding of chemically synthesized protein chains suggests that the native state is thermodynamically the most stable under "biological" conditions.

On the other hand, a chain has zillions of possible conformations (at least 2^{100} for a 100-residue chain, since at least two conformations are possible for each residue), and the protein can "feel" the right stable structure only if it is achieved exactly, since even a 1Å deviation can strongly increase the chain energy in the closely packed globule. Thus, the chain needs at least ~2^{100} picoseconds, or ~10^{10} years to sample all possible conformations in its search for the most stable fold.

Then, a question arises: how can the chain find its most stable structure within a "biological" time (minutes) at all?

The paradox is that, on the one hand, the achievement of the same (native) state by a variety of processes is (in physics) clear-cut evidence of its stability. On the other hand, Levinthal's estimate shows that the protein simply does not have enough time to prove that the native structure is the most stable among all possible structures!

In order to solve this paradox, Levinthal suggested the existence of specific folding pathways, and hypothesized that the native fold is simply an end of the protein-specific pathway rather than the most stable fold of its chain. Should this pathway be narrow, only a small part of the conformational space would be sampled, and the paradox would be avoided. In other words, Levinthal suggested that the native protein structure is under kinetic rather than under thermodynamic control, *i.e.*, that it corresponds not to the global, but rather to the easily accessible, free energy minimum.

2.4.2. Folding Pathways and Folding Intermediates

The question as to whether the protein structure is under kinetic or thermodynamic control is not a purely speculative question. It is raised again and again when one faces practical problems of protein physics and engineering. For example: when trying to predict protein structure from a sequence, for what must we look? The most stable, or the most rapidly folding structure? When designing a *de novo* protein, what must we do? Should we maximize the stability of the desired fold or create a rapid pathway to this fold?

A discussion on protein folding mechanisms started immediately after discovery of spontaneous folding. It seems that the first proposed hypothesis was by Phillips, who suggested that the folding nucleus is formed by the N-end of the nascent protein chain, and that the remaining part of the chain wraps around it [37]. This appealing hypothesis is present in some works up to now. However, it has been refuted experimentally, as far as single-domain proteins are concerned. The elegant works by Goldenberg & Creighton have shown that the N-terminus has no special role in *in vitro* folding: it is possible to glue the ends of the chain of a small protein with a peptide bond, and nevertheless it folds into the correct 3D structure [38]. Moreover, it is possible to cut this circular chain to make a new N-end at the former middle of the chain; the protein folds, nevertheless, to the former native structure. Nowadays, protein engineering routinely produces circularly permuted proteins.

In an effort to solve the folding problem, Ptitsyn proposed a model of stepwise protein folding [39] (Fig. **3**). Later given the name "framework model," this hypothesis stimulated investigation of folding intermediates. It postulated a stepwise involvement of different interactions in the protein structure formation, and it stressed the importance of rapidly folded α-helices and β-hairpins at the initial folding steps, gluing of these helices and hairpins into a native-like globule, and crystallization of the final structure within this globule at the last step of folding.

The cornerstone of this concept was the then hypothetical and now well known folding intermediate, the "molten globule", which was later discovered and studied first as the equilibrium state of a "weakly denatured" protein [19,40] and then as a kinetic folding intermediate [41].

Experiment shows that different properties of the native protein have two quite different rates of restoring. Nearly-native volume and secondary structure restore within a fraction of second, while side-chain order, tertiary structure and biochemical activity often take minutes to restore. This is evidence for the accumulation of some "intermediate" state of the protein molecule (as shown, the molten globule [21]) at the beginning of the folding process.

The molten globule is an early folding intermediate in the *in vitro* folding of many proteins at physiological conditions [21,42,43]. It takes a few milliseconds to form, while the complete restoration of native properties of a 100 – 300 residue chain can take seconds for some proteins and hours for others. Thus, the rate-limiting folding step concerns formation of the native "solid" protein from the molten globule rather than formation of the molten globule from the coil.

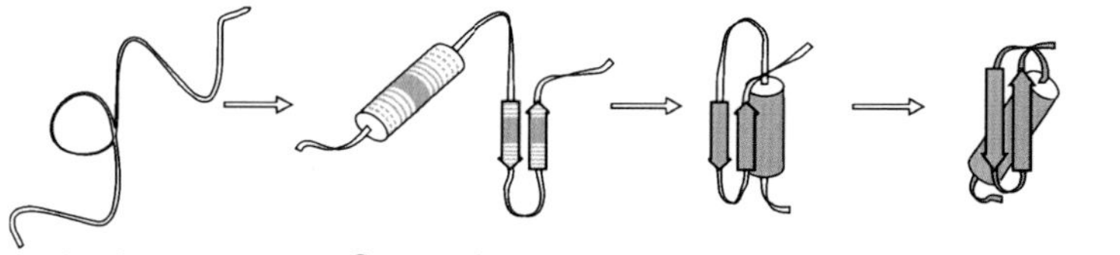

Figure 3: Framework model of stepwise folding [39]. The secondary structures are shown as cylinders (α-helices) and arrows (β-strands). Both predicted intermediates have already been observed; the first is now known as the "pre-molten globule" and the other as the "molten globule" [21].

The molten globule is compact due to remainders of hydrophobic interactions of the native protein [23,24]. However, these "remainders" are not as strong: their strength amounts to 15 – 50% of that in the protein's native state for some (especially evolutionary conserved) residues, and practically to 0% for other (non-conserved) residues [44,45].

The molten globule is not the only intermediate observed in protein folding. The "pre-molten" globule (that also fits the "framework model") was observed [46] to precede the molten globule formation. As a kinetic intermediate it was discovered with the use of ultra-fast (sub-millisecond) measuring techniques [47,48]. In addition, proteins with disulfide bonds allow for the trapping of various intermediates indicative of the order of S-S bond formation, *etc.* [12].

The "kinetic control" hypothesis initiated very intensive studies of folding intermediates. Actually, it was clear almost from the very beginning that the metastable intermediates are not obligatory for folding (since the protein can fold also near the point of equilibrium between the native and denatured states, where the transition is of the "all-or-none" type, which excludes any metastable intermediates). The idea was, though, that the intermediates, if trapped, would help to trace the folding pathway, like intermediates in a complicated (bio)chemical reaction trace its pathway. This was, as it is now called, "chemical logic". However, this logic worked only in part when it came to the protein folding. The intermediates (like molten globules) were found for many proteins, but the main question as to how the protein chain can rapidly find its native structure among zillions of alternatives remained unanswered.

2.4.3. "Two-State" and "Multi-State" Protein Folding

Progress in the understanding of protein folding [49,50] has been achieved just by investigation of those proteins, which fold without "unnecessary complications" (previously widely used to trace the folding pathway): without accumulation of any intermediates at the folding pathways, without cis-trans proline isomerization, and without S-S bond formation. The folding (and the unfolding) kinetics look very simple in this case: all the properties of the native (or denatured) protein are restored synchronically, following the single-exponential kinetics [51]. For some proteins, this simplicity is observed in a wide range of conditions, including the denaturant-free water ("biological zone" in Fig. **4**), the zone of the reversible thermodynamic transition between two phases (the native and the denatured state) and the unfolding zone; these proteins obtained a name of "two-state proteins". For the other, "multi-state" proteins, the two-state folding occurs only in the transition zone, if any (see [44,45]), while the unfolding demonstrates a "two-state" manner [52] (Fig. **4**). Usually, the complicated folding demonstrates three phases, and the corresponding proteins obtained a name of "three-state proteins" [49-53]. Thus, the most universal features of folding (and unfolding) can be observed just in and around the transition zone, while the moving of this zone towards the "biological" conditions reveals individualities of various proteins (which are the "unnecessary complications", when we try to understand the basics of protein folding).

The above statement looks, in a sense, paradoxical. Indeed, what can we get from investigation of folding (or unfolding) in the transition zone, where we cannot accumulate any transition intermediates? The answer is: just here we can most readily, though indirectly, observe the folding transition state, whose stability (or, more exactly, instability) determines the folding (and unfolding) rate [49-52,54-

56]. The transition state corresponds to the free energy maximum on the folding/unfolding pathway, – or, it is better to say, to the free energy saddle point on the network of these pathways (see Fig. **10** below). The folded part of the transition state is called the "folding nucleus", and the folding pathway *via* formation of a nucleus (which often consists of amino acid residues remote in protein chain [57,58]) obtained a name of "nucleation-condensation" mechanism of folding [59].

2.4.4. Folding Nucleus

The "folding nucleus" plays a key role in protein folding: its instability determines the folding and unfolding rate-limiting steps. It should be stressed that the folding nucleus is **not** the molten globule, although some of their characteristics may be similar [56]: the nucleus corresponds to the free energy maximum, while the molten globule which is observed as a folding intermediate corresponds to a local free energy minimum [21]. It has been shown that the nucleus looks like part of the 3D structure of the native protein [52,56].

So far, there is only one, very difficult experimental method to identify the folding nuclei in proteins: to find residues whose mutations affect the folding rate by changing the transition state stability as strongly as that of the native protein [52,56] (Figs. **5**, **6**). This method is called "ϕ-analysis". A similar method, called "ψ-analysis" [60] (it includes mutation of two close in space protein residues) is used to measure interactions in transition states.

The participation of a residue in the folding nucleus is expressed by the residue's ϕ value. ϕ is defined as $\Delta \ln k_f / \Delta \ln K$, where k_f is the folding rate constant at given conditions (usually, in denaturant-free water), $K = k_f / k_u$ is the folding-unfolding equilibrium constant at the same conditions, and Δ means the mutation-induced shift of the corresponding value. According to the model of a native-like folding nucleus [52,56], ϕ=1 means that the residue has its native conformation and environment already in the transition state (*i.e.*, that this residue is in the folding nucleus), while ϕ=0 means that the residue remains unfolded in the transition state. The values $\phi \approx 0.5$ (which are observed quite often) are ambiguous: either the residue is at the surface of the nucleus, or it is in one of the alternative nuclei, belonging to a different folding pathway. It is noteworthy that the values $\phi < 0$ and ϕ>1 (which would be inconsistent with the model of a native-like folding nucleus) are extremely rare and never concern a residue with a reliably measured ΔlnK.

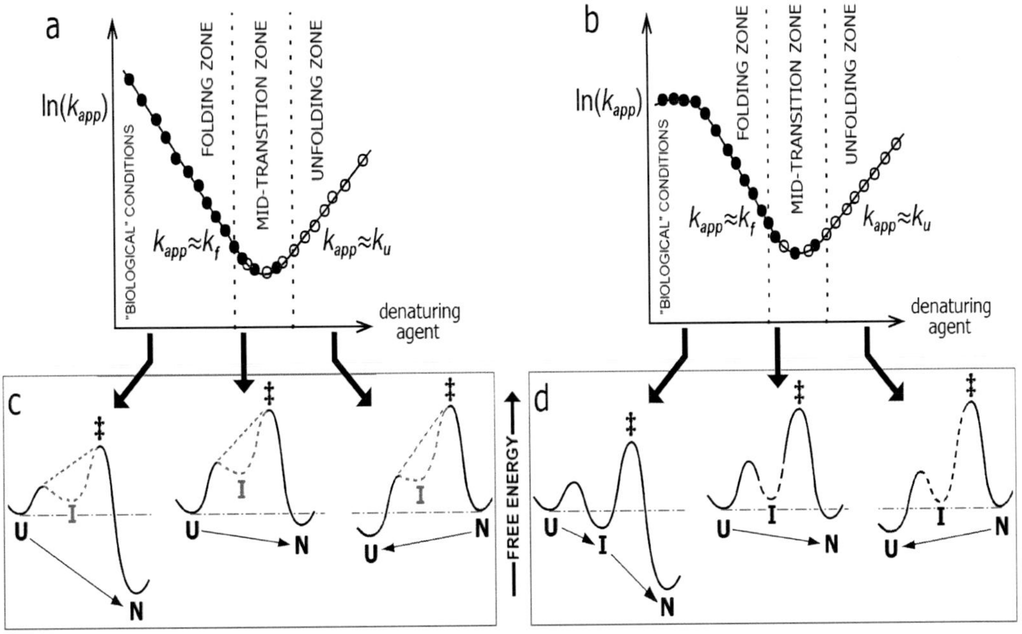

Figure 4: (a, b) Typical appearance of a "chevron plot" presenting an apparent rate of the folding/unfolding process (k_{app}) *vs.* the denaturant concentration (or the temperature). The closed circles correspond to folding which occurs when the protein is transferred from the unfolding to renaturing conditions. The open circles correspond to unfolding that occurs when it is transferred from the native to denaturing conditions. Note that circles of both kinds overlap at mid-transition. (a) Typical plot for a protein having the *two-state* folding throughout the whole range of experimental conditions. For the two-state folding, different characteristics of the protein change with equal rate k_{app}, and $k_{app} = k_f + k_u$, where k_f is the folding rate and k_u is the unfolding rate: thus, $k_{app} \approx k_f$ in the folding zone (where $k_f \gg k_u$), $k_{app} \approx k_u$ in the unfolding zone (where $k_f \ll k_u$) and $k_f \approx k_u \approx k_{app}/2$ at the mid-transition [52]. (b) Typical plot for the rate-limiting step of folding and unfolding of a "*multi-state*" folding protein: such a protein has an apparent two-state folding only close to the mid-transition (*i.e.*, the point of thermodynamic equilibrium between the native and denatured states), but not at the "biological" conditions where a multi-state folding occurs (and some state(s) arise(s) and/or decay(s) much faster than the complete transition occurs). The schemes at the bottom of the Figure show the free energy changes along the folding pathways of the two-state (c) and three-state (d) proteins. N is the native state, U the unfolded, I the metastable (molten globule-like) folding intermediate, and ‡ the main, most unstable transition state (as a rule, ‡ is situated directly before the N state) [49]. The thin arrows show the directions of the processes, and the dash-dot lines show the free energy levels corresponding to the beginnings of the processes. The dotted lines in (c, d) correspond to those intermediates, which are either invisible in kinetic experiments (this happens when I is either unstable relative to the initial state of the process, or is situated after ‡ along the process pathway [12,21]), or absent entirely (the latter possibility is shown by the alternative dotted line).

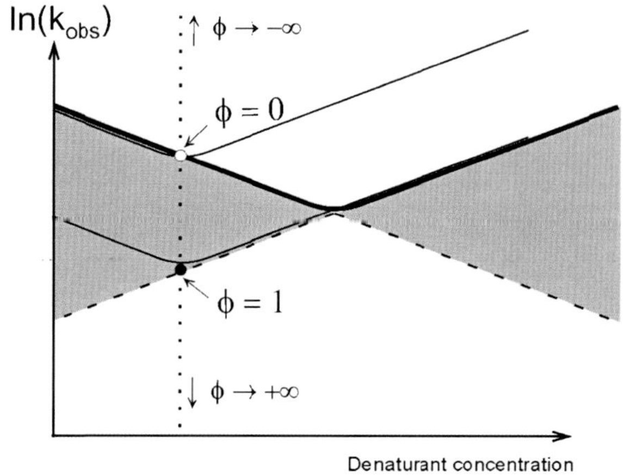

Figure 5: Identification of involvement of a residue into the folding nucleus using ϕ-analysis of site-directed mutations (a scheme). The wild type chevron plot is drawn in bold. The dashed lines denote extrapolations of processes to conditions where they are non-observable in experiment (folding at high denaturant concentration and unfolding at low denaturant concentration). (Closed circle) The mid-transition point for chevron plot of a mutant protein if the mutated residue has its native conformation and environment (*i.e.*, its native interactions) already in the folding nucleus; in this case ϕ=1. (Open circle) The mid-transition point for chevron plot of a mutant protein if the mutated residue remains denatured in the transition state; in this case ϕ=0. If the mid-transition point moves from open circle to the closed circle, the corresponding ϕ-value changes from 0 to 1. If the mid-transition point moves up from open circle, then $\phi \to -\infty$; if the mid-transition point moves down from closed circle, then $\phi \to +\infty$. The grey region corresponds to those positions of mid-transition points when $0 \leq \phi \leq 1$.

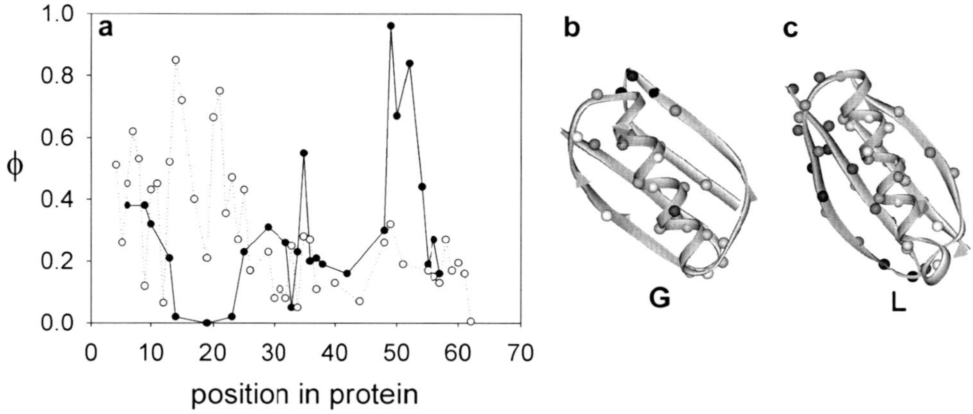

Figure 6: (a) Profiles of experimental ϕ-values obtained for B1 domains of proteins G (filled circles) and L (open circles). (b,c) Schemes of three-dimensional structures of these proteins [61,62], colored according to the ϕ values of the amino-acid residues, from white (ϕ=0) to black (ϕ=1). The experimentally studied residues are shown as beads against the background of the native chain fold. Adapted from [63,64]. Sequence identity of G and L proteins is 15% [65].

The major assumptions underlying the ϕ-analysis of the folding nucleus by point mutations [56] are that neither the folding pathway, nor the nucleus, nor the structure of the folded state, nor the unfolded state ensemble is changed as a result of the mutations. Experimentally, this is proved to be usually (there are exceptions [66]) correct when the mutated residue is not larger than the initial one, and when the mutation is not connected with introduction of charges inside the globule; the proof is done by double mutations [67].

The data obtained on the structure of nuclei can be summarized as follows.

(1) In some proteins, the nucleus is situated in the center, in the hydrophobic core [57,68-70]; in some, it is on the boundary of the globule [71-73].

(2) In some proteins the nucleus is stabilized by hydrophobic interactions [57,68,72]; in some it includes hydrogen bonds and salt bridges [71,73].

(3) The measurements of the nuclei's solvent-accessible surface area are also rather informative [43]. The ratio of accessible surfaces of the protein's transition and native states can be estimated from the dependence of k_f and $K = k_f/k_u$ (reflecting the transition state and the native protein free energies) on the denaturant concentration C: $\beta_T = (\delta \ln k_f/\delta C)/(\delta \ln K/\delta C)$ [74]. When β_T is close to 1, the solvent-accessible area of the nucleus is close to that of the native protein; when β_T is close to 0, the nucleus is close to that of denatured protein. As a rule, the observed values of β_T are close to 0.6 – 0.8 for small proteins. At the same time, the average value of ϕ (which shows what fraction of the residue's environment already exists in the transition state) is usually about 0.3 – 0.4. The main difference between ϕ-values and β_T values is that ϕ-values report on the side chain only, whereas the β_T values report on both side chains and main chain. However, non-native interactions in the transition state can increase the β_T values as compared to the ϕ-values, and moreover, they may give rise to ϕ-values that are negative or larger than unity (cf. [75]).

(4) Proteins with different sequences but similar 3D structures often have similar folding nuclei [71,76-78]. However, there are some exceptions [79] (Fig. **6**). It also has been shown that a circular permutation, which changes the protein topology, sometimes changes [80-82], and sometimes does not change [83], the position of the nucleus within the protein. The observed abundance of ϕ values of about 0.5 and the observed sensitivity of nuclei to mutations, together with the results of computer simulations [58,84], led to a conclusion that a "nucleus" is an ensemble of structures rather that a single structure.

Summing up the experimental data, Grantcharova *et al.* conclude that mutations, both artificial and natural, can radically change folding pathways (create and destroy folding intermediates, transforming two- into multi-state folding proteins and *vice versa*, shift the folding nuclei at the opposite side of the molecule, *etc.*) — without any considerable variation of three-dimensional structures of native proteins [85]. This means that the native structure is a subject of much more severe natural selection than that of the folding nucleus and than folding pathways, – at least when we speak about relatively small proteins, which usually fold *in vitro* much faster than they are synthesized by a ribosome.

3. THEORY OF PROTEIN FOLDING

All of the experimental data that we have discussed, although exciting, cannot answer the main question about how a protein manages to find its native, apparently most stable, structure among zillions of others within those minutes or seconds that are assigned for its folding. The explanation for this comes from physical theory.

3.1. Solution of the Levinthal Paradox

The difficulty of this "Levinthal problem" is that it cannot be solved by a direct experiment. Indeed, suppose that the protein has some structure that is more stable than the native one but folds very slowly. How can we find it if the protein does not do so itself? Shall we wait for $\sim 10^{10}$ years?

However, is there a real contradiction between "the most stable" and the "rapidly folding" structure? Maybe, the stable structure *automatically* forms a focus for the "rapid" folding pathways, and therefore it is *automatically* capable of fast folding?

Before considering *kinetic* aspects of protein folding, let us recall some basic facts concerning protein *thermodynamics* (as above, we will consider single-domain globular proteins only). This will help us to understand what chains and what folding conditions we have to consider. The facts are as follows:

- Protein folding and unfolding are usually reversible "all-or-none" transitions: only the native and denatured states of the chain are present (close to the denaturation point) in a visible quantity, while the other states are unstable and therefore virtually absent. An "all-or-none" folding phase transition requires the amino acid sequence that provides a large energy gap between the native and the other folds.

- The denatured state, at least that of small proteins unfolded by a strong denaturant, is often the random coil.

- Even under physiological conditions the native state of a protein is only by a few kcal/mol more stable than its unfolded state (and these states have equal stability at mid-transition, naturally).

Thus, to solve the "Levinthal paradox" and to show that the most stable chain fold can be found within a reasonable time, we could, as a first approximation, consider only the rate of the "all-or-none" transition between the coil and the most stable structure. For this, we can consider the case when only one, *the most stable*, chain fold is in (or close to) thermodynamic equilibrium with the coil state, and all other folds of the chain are unstable. Here the analysis is the simplest: it must not consider accumulating intermediates. True, the maximal folding rate is achieved when the native fold is much more stable than the coil (Fig. **4**), and then the observable intermediates often arise. But let us consider the situation when the folding is not the fastest but the simplest…

Since the "all-or-none" transition requires a large energy gap between the most stable structure and the other ones, *we will assume that the considered amino acid sequence provides such a gap.*

We are going to show that the "gap condition" provides a rapid folding pathway to the global energy minimum, to estimate the rate of folding, and to prove that the most stable structure of a normal size domain can fold within seconds or minutes [2,86,87].

To prove that the most stable chain structure is capable of rapid folding, it is sufficient to prove that at least one rapid folding pathway leads to this structure. Additional pathways can only accelerate the folding since the rates of parallel reactions are additive. (One can imagine water leaking from a full to an empty pool through cracks in the wall between them: when the cracks cannot absorb all the water, each additional crack accelerates filling of the empty pool. And, by definition of the "all-or-none" transition, all semi- and mis-folded forms together are too unstable to absorb a significant fraction of the folding chains and trap them).

A rapid pathway must include not too many steps, but, first of all, it must not require overcoming of a free energy barrier that is too high.

An L-residue chain can attain its lowest-energy fold in L steps, each adding one fixed residue to the growing structure (Fig. **7**). *If* the free energy went downhill along the entire pathway, a 100-residue chain would fold in ~ 100 – 1000 ns, since the growth of a structure (*e.g.*, an α-helix) by one residue (τ) is known to take ≈10 nanoseconds [88]. Protein folding takes much more than 1 μs only because of the free energy barrier, since most of the folding time is spent on climbing up this barrier and falling back, rather than on moving along the folding pathway.

According to the conventional transition state or Kramers' theory [89-91], characteristic time (≡1/k) of the multi-step process is estimated as

$$TIME \sim \tau \times \exp(+\Delta F^{\ddagger}/RT). \qquad \textbf{(1)}$$

Here T is the temperature, R the gas constant, τ the time of one step (~ns in polypeptide chains [88]), and $\Delta F^{\ddagger}$ the free energy of transition state relative to the initial one, *i.e.*, the barrier height.

Our main question is: how high is the free energy barrier $\Delta F^{\ddagger}$ for the pathway leading from the unfolded to the lowest-energy structure?

If the fold-stabilizing contacts start to arise only when the chain comes very close to its final structure (that is, if the chain has to lose almost all its entropy *before* the energy starts to decrease), the initial free energy increase would form a very high free energy barrier (proportional to the *total* chain entropy lost). The Levinthal paradox claiming that the lowest-energy fold cannot be found within any reasonable time, since this involves exhaustive sampling of all chain conformations, originates exactly from this "golf course" picture of the energy landscape (loss of the entire entropy *before* the energy gain).

A note: In analysis of the protein folding problem, the term "energy" conventionally means "Gibbs free energy of interactions" (since hydrophobic and other solvent-mediated forces are connected with the solvent entropy), while the term "entropy" is reserved for only the chain's conformational entropy and does not include the solvent entropy. This terminology is commonly used to concentrate on the main problem of sampling the protein chain conformations [2,3], and it will be used in this paper as well.

However, this paradox can be avoided if there is a folding pathway where the entropy decrease is almost immediately covered by the energy decrease [86,87] (as in the usual first order phase transitions).

Let us consider a *sequential* (Fig. **7**) folding pathway.

At each step of this process, one residue leaves the coil and takes its final position in the lowest-energy 3D structure. This pathway looks a bit artificial, but it is exactly the pathway of unfolding of the lowest-energy structure going in the opposite direction. The detailed balance law [92] reads that direct and reverse reactions must follow the same pathway under the same conditions (and we already agreed to consider the mid-point of the folding-unfolding equilibrium). The advantage of considering this folding scenario is that it obviously exists (though the others are also not excluded); second, it allows us to consider only those residue-residue contacts which exist in the native protein [93]. Thus, we can replace a difficult analysis of folding by a simpler analysis of unfolding, and consider the folding nucleus instead of the nucleus of unfolding: these two nuclei coincide at the thermodynamic mid-transition conditions!

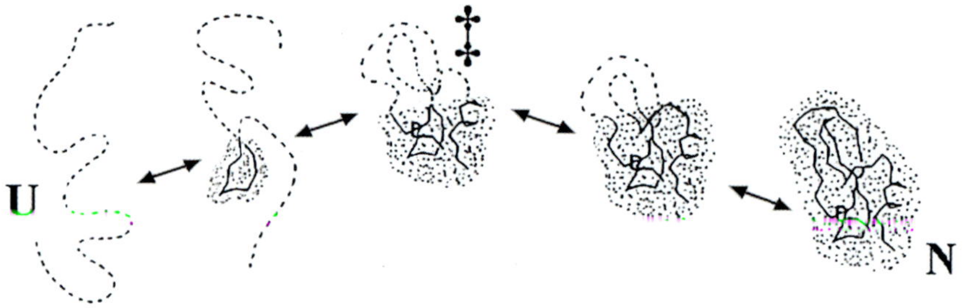

Figure 7: Sequential folding (and unfolding) pathway [86]. U is the unfolded state, N the native state, ‡ the transition state. The folded part (dotted) is native-like. The bold line shows the backbone fixed in this part; the fixed side chains are not shown for the sake of simplicity (the volume that they occupy is dotted). The dashed line shows the unfolded chain.

Since, at the equilibrium point, the free energies of the native and unfolded phases are equal, the additional free energy $\Delta F^{\ddagger}$ of the nucleus is due only to the boundary between the native and the unfolded phases. The largest boundary which will be met at the optimal phase transition pathway, *i.e.*, when the boundary between phases moves along the longest axes of the globule (Fig. 7), includes not more than $\approx L^{2/3}$ out of L residues of the chain. This corresponds to a folding nucleus embracing about half of the globule. The energy of this boundary can be estimated [86] as $\approx \frac{1}{3}\varepsilon L^{2/3}$, where ε is the protein denaturation energy per residue and ≈⅓ is a fraction of the residue's contacts that are lost at a 2-dimensional surface in the 3-dimensional space. The ε value has been experimentally estimated as ≈1 kcal/mol [13], or $\approx 2RT$ at room temperature. Besides, depending on the protein topology and the boundary position, the surface of the nucleus may or may not be covered by unfolded closed loops, whose Flory entropy creates an additional surface tension that adds to the conventional surface energy of the boundary. At the very maximum (when all loops have equal length), this entropic term is

$$T\Delta S^{\ddagger,\mathrm{Flory}} = -\,L^{2/3}\cdot{}^{1}/_{6}\cdot{}^{5}/_{2}\cdot\ln(3L^{1/3}). \qquad \textbf{(2)}$$

Here $L^{2/3}$ is the number of surface amino acid residues, the multiplier $^{1}/_{6}$ reflects that only 1 out of 6 possible directions of the surface residue is consistent with beginning of a loop, the multiplier $^{5}/_{2}$ is used instead of the Flory multiplier $^{3}/_{2}$ because each loop avoids the space occupied by the globule, and $3L^{1/3}$ is the average loop length, when *L/2* non-globular residues are divided into $^{1}/_{6}L^{2/3}$ equal-length loops. However, in a more typical case of random division of *L/2* residues into $^1/_6L^{2/3}$ loops, the term $-T\Delta S^{\ddagger,\mathrm{Flory}}$ does not exceed [86] $RT{\cdot}L^{2/3}$ (which is, however, numerically very close to the estimate (2) when $L<10^3$).

Thus, the free energy barrier for folding (and unfolding) in the mid-transition can be estimated as

$$\Delta F^{\ddagger} \approx (1 \pm 0.5)\ RT{\cdot}L^{2/3}. \quad \mathbf{(3)}$$

$\Delta F^{\ddagger}$ is $\approx 1.5L^{2/3}RT$, when the boundary is densely covered by loops, and $\approx 0.5L^{2/3}RT$, when the boundary is free of them [86]. Since a characteristic time of rearrangement of one residue τ is $\approx$10 ns [88], it takes ~10ns×exp($1.5L^{2/3}$) to overcome the free energy barrier of nucleation in the first case, and only ~10ns×exp($0.5L^{2/3}$) in the second. This range is exactly consistent with the observed times of protein folding near the mid-transition (Fig. **8**). It has also been estimated that knotting of a 100-residue chain can increase the folding time by no more than two times, and of a 400-residue chain by an order of magnitude at most [94].

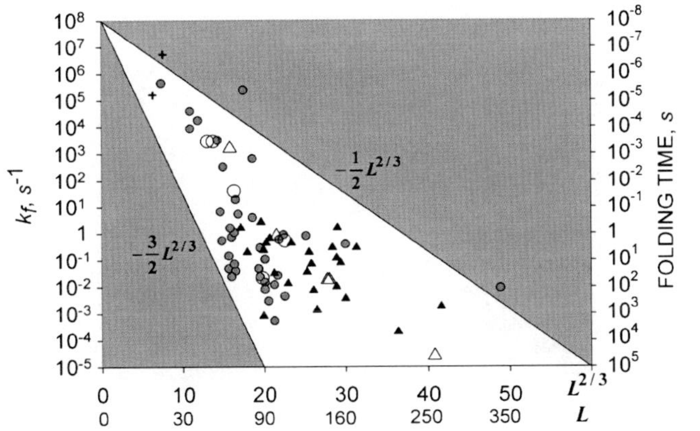

Figure 8: Observed folding rate (and time) at the point of equilibrium between the unfolded and the native states *vs.* $L^{2/3}$ (L being the number of residues in the chain). The circles and triangles correspond to two-state and multi-state folding proteins, respectively. All these proteins do not contain S-S bonds or large ligands. Symbols + correspond to the α-helical and β-hairpin peptides. The theoretically predicted region of folding rates (white) is between the lines corresponding to *TIME* = 10ns×exp($0.5L^{2/3}$) and *TIME* = 10ns×exp($1.5L^{2/3}$); this region is in excellent agreement with experimental results, but the folding time *vs.* $L^{2/3}$ correlation is only 0.65. The experimental folding rate errors do not exceed the size of used symbols. Adapted from [95,96]. The data marked by black and grey symbols are taken from [96,97]; the data marked by larger open symbols are obtained after 2007, *i.e.*, after publication of [96]. The database is available at http://phys.protres.ru/resources/k-fold.doc

The reason for the obtained "non-Levinthal" estimate of achievement of the lowest-energy structure,

$$TIME \sim \exp[(1\pm0.5)L^{2/3}]\times10\text{ns}, \quad \textbf{(4)}$$

is that the entropy decrease is almost immediately compensated for by the energy gain along the sequential folding pathway (cf. [98,99]), and the free energy barrier occurs due to the phase surface effects only.

It is noteworthy that the sequential folding pathway does not require any rearrangement of the dense globular part (which could take a lot of time): all rearrangements occur in the swollen (coil) phase and are therefore rapid.

A few notes in conclusion of this chapter:

- It is noteworthy that numerous computer simulations of folding never encounter any kinetically inaccessible structures of 3D protein models [84].

- It is remarkable that a subsequent mathematical estimate [100] of the maximum time necessary to find the most stable structure of a chain molecule in a *d*-dimensional space follows the equation $\ln(TIME) \sim L^{(1-1/d)}\cdot\ln(L)$, in accordance with the above given estimates (2) – (4), which have been obtained from simple physical considerations.

- Computer experiments of Sali and co-workers [101] established that the only requirement for fast folding was a sufficient energy gap between the native state and the next lowest energy state. A successful prediction (Fig. **8**) of the folding rate region by a theory of Finkelstein and Badretdinov [86], that ignores any special construction of the folding nucleus, shows that proteins have only evolved for having sufficient stability of their native folds (and therefore function) and not folding rates (see also a paper [102] on this).

- For the overwhelming majority of proteins, the native state is the only stable folded structure; however, there is at least one exception, *i.e.*, serpin, whose native state is metastable [103] (see Fig. **11b** for the case when one "misfolded" structure is close to the most stable structure N and divided by a large energy gap from the other misfolded structures M).

3.2. Theories of Protein Folding Rates

The theory described above not only "demystifies" (as it was written in [84]) "the protein folding problem cast in terms of the Levinthal paradox", but also opens a way to the creation of more detailed theories of protein folding rates.

These theories have to take into account such factors as the non-uniform distribution of strongly and weakly interacting amino acid residues within the globule, the folding pattern ("topology") formed by the native fold of the protein chain and folding under strongly non-equilibrium conditions.

The non-uniform distribution of weak and strong interactions between the nucleus and the remaining part of the protein contributes to $\Delta F^{\ddagger}$ only a term of less than [86] $\sim L^{1/2}RT$. This effect is of secondary importance when the folding takes place close to the mid-transition where $\Delta F^{\ddagger} \sim L^{2/3}RT$. However, the non-uniformity effect seems to be important when the native state is much more stable than the denatured one [84,104-106]. Besides, it is of crucial importance for a theory of those relatively small changes in the folding rates which are caused by mutations of a protein's residues.

When a protein folds under the non-equilibrium conditions, the transition state free energy becomes smaller than in the case of equilibrium conditions (see Fig. **11** below). It was shown that this free energy $\Delta F^{\ddagger}$ scales with the chain length L as $\Delta F^{\ddagger} \sim L^{1/2}RT$ when the native state is moderately more stable than the denatured one [104,106], and as $\Delta F^{\ddagger} \sim (4 \div 6)\ln(L) \cdot RT$ when the native state is much more stable than the denatured one [105,106]. All of these length-based dependencies are consistent with currently available experimental data [107,108] for protein folds of extreme stability.

A review [84] of Shakhnovich gives an excellent overview of effects that occur at non-equilibrium folding.

The capillarity model [86] gave rise to the hypothesis that protein folding rates are determined by the average "entropy capacity" (the entropy capacity of an amino acid residue is defined as the number of contacts divided by the number of degrees of freedom; thus, this value is, in a sense, reciprocal to the expected melting temperature) [109]. It has been shown [110] that entropy capacity correlates with folding rates for alpha-helical proteins (correlation coefficient is 0.79) and proteins with mixed (α/β) secondary structure (correlation coefficient is 0.84). For 84 proteins of all classes, the calculations based on the "entropy capacity" correlates with experimentally measured folding rates at the level of about 73% at the mid-transition point [111,112].

Statistical analysis showed that, among proteins of the same size, α/β proteins have, on the average, a greater number of contacts per residue compared to other protein classes [113]. This difference is due to their more compact (more "spherical") structure of α/β proteins, rather than to tighter packing. The relationship between the average number of contacts per residue and folding rates in globular proteins according to general protein structural class (all-α, all-β, α/β, α+β) has been examined. The analysis demonstrates that α/β proteins have both the greatest number of contacts and the slowest folding rates in comparison to proteins from the other structural classes [113].

The influence of the protein chain topology upon the folding time has been estimated using a "contact order" (CO) parameter [114,115] and other CO-like parameters [116-121]. The CO is equal to the average chain separation of contacting (in the native fold) residues, divided by the chain length. The CO is low for a structure rich in local contact, and high for a structure rich in contacts between remote chain regions. Specifically, CO is low for α-helices and high for β-structure; this may explain the observed 50-fold difference in folding rates of fast-folding α-helices and slow-folding β-hairpins [84,122]. However, modern data show that CO works well only for two-state folding proteins of an approximately equal size [123]. It has been shown also that the fraction of local contacts in protein structure gives a better folding rate prediction [124].

A high CO value reflects the existence of many long closed loops in the fold of a complicated topology; thus, CO is roughly proportional to the "1±0.5" multiplier in Equations (3), (4), see [125]. When CO is taken into account in addition to the folding-rate on the chain-length dependence shown in Fig. **8**, the correlation of theory with experiment rises by an additional 10% over that presented in Fig. **8** and reaches 0.74 [125]. The correlation is nearly the same for two- and multi-state proteins, in contrast to results obtained for "pure" CO in [114], which seemed to be good for two-state proteins only. However, with later obtained data on folding rates for large proteins with two-state kinetics [126] and small proteins and peptides, the prediction strength of CO, even for two-state folders, decreases to 0.41 only [123]. The most striking example [126] is presented by the 338-residue protein VlsE (PDB entry 1L8W) that has a very low CO but folds slowly. On the contrary, the absolute contact order, AbsCO = $L \times$ CO, which scales as $L^{2/3}$ [125], is good for predicting of the folding rate of VlsE and other proteins, having both two-state and multi-state folding [125].

The need for simultaneously taking into account both CO and chain length in protein folding rate estimation also follows from simplified off-lattice folding simulations [127], as well as from analysis of other parameters used to predict the protein folding rate [123].

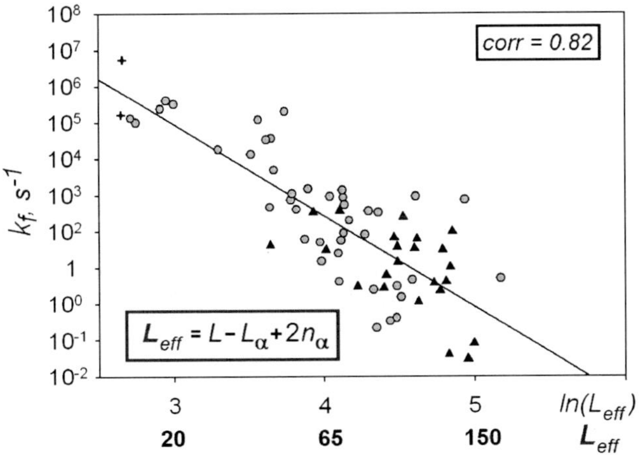

Figure 9: Correlation between observed protein folding rate in denaturant-free water and the number of amino acid residues in the protein chain except for the number of α-helical hydrogen bonds in this chain. The folding rate errors do not exceed the size of used symbols. Adapted from [97].

Interestingly, an even higher correlation (0.82) is observed between the length of an α-helix-free part of a protein chain and the observed protein folding rate in denaturant-free water [97]; in this case, again, the correlation is nearly the same for two- and multi-state proteins (Fig. **9**). It is noteworthy that this highly accurate prediction does not require any knowledge of the protein's tertiary structure: it can be obtained from the amino acid sequence alone, since one can predict [128] α-helices from the sequence.

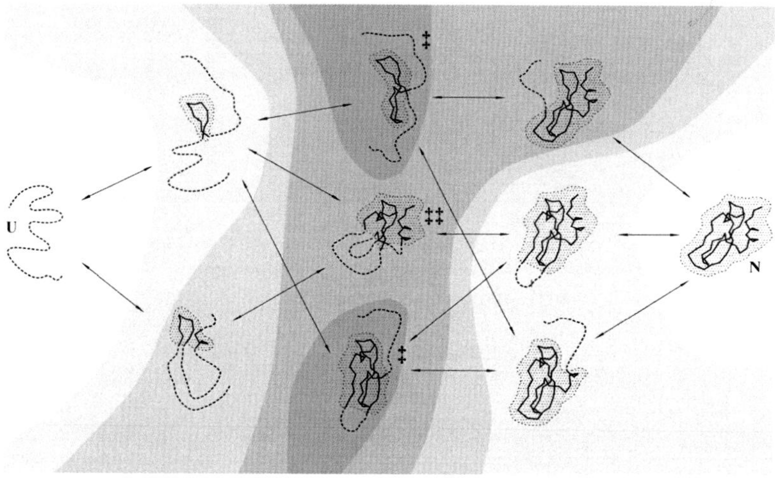

Figure 10: Free energy landscape of a protein chain; the darker the color, the higher the free energy. Arrows show possible transitions between semi-folded microstates (a "transition" is either an addition of one "chain link", *i.e.*, a chain segment including 1 – 5 amino acid residues, to the growing globule or a removal of one chain link from the globule). These transitions are parts of various folding-unfolding pathways, one of which is shown in Fig. 7. Transition states are marked with ‡. The main transition state, *i.e.*, the one with the lowest free energy, is marked with ‡‡.

A growing mass of experimental data has stimulated many studies where experimental folding rates are fitted by a function of secondary structure [129,130] or amino acid sequence, *etc.* [131-135]. We do not discuss them since they do not follow from any folding theory or folding mechanism, while the interpretation of obtained results suggested by the authors, if any, has obscure physical sense.

A more detailed and physically strict scheme to estimate protein folding rates [136-142] can be obtained from the analysis of the networks of folding pathways, or rather, the networks of pathways of unfolding of native protein structures (Fig. **10**). The free energy of each microstate (a network used contains ~10^6 microstates) can be estimated [136-140] from the energy of contacts within the folded region and the entropy of the unfolded region (the latter includes the free energy deficit caused by the Flory's entropy of closed unfolded loops as well). In addition to predicting protein folding/unfolding rates at various conditions, a theory based on this scheme is also able to find approximate localization of the protein folding nuclei (see below).

The theoretical results described show, in accordance with experiment, that the most stable protein fold of a 100-residue protein must be found within minutes, maximum hours. They explain also why very large proteins should consist of separately folding domains: otherwise, the chains of more than ~400 residues would fold too slowly (within many days or more) even when the protein chain topology is simple.

3.3. Protein Folding Under Non-Equilibrium Conditions

So far, we only have considered the folding rate close to mid-transition, where only one (the "native") fold competes with the coil, and all other globular forms, even taken together, are unstable.

If we have an environment that stabilizes the native protein fold, it stabilizes the other globular and semi-folded structures of the chain as well. At first, this stabilization only increases the folding rate (see the rise of the left chevron limb from the mid-transition in Figs. **4**, **5**), since the folding nucleus is also stabilized, and the competing (with the native fold) misfolded structures are still unstable (relative to the initial unfolded state as well) due to the energy gap between them and the native fold (Fig. **11a**).

It has been estimated [86, 143] that each additional 1 kcal/mol in stability of the native (relative to the unfolded) state decreases the barrier height by about ½ kcal/mol (since the critical nucleus includes about half of the protein; though, experimentally, this free energy change is not ½ but rather ≈0.7 kcal/mol [74], seemingly due to the presence of the ignored by our theory non-native contacts in the transition state). The decreased instability of the nucleus accelerates folding at room temperature by about 2.5 times for each additional 1 kcal/mol of native state stability.

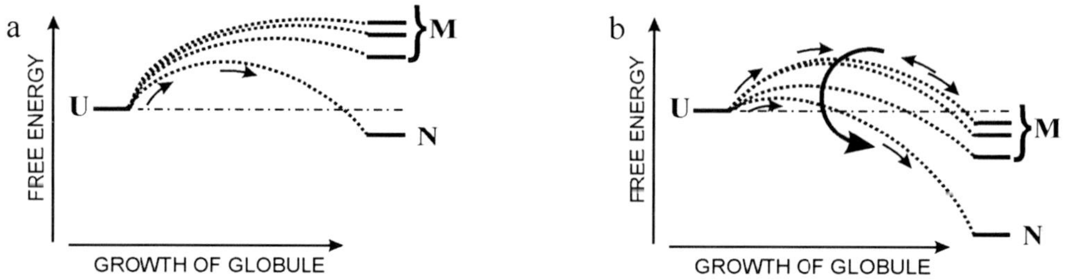

Figure 11: Folding under conditions when (a) the most stable fold N is only a little more stable than the unfolded chain U, and (b) when N is much more stable than U, and some mis- or semi-folded structures (shown by numerous free energy levels M) are also more stable than U. Dotted lines schematically show the free energy changes along the folding pathways leading to different structures; maxima of these lines correspond to the transition states on these pathways. The dash-dot lines show the free energy level of state U. Adapted from [86].

However, this acceleration proceeds only up to a certain limit: the maximum folding rate is achieved when the "misfolded" states become as stable as the initial unfolded state, *i.e.*, when the globular folding intermediates become stable (see the plateau at the top of the left chevron limb in Fig. **11b**). The further increase in stability of the folded states leads to a rapid misfolding followed by a relatively slow conversion to the native state (Fig. **11b**). This is a rather complicated process; it includes differences in coming to the native state by the protein main chain and side chains [84], own "folding nucleus" for each step, *etc.* All of these complications may be considered "artifacts" created by extreme folding conditions (cf. [144]), which are avoided (both in experimental and in theoretical studies) when the protein folding is considered close to mid-transition.

The above describes a situation when the native state N is moderately more stable than the unfolded state U. The case when N is much more stable (by tens kilocalories per mol) than U seems to be not feasible for a "wet" experiment; however, it is a theoretically interesting case, and it has been considered in [84,104,106].

3.4. Nucleation-Based Prediction of the Protein Folding Rates

A progress in prediction of protein folding rates was achieved by analysis of multidimensional networks of protein folding-unfolding trajectories made by a variety of algorithms [109,136,137,145,146]. All these approaches consider only native interactions (as in the "Gō model" [147]) and a trade-off between formation of attracting interactions and the loss of conformational entropy during protein folding. As in [86, 148], they simulate unfolding of known 3D protein structures rather than their folding. Table **1** shows that correlation coefficients between theoretically estimated and experimentally found folding rates are, independently of the number of examined proteins, about 0.70 - 0.80 for methods that do not use many adjustable parameters.

Table 1: Nucleation-based methods for calculation of folding rates of protein with known 3D structures

Method	**Condition of Modeling**	**Number of Proteins**	**Number of Adjustable Parameters**	**Correlation Coefficient**	**References**
Diffusion equations on the one-dimension profile	water	22	many	0.87	Munoz, Eaton, 1999 [141]
Kinetic equations	mid-transition	45	none	0.78	Ivankov, Finkelstein, 2001 [136]
Recursive search	water	37	a few	0.67	Alm *et al.*, 2002 [142]
Dynamic programming	mid-transition	17	one	0.73	Garbuzynskiy *et al.*, 2004 [138]
Monte Carlo	mid-transition	10	none	0.70	Galzitskaya *et al.*, 2005 [140]
Monte Carlo	mid-transition	67	none	0.82	Galzitskaya, 2010 [112]
Kinetic equations	mid-transition	82	none	0.79	Ivankov, Finkelstein, 2010 [145]
Kinetic equations	water	82	none	0.70	Ivankov, Finkelstein, 2010 [146]
Dynamic programming	mid-transition	82	none	0.75	Galzitskaya, Glyakina, 2012 [111]

Comparison of results of two recent methods based on the nucleation mechanism (the method of dynamic programming and solution of kinetic equations) is presented in Fig. **12**. One can see that the both methods are in a good agreement with each other.

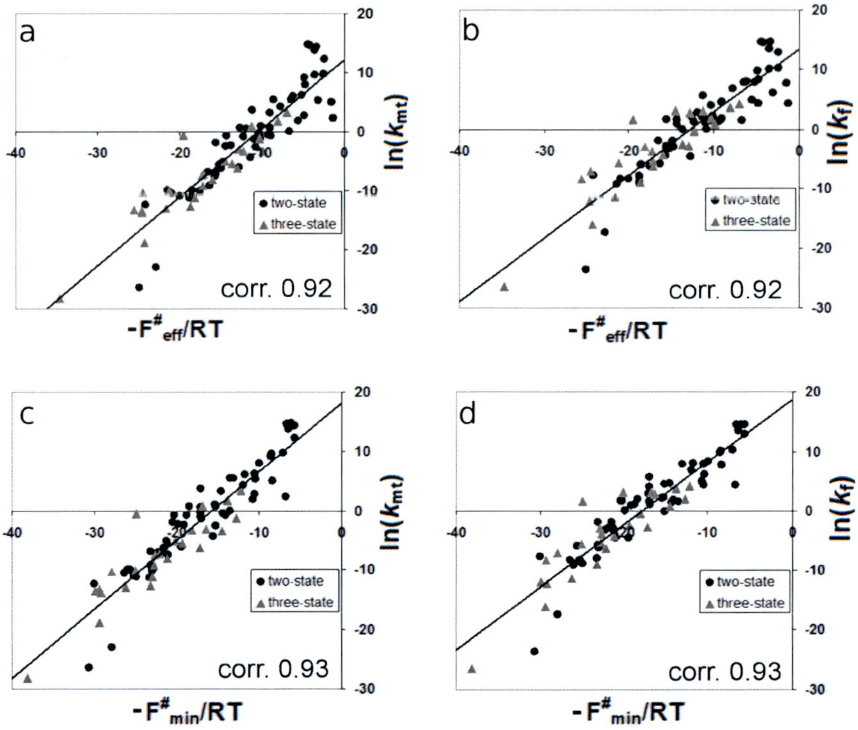

Figure 12: Correlations of the estimated folding rates at mid-transition (k_{mt}, [145]) for 82 proteins (a, c) and in water (k_f, [146]) for 81 proteins (b, d) with the computed free energy $F^{\#}_{eff}/RT$ of the transition state ensemble [111] and with the minimal free energy $F^{\#}_{min}/RT$ [111] of the sole transition state. The two-state folders are marked by circles and multi-state folders by triangles.

3.5. Recent Advances in All-Atom Simulations of Protein Folding

Molecular dynamics simulations (MDS) are widely used to investigate protein chain motions at an atomic level. However, until recently they have been done to the time insufficient to model the folding of even small proteins.

Currently, many millisecond-long MDS become feasible due to recent advances in computational power and the progress in elaboration of accurate force fields. It is very important that modern MDS of protein folding can be made in or close to the conditions of thermodynamic equilibrium between the native and denatured states of protein chains [149-152], which, as it has been stressed in this paper, are of key importance for understanding of protein folding. A great success is that these MDS demonstrate us how a protein, under these conditions, folds to its correct native structure, then unfolds, and then folds again to the same structure following some another folding pathway. Strikingly, the folding/unfolding times obtained by MDS [149] are rather close to the experimental ones. This is remarkable, because the force field parameters are fitted for a good description of energy minima rather than for description of energy maxima (and it is just the latter that are responsible for kinetics).

An excellent review of the modern MDS one can find in [153].

3.6. The Search for the Protein Folding Nucleus

The understanding of the nucleation mechanism has a long, contradictory and still unfinished story. The pioneering experimental results [52,56,67] suggested that the folding nucleus is small and specific (*i.e.*, it consists of only a few definite residues). This idea was reinforced by molecular dynamic simulations of protein unfolding (held at a temperature extremely favorable for unfolding) [148] and especially by very accurate and careful analysis of *in silico* folding of simplified lattice protein chains [58], held at a temperature extremely favorable for folding. Later, it was hypothesized that the folding nucleus consists of a *small* set of highly conserved residues having no obvious functional role [154] (however, recent experiments [45] do not support this hypothesis). On the contrary, the model of sequential protein folding at the mid-transition temperature [86] suggested to solve the Levinthal paradox led to a conclusion that the nucleus is large (including ≈ ⅓ – ½ of the protein molecule) and may be non-specific; for a long time, this was considered a drawback that resulted from oversimplification inherent to this model. However, current estimates of the average ϕ values show that ≈ ⅓ of the interactions existing in native-state proteins are present in the "folding nucleus" and argue against the existence of a few specific key residues which are *only* important during the kinetics of protein folding [155] (although Fersht and Sato [156] mentioned that the above result was obtained due to undeservedly massive discarding of experimental ϕ-values with small $\Delta \ln K$, *i.e.*, those where experimental errors in ϕ values can be more significant [155]). The latest statistical analysis also shows no indication that residues with increased ϕ-values are significantly more conserved than the others [157]. Ref. [84] argues the opposite, however, and states that the transition state ensemble includes a variety of large nuclei, but that all this ensemble of nuclei includes a small "obligatory" part, common for all nuclei. Thus, theoretical and experimental investigation of the nucleation mechanism is far from being completed.

As regards the theoretical search for folding/unfolding nuclei in proteins, several different approaches have been suggested.

The idea of specific nuclei, reinforced by the lattice simulations of protein folding [58], generated an evolutionary approach to prediction of the nuclei. It is based on the search for a set of highly conserved residues having no obvious functional role [154,158-160]. It should be mentioned that this approach, at best, can give only the common part of the nuclei existing in homologous proteins. Moreover, some observations show that the residue conservation across the homologous proteins correlates with deep immersion into the hydrophobic core of a protein [160] rather than into the folding nucleus [161]. It should be noted that there is some correlation between the nuclei (the regions of high ϕ values) and the hydrophobic cores and secondary structures [162-165], but it is rather low on average [161,163]. On the contrary, the positions of the folding nuclei demonstrate [166] a considerable correlation with the so-called “root structural motifs”, which are supersecondary structures used to classify protein folds [167].

Exciting progress has been made using experimental constraints to obtain the folding nucleus at atomic resolution (or rather, to visualize a possible shape for the folding nucleus that is consistent with the available, but sparse experimental data). Review of these works, excellently done in [84], is outside the scope of our paper.

The most direct approach to the theoretical search for the nucleus generates a plausible transition state using the all-atom molecular dynamic simulations of protein unfolding [148,168-170]. According to these simulations, held for very few small proteins at highly denaturing conditions, the unfolding is hierarchic [171-173] (at least when it occurs very far from equilibrium): tertiary interactions break early, whereas secondary structures remain longer. The repeated trajectories show a statistical distribution around the experimentally found transition states and demonstrate a broad ensemble of the TS structures. However, these simulations usually need extremely denaturing conditions (500 K, *etc.*) to be completed. Therefore, the transition states found for such an extreme *un*folding can be, in principle, rather different from those existing for folding [144]. Some other molecular dynamic simulations of unfolding of small proteins [174-176] have been performed at more realistic, although also highly denaturing, conditions. They have been performed at temperatures accessible for "wet" experiments (350°K), as well as for simulations at current supercomputers. They gave transition state structures, which are consistent with experiment [176]; however, these simulations took enormous time and were performed for very small proteins only.

Now, when many millisecond-long folding simulations become feasible, at least for small proteins [149-152], we can see how the folding nucleus is formed at the abrupt rate-limiting step of transition from the unfolded to the folded state. We see that some part of the protein chain attains the native-like structure, while the other parts of the chain remain unfolded and have different conformations at different runs of MDS. Thus, these results do not alter the presented general picture of protein folding, but they supported it and provided us many interesting details, especially on rearrangement of the protein chain just before and in the course of the rate-limiting steps of folding (and unfolding).

Other, more traditional approaches are based on a simplified modelling of the protein folding landscapes and trajectories [177-179]. The lattice simulations [180], the above described off lattice simulations, as well as experiment [66] show that the transition state is an ensemble rather than a single conformation, and that it can be described by an order parameter, such as the fraction of native contacts. The transition state search is based on the projection of the folding trajectories onto a single reaction coordinate (usually, the fraction of the native contacts) and the investigation of barriers at the obtained free-energy profiles [181-183]. Although such a projection is not a rigorous procedure (*e.g.*, the structure which is "nearly native" in terms of contacts can contain an additional knot of the chain and needs complete unfolding before coming to the native state), these studies were able to outline, though crudely, the folding nuclei of some small proteins.

Important progress is also being made in the prediction of folding nuclei by starting from the known 3D structures of native proteins. This progress is due to the analysis of multidimensional networks (see Fig. **10**) of the protein folding-unfolding trajectories done by various algorithms [137,145,163,184]. All these approaches (applied also to the studies of the folding rates [86,141], see the end of section 3.2) use different approximations and algorithms, consider only the attractive native interactions (the "Gō model" [147]) to reduce the energy frustrations and heterogeneity of interactions, and model the trade-off between the formation of attractive interactions and the loss of conformational entropy during protein folding. These works also simulate unfolding of known 3D protein structures rather than their folding, but the unfolding is considered close to the mid-transition point, where folding and unfolding pathways coincide according to the detailed balance principle. These works allowed the authors to outline the folding nuclei in more detail. Despite the relative simplicity of these models, they give a promising (~50%) correlation with experiment [138, 142,185,186].

It is noteworthy that this correlation is considerably worse than those typical for prediction of protein folding rates. The first obvious reason is that the observed ϕ values being predicted are restricted to the narrow region of 0 – 1 with an experimental error of ~±0.1, while the observed folding rates (determined with a relatively small experimental error) belong to the wide range of 10^7 s^{-1} – 10^{-4} s^{-1}. A more important reason is that the folding nucleus is not as stable to the action of mutations (and thus to the unavoidable errors in energy estimates used to outline them) as a 3D protein structure (see Fig. **6**), and it would be strange to obtain a perfect prediction of the folding nuclei with the same force fields which are still not able to predict the mutation-stable 3D native structure of a protein [84,187].

4. CONCLUDING REMARKS

Actually, protein folding resembles crystallization (where a kind of a Levinthal paradox also exists, since the decrease in the number of configurations during crystallization is huge, but is solved by a mechanism of nucleation) [143]. When crystallization occurs close to the freezing temperature, a perfect, large monocrystal (the lowest-energy structure, an analog of the "native structure") arises, although extremely slowly. As the temperature decreases a little, the monocrystal grows faster; and a further temperature decrease leads to a rapid formation of many chips rather than a perfect large monocrystal [90].

The scheme given above of entropy-by-energy compensation along the folding pathway and the conclusion that it leads to nucleation that can solve the Levinthal paradox are applicable to the formation of the native protein structure not only from the coil but also from the molten globule or from another intermediate. However, for these scenarios, all estimates would be much more cumbersome, while these processes (a variety of which are discussed in Refs. [23, 38, 39, 41, 50, 52, 81,95,101-104,114,115] do not show (in experiment) drastic advantages in the folding rate. Therefore, here we will not go beyond the simplest case of the coil-to-native globule transition.

Finally, it should be noted that a hierarchic scheme of protein folding [35,37,39,188], as well as many simplified "protein folding funnel" models [189,190], which ignore a separation of "native" and "unfolded" phases during the folding, do not solve the Levinthal paradox since they cannot provide a *simultaneous* [191,192] explanation for all of the three major observations about protein folding: (i) spontaneous, non-assisted folding of a unique native structure within non-astronomical time, (ii) the fact that the same native structure can be achieved at very different conditions (including the thermodynamic mid-transition), from different structural states and with very different folding rates, and (iii) co-existence, to a visible quantity, of only the native and unfolded molecules during folding of single-domain proteins near the thermodynamic mid-transition between the native and denatured states.

On the contrary, a nucleation-condensation mechanism can account for all these major features simultaneously and thus resolves the Levinthal paradox and opens a way to a theory of protein folding nuclei and protein folding/unfolding rates.

ACKNOWLEDGEMENTS

We thank A.V. Glyakina for assistance in calculations and E.V. Serebrova for assistance in manuscript preparation. This work was supported by the RFBR (grant numbers 11-04-00763a, 10-04-00162a, 13-04-00253a), by the Russian Academy of Sciences (Molecular and Cell Biology program (grants 01200957492, 01201358029, 01200959110, 01201353567) and "Fundamental Sciences to Medicine" program), by the Russian Young Scientists' grant (#MK-5540.2011.4), by MES RK and by the "Dynasty Foundation".

CONFLICT OF INTEREST

The authors confirm that this chapter contents have no conflict of interest.

DISCLOSURE

The chapter submitted for eBook series titled "Advances in Protein and Peptide Sciences Volume 1" is an update of our article published in Current Protein and Peptide Science (Invited Review), Volume 10, Number 3 (June), Pages 196-215, 2009 (PMID: 19519451), with additional text and references.

REFERENCES

[1] Stryer, L. *Biochemistry*, 4th ed.; W.H. Freeman & Co: New York, **1995**.

[2] Finkelstein, A.V.; Ptitsyn, O.B. *Protein physics. A course of lectures*; Academic Press, An Imprint of Elsevier Science: Amsterdam – Boston – London – NY – Oxford – Paris – San Diego – San Francisco – Singapore – Sydney – Tokyo, **2002**.

[3] Branden, C.; Tooze, J. *Introduction to protein structure*; Garland Publ Inc: New York, **1999**.

[4] Uversky, V.N.; Gillespie, J.R.; Fink, A.L. Why are "natively unfolded" proteins unstructured under physiologic conditions? *Proteins*, **2000**, *41*(3), 415-427.

[5] Xue, B.; Dunker, A.K.; Uversky, V.N. Orderly order in protein intrinsic disorder distribution: disorder in 3500 proteomes from viruses and the three domains of life. *J. Biomol. Struct. Dyn.*, **2012**, *30*(2), 137-149.

[6] Kendrew, J.C.; Bodo, G.; Dintzis, H.M.; Parrish, H.; Wyckoff, H.; Phillips, D.C. A three-dimensional model of the myoglobin molecule obtained by x-ray analysis. *Nature*, **1958**, *181*(4610), 662-666.

[7] Perutz, M.F.; Rossmann, M.; Cullis, G.A.F.; Muirhead, G.; Will, G.; North, T. Structure of haemoglobin: a three-dimensional Fourier synthesis at 5.5-A. resolution, obtained by X-ray analysis. *Nature*, **1960**, *185*(4711), 416-422.

[8] Wüthrich, K. *NMR of proteins and nucleic acids*. John Wiley & Sons: New York, **1986**.

[9] Anfinsen, C.B.; Haber, E.; Sela, M.; White, F.H. The kinetics of formation of native ribonuclease during oxidation of the reduced polypeptide chain. *Proc. Natl. Acad. Sci. USA*, **1961**, *47*, 1309-1314.

[10] Gutte, B.; Merrifield, R.B. The total synthesis of an enzyme with ribonuclease A activity. *J. Amer. Chem. Soc.*, **1969**, *91*(2), 501-502.

[11] Privalov, P.L.; Khechinashvili, N.N. A thermodynamic approach to the problem of stabilization of globular protein structure: a calorimetric study. *J. Mol. Biol.*, **1974**, *86*(3), 665-684.

[12] Creighton, T.E. *Proteins*, 2nd ed.; W. H. Freeman & Co.: New York, **1991**.

[13] Privalov, P.L. Stability of proteins: small globular proteins. *Adv. Protein Chem.*, **1979**, *33*, 167-241.

[14] Nojima, H.; Ikai, A.; Oshima, T.; Noda, H. Reversible thermal unfolding of thermostable phosphoglycerate kinase. Thermostability associated with mean zero enthalpy change. *J. Mol. Biol.*, **1977**, *116*(3), 429-442.

[15] Privalov, P.L. Stability of proteins. Proteins which do not present a single cooperative system. *Adv. Protein Chem.*, **1982**, *35*, 1-104.

[16] Tanford, C.; Kawahara, K.; Lapanje, S.; Hooker, T.M., Jr.; Zarlengo, M.H.; Salahuddin, A.; Aune, K.C.; Takagi, T. Proteins as random coils. 3. Optical rotatory dispersion in 6 M guanidine hydrochloride. *J. Am. Chem. Soc.*, **1967**, *89*(19), 5023-5029.

[17] Lapanje, S.; Tanford, C. Proteins as random coils. IV. Osmotic pressures, second virial coefficients, and unperturbed dimensions in 6 M guanidine hydrochloride. *J. Am. Chem. Soc.*, **1967**, *89*(19), 5030-5033.

[18] Tanford, C. Protein denaturation. *Adv. Prot. Chem.*, **1968**, *23*, 121-282.

[19] Kuwajima, K.; Sugai, S. Equilibrium and kinetics of the thermal unfolding of alpha-lactalbumin. The relation to its folding mechanism. *Biophys Chem.*, **1978**, *8*(3), 247-254.

[20] Dolgikh, D.A.; Abaturov, L.V.; Bolotina, I.A.; Brazhnikov, E.V.; Bychkova, V.E.; Gilmanshin, R.I.; Lebedev, Yu.O.; Semisotnov, G.V.; Tiktopulo, E.I.; Ptitsyn, O.B. Compact state of a protein molecule with pronounced small-scale mobility: bovine alpha-lactalbumin. *Eur. Biophys. J.*, **1985**, *13*(2), 109-121.

[21] Ptitsyn, O.B. Molten globule and protein folding. *Adv. Prot. Chem.*, **1995**, *47*, 83-229.

[22] Dobson, C.M. Protein folding. Solid evidence for molten globules. *Curr. Biol.*, **1994**, *4*(7), 636-640.

[23] Shakhnovich, E.I.; Finkelstein, A.V. Theory of cooperative transitions in protein molecules. I. Why denaturation of globular protein is the first order phase transition. *Biopolymers*, **1989**, *28*(10), 1667-1680.

[24] Finkelstein, A.V.; Shakhnovich, E.I. Theory of cooperative transitions in protein molecules. II. Phase diagram for protein molecule in solution. *Biopolymers*, **1989**, *28*(10), 1681-1694.

[25] Shakhnovich, E.I.; Gutin, A.M. Implications of thermodynamics of protein folding for evolution of primary sequences. *Nature*, **1990**, *346*(6286), 773-775.

[26] (a) Goldstein, R.A.; Luthey-Schulten, Z.A.; Wolynes, P.G. Optimal protein-folding codes from spin-glass theory. *Proc. Natl. Acad. Sci. USA*, **1992**, *89*(11), 4918-4922; (b) Goldstein, R.A.; Luthey-Schulten, Z.A.; Wolynes, P.G. Protein tertiary structure recognition using optimized Hamiltonians with local interactions. *Proc. Natl. Acad. Sci. USA*, **1992**, *89*(19), 9029-9033.

[27] Finkelstein, A.V.; Gutin, A.M.; Badretdinov, A.Y. Boltzmann-like statistics of protein architectures. Origins and consequences. In: *Subcellular Biochemistry*; Biswas, B.B., Roy, S., Eds.; Plenum Press: New York, **1995**; Vol. *24*, pp. 1-26.

[28] Kolb, V.A.; Makeev, E.V.; Spirin, A.S. Folding of firefly luciferase during translation in a cell-free system. *EMBO J.*, **1994**, *13*(15), 3631-3637.

[29] Ellis, R.J.; Hartl, F.U. Principles of protein folding in the cellular environment. *Curr. Opin. Struct. Biol.*, **1999**, *9*(1), 102-110.

[30] Hardesty, B.; Tsalkova, T.; Kramer, G. Co-translational folding. *Curr. Opin. Struct. Biol.*, **1994**, *9*(1), 111-114.

[31] Eichmann, C.; Preissler, S.; Riek, R.; Deuerling, E. Cotranslational structure acquisition of nascent polypeptides monitored by NMR spectroscopy. *Proc. Natl. Acad. Sci. USA*, **2010**, *107*(20), 9111-9116.

[32] Han, Y.; David, A.; Liu, B.; Magadan, J.G.; Bennink, J.R.; Yewdell, J.W.; Qian S.-B. Monitoring cotranslational protein folding in mammalian cells at codon resolution. *Proc. Natl. Acad. Sci. USA*, **2012**, *109*(31), 12467-12472.

[33] Fändrich, M.; Forge, V.; Buder, K.; Kittler, M.; Dobson, C.M.; Diekmann, S. Myoglobin forms amyloid fibrils by association of unfolded polypeptide segments. *Proc. Natl. Acad. Sci. USA*, **2003**, *100*(26), 15463-15468.

[34] Lührs, T.; Ritter, C.; Adrian, M.; Riek-Loher, D.; Bohrmann, B.; Döbeli, H.; Schubert, D.; Riek, R. 3D structure of Alzheimer's amyloid-beta(1-42) fibrils. *Proc. Natl. Acad. Sci. USA*, **2005**, *102*(48), 17342-17347.

[35] Levinthal, C. Are there pathways for protein folding? *J. Chim. Phys. Chim. Biol.*, **1968**, *65*, 44-45.

[36] Levinthal, C. *How to fold graciously*. In: Mössbauer spectroscopy in biological systems: Proceedings of a meeting held at Allerton House, Monticello, Illinois. Champaign; DeBrunner, J.T.P., Munck, E., Eds.; University of Illinois Press: IL, **1969**, pp. 22-24.

[37] Phillips, D.C. The three-dimensional structure of an enzyme molecule. *Sci. Am.*, **1966**, *215*(5), 78-90.

[38] (a) Goldenberg, D.P.; Creighton, T.E. Circular and circularly permuted forms of bovine pancreatic trypsin inhibitor. *J. Mol. Biol.*, **1983**, *165*(2), 407-413; (b) Goldenberg, D.P.; Creighton, T.E. Folding pathway of a circular form of bovine pancreatic trypsin inhibitor. *J. Mol. Biol.*, **1984**, *179*(3), 527-545.

[39] Ptitsyn, O.B. Stages in the mechanism of self-organization of protein molecules. *Dokl. Akad. Nauk SSSR (Moscow)*, **1973**, *210*(5), 1213-1215.

[40] Dolgikh, D.A.; Gilmanshin, R.I.; Brazhnikov, E.V.; Bychkova, V.E.; Semisotnov, G.V.; Venyaminov, S.Yu.; Ptitsyn, O.B. Alpha-Lactalbumin: compact state with fluctuating tertiary structure? *FEBS Lett.*, **1981**, *136*(2), 311-315.

[41] Dolgikh, D.A.; Kolomiets, A.P.; Bolotina, I.A.; Ptitsyn, O.B. 'Molten-globule' state accumulates in carbonic anhydrase folding. *FEBS Lett.*, **1984**, *165*(1), 88-92.

[42] Dyson, J.H.; Wright, P.E. Insights into protein folding from NMR. *Annu. Rev. Phys. Chem.*, **1996**, *47*, 369-395.

[43] Jackson, S.E. How do small single-domain proteins fold? *Fold. Des.*, **1998**, *3*(4), R81-R91.

[44] Samatova, E.N.; Katina, N.S.; Balobanov, V.A.; Melnik, B.S.; Dolgikh, D.A.; Bychkova, V.E.; Finkelstein, A.V. How strong are side chain interactions in the folding intermediate? *Protein Sci.*, **2009**, *18*(10), 2152-2159.

[45] Samatova, E.N.; Melnik, B.S.; Balobanov, V.A.; Katina, N.S.; Dolgikh, D.A.; Semisotnov, G.V.; Finkelstein, A.V.; Bychkova, V.E. Folding intermediate and folding nucleus for I—N and U—N transitions in apomyoglobin: different contributions of conserved and non-conserved residues. *Biophys. J.*, **2010**, *98*(8), 1694-1702.

[46] Uversky, V.N.; Semisotnov, G.V.; Pain, R.H.; Ptitsyn, O.B. 'All-or-none' mechanism of the molten globule unfolding. *FEBS Lett.*, **1992**, *314*(1), 89-92.

[47] Jones, C.M.; Henry, E.R.; Hu, Y.; Chan, C.K.; Luck, S.D.; Bhuyan, A.; Roder, H.; Hofrichter, J.; Eaton, W.A. Fast events in protein folding initiated by nanosecond laser photolysis. *Proc. Natl. Acad. Sci. USA*, **1993**, *90*(24), 11860-11864.

[48] Shastry, M.C.R.; Roder, H. Evidence for barrier-limited protein folding kinetics on the microsecond time scale. *Nat. Struct. Biol.*, **1998**, *5*(5), 385-392.

[49] Fersht, A.R. Characterizing transition states in protein folding: an essential step in the puzzle. *Curr. Opin. Struct. Biol.*, **1995**, *5*(1), 79-84.

[50] Dobson, C.M.; Karplus, M. The fundamentals of protein folding: bringing together theory and experiment. *Curr. Opin. Struct. Biol.*, **1999**, *9*(1), 92-101.

[51] Kragelund, B.B.; Robinson, C.V.; Knudsen, J.; Dobson, C.M.; Poulsen, F.M. Folding of a four-helix bundle: studies of acyl-coenzyme A binding protein. *Biochemistry*, **1995**, *34*(21), 7217-7224.

[52] Matouschek, A.; Kellis, J.T., Jr; Serrano, L.; Bycroft, M.; Fersht, A.R. Transient folding intermediates characterized by protein engineering. *Nature*, **1990**, *346*(6283), 440-445.

[53] Maxwell, K.L.; Wildes, D.; Zarrine-Afsar, A.; De Los Rios, M.A.; Brown, A.G.; Friel, C.T.; Hedberg, L.; Horng, J.C.; Bona, D.; Miller, E.J.; Vallée-Bélisle, A.; Main, E.R.; Bemporad, F.; Qiu, L.; Teilum, K.; Vu, N.D.; Edwards, A.M.; Ruczinski, I.; Poulsen, F.M.; Kragelund, B.B.; Michnick, S.W.; Chiti, F.; Bai, Y.; Hagen, S.J.; Serrano, L.; Oliveberg, M.; Raleigh, D.P.; Wittung-Stafshede, P.; Radford, S.E.; Jackson, S.E.; Sosnick, T.R.; Marqusee, S.; Davidson, A.R.; Plaxco, K.W. Protein folding: defining a "standard" set of experimental conditions and a preliminary kinetic data set of two-state proteins. *Protein Sci.*, **2005**, *14*(3), 602-616.

[54] Segawa, S.-I.; Sugihara, M. Characterization of the transition state of lysozyme unfolding. I. Effect of protein-solvent interactions on the transition state. *Biopolymers*, **1984**, *23*(11 pt. 2), 2473-2488.

[55] Fersht, A.R. Nucleation mechanisms in protein folding. *Curr. Opin. Struct. Biol.*, **1997**, *7*(1), 3-9.

[56] Matouschek, A.; Kellis, J.T., Jr.; Serrano, L.; Fersht, A.R. Mapping the transition state and pathway of protein folding by protein engineering. *Nature*, **1989**, *340*(6229), 122-126.

[57] Itzhaki, L.S.; Otzen, D.E.; Fersht, A.R. The structure of the transition state for folding of chymotrypsin inhibitor 2 analysed by protein engineering methods: evidence for a nucleation-condensation mechanism for protein folding. *J. Mol. Biol.*, **1995**, *254*(2), 260-288.

[58] Abkevich, V.I.; Gutin, A.M.; Shakhnovich, E.I. Specific nucleus as the transition state for protein folding: evidence from the lattice model. *Biochemistry*, **1994**, *33*(33), 10026-10036.

[59] Wetlaufer, D.B. Nucleation, rapid folding, and globular intrachain regions in proteins. *Proc. Natl. Acad. Sci. USA*, **1973**, *70*(3), 697-701.

[60] Sosnick, T.R.; Dothager, R.S.; Krantz, B.A. Differences in the folding transition state of ubiquitin indicated by phi and psi analyses. *Proc. Natl. Acad. Sci. USA*, **2004**, *101*(50), 17377-17382.

[61] Gallagher, T.; Alexander, P.; Bryan, P.; Gilliland, G.L. Two crystal structures of the B1 immunoglobulin-binding domain of streptococcal protein G and comparison with NMR. *Biochemistry*, **1994**, *33*(15), 4721-4729.

[62] Wikström, M.; Drakenberg, T.; Forsén, S.; Sjöbring, U.; Björck, L. Three-dimensional solution structure of an immunoglobulin light chain-binding domain of protein L. Comparison with the IgG-binding domains of protein G. *Biochemistry*, **1994**, *33*(47), 14011-14017.

[63] Galzitskaya, O.V. Sensitivity of the folding pathway to the details of amino acid sequence. *Mol. Biol. (Moscow)*, **2002**, *36*(3), 386-390.

[64] Finkelstein, A.V.; Galzitskaya, O.V. Physics of protein folding. *Phys. Life Rev.*, **2004**, *1*(1), 23-56.

[65] McCallister, E.L.; Alm, E.; Baker, D. Critical role of beta-hairpin formation in protein G folding. *Nat. Struct. Biol.*, **2000**, *7*(8), 669-673.

[66] Burton, R.E.; Huang, G.S.; Daugherty, M.A.; Calderoni, T.L.; Oas, T.G. The energy landscape of a fast-folding protein mapped by Ala-->Gly substitutions. *Nat. Struct. Biol.*, **1997**, *4*(4), 305-310.

[67] Fersht, A.R.; Matouschek, A.; Serrano, L. The folding of an enzyme. I. Theory of protein engineering analysis of stability and pathway of protein folding. *J. Mol. Biol.*, **1992**, *224*(3), 771-782.

[68] Fulton, K.; Main, E.; Daggett, V.; Jackson, S.E. Mapping the interactions present in the transition state for unfolding/folding of FKBP12. *J. Mol. Biol.*, **1992**, *291*(2), 445-461.

[69] Chiti, F.; Taddei, N.; White, P.; Bucciantini, M.; Magherini, F.; Stefani, M.; Dobson, C. Mutational analysis of acylphosphatase suggests the importance of topology and contact order in protein folding. *Nat. Struct. Biol.*, **1999**, *6*(11), 1005-1009.

[70] Villegas, V.; Martinez, J.C.; Aviles, F.X.; Serrano, L. Structure of the transition state in the folding process of human procarboxypeptidase A2 activation domain. *J. Mol. Biol.*, **1998**, *283*(5), 1027-1036.

[71] López-Hernándes, E.; Serrano, L. Structure of the transition state for folding of the 129 aa protein CheY resembles that of a smaller protein, CI-2. *Fold. Des.*, **1996**, *1*(1), 43-55.

[72] Kragelund, B.B.; Osmark, P.; Neergaard, T.B.; Schiodt, J.; Kristiansen, K.; Knudsen, J.; Poulsen, F.M. The formation of a native-like structure containing eight conserved hydrophobic residues is rate limiting in two-state protein folding of ACBP. *Nat. Struct. Biol.*, **1999**, *6*(6), 594-601.

[73] Grantcharova, V.P.; Riddle, D.S.; Santiago, J.V.; Baker, D. Important role of hydrogen bonds in the structurally polarized transition state for folding of the src SH3 domain. *Nat. Struct. Biol.*, **1998**, *5*(8), 714-720.

[74] Matouschek, A.; Otzen, D.E.; Itzhaki, L.S.; Jackson, S.E.; Fersht, A.R. Movement of the position of the transition state in protein folding. *Biochemistry*, **1995**, *34*, 13656-13662.

[75] Li, L.; Mirny, L.A.; Shakhnovich, E.I. Kinetics, thermodynamics and evolution of non-native interactions in a protein folding nucleus. *Nat. Struct. Biol.*, **2000**, *7*(4), 336-342.

[76] Martinez, J.C.; Serrano, L. The folding transition state between SH3 domains is conformationally restricted and evolutionarily conserved. *Nat. Struct. Biol.*, **1999**, *6*(11), 1010-1016.

[77] Riddle, D.S.; Grantcharova, V.P.; Santiago, J.V.; Alm, E.; Ruczinski, I.; Baker, D. Experiment and theory highlight role of native state topology in SH3 folding. *Nat. Struct. Biol.*, **1999**, *6*(11), 1016-1024.

[78] Perl, D.; Welker, C.; Schindler, T.; Schroder, K.; Marahiel, M.A.; Jaenicke, R.; Schmid, F.X. Conservation of rapid two-state folding in mesophilic, thermophilic and hyperthermophilic cold shock proteins. *Nat. Struct. Biol.*, **1998**, *5*(3), 229-235.

[79] Steensma, E.; van Mierlo, C.P. Structural characterisation of apoflavodoxin shows that the location of the stable nucleus differs among proteins with a flavodoxin-like topology. *J. Mol. Biol.*, **1998**, *282*(3), 653-666.

[80] Viguera, A.R.; Serrano, L.; Wilmanns, M. Different folding transition states may result in the same native structure. *Nat. Struct. Biol.*, **1996**, *3*(10), 874-880.

[81] Lindberg, M.O.; Tangrot, J.; Otzen, D.E.; Dolgikh, D.A.; Finkelstein, A.V.; Oliveberg, M. Folding of circular permutants with decreased contact order: general trend balanced by protein stability. *J. Mol. Biol.*, **2001**, *314*(4), 891-900.

[82] Lindberg, M.O.; Haglund, E.; Hubner, I.A.; Shakhnovich, E.I.; Oliveberg, M. Identification of the minimal protein-folding nucleus through loop-entropy perturbations. *Proc. Natl. Acad. Sci. USA*, **2006**, *103*(11), 4083-4088.

[83] Otzen, D.E.; Fersht, A.R. Folding of circular and permuted chymotrypsin inhibitor 2: retention of the folding nucleus. *Biochemistry*, **1998**, *37*(22), 8139-8146.

[84] Shakhnovich, E. Protein folding thermodynamics and dynamics: where physics, chemistry, and biology meet. *Chem. Rev.*, **2006**, *106*(5), 1559-1588.

[85] Grantcharova, V.; Alm, E.J.; Baker, D.; Horwich, A.L. Mechanisms of protein folding. *Curr. Opin. Struct. Biol.*, **2001**, *11*(1), 70-82.

[86] Finkelstein, A.V.; Badretdinov, A.Ya. Rate of protein folding near the point of thermodynamic equilibrium between the coil and the most stable chain fold. *Fold. Des.*, **1997**, *2*(2), 115-121.

[87] Finkelstein, A.V.; Badretdinov, A.Y. Physical reason for fast folding of the stable spatial structure of proteins: A solution of the Levinthal paradox. *Mol. Biol. (Moscow)*, **1997**, *31*(5), 391-398.

[88] Zana, R. On the rate determining step for helix propagation in the helix-coil transition of polypeptides in solution. *Biopolymers*, **1975**, *14*, 2425-2428.

[89] Kramers, H.A. Brownian motion in a field of force and the diffusion model of chemical reactions. *Physica*, **1940**, *7*, 284-304.

[90] Ubbelohde, A.R. *Melting of Crystal Structure*; Clarendon Press: Oxford, **1965**.

[91] Moore, J.W.; Pearson, R.G. *Kinetics and Mechanism*; John Wiley & Co: New York, **1981**.

[92] Lifshiz, E.M.; Pitaevskii, L.P. *Physical Kinetics*; Pergamon Press: London, **1981**.

[93] Gō, N. Theory of reversible denaturation of globular proteins. *Int. J. Pept. Prot. Res.*, **1975**, *7*(4), 313-323.

[94] Finkelstein, A.V.; Badretdinov, A.Ya. Influence of chain knotting on the rate of folding. ADDENDUM to: Rate of protein folding near the point of thermodynamic equilibrium between the coil and the most stable chain fold. *Fold. Des.*, **1998**, *3*(1), 67-68.

[95] Finkelstein, A.V.; Ivankov, D.N.; Galzitskaya, O.V. Prediction of folding rates and folding nuclei in globular proteins, based on theory of their folding. *Progr. Biol. Chem. (Moscow)*, **2005**, *45*(1), 3-36.

[96] Finkelstein, A.V.; Ivankov, D.N.; Garbuzynskiy, S.O.; Galzitskaya, O.V. Understanding the folding rates and folding nuclei of globular proteins. *Curr. Protein Pept. Sci.*, **2007**, *8*(6), 521-536.

[97] Ivankov, D.N.; Finkelstein, A.V. Prediction of protein folding rates from the amino-acid sequence-predicted secondary structure. *Proc. Natl. Acad. Sci. USA,* **2004**, *101*(24), 8942-8944.

[98] Gō, N. Theoretical studies of protein folding. *Ann. Rev. Biophys. Bioeng.*, **1983**, *12*, 183-210.

[99] Leopold, P.E.; Montal, M.; Onuchic, J.N. Protein folding funnels: a kinetic approach to the sequence-structure relationship. *Proc. Natl. Acad. Sci. USA,* **1992**, *89*(11), 8721-8725.

[100] Fu, B.; Wang, W. A $2^{O(n^{1-\frac{1}{d}} \log n)}$ time algorithm for d-dimensional protein folding in the HP-model. In: *Lecture Notes in Computer Science*; Díaz, J., Karhumäki, J., Lepistö, A., Sannella, D., Eds.; Springer: Berlin, **2004**; Vol. *3142*, pp. 630-644.

[101] Sali, A.; Shakhnovich, E.; Karplus, M. Kinetics of protein folding. A lattice model study of the requirements for folding to the native state. *J. Mol. Biol.*, **1994**, *235*(5), 1614-1636.

[102] Larson, S.M.; Ruczinski, I.; Davidson, A.R.; Baker, D.; Plaxco, K.W. Residues participating in the protein folding nucleus do not exhibit preferential evolutionary conservation. *J. Mol. Biol.*, **2002**, *316*(2), 225-233.

[103] Tsutsui, Y.; Cruz, R.D.; Wintrode, P.L. Folding mechanism of the metastable serpin α1-antitrypsin. *Proc. Natl. Acad. Sci. USA,* **2012**, *109*(12), 4467-4472.

[104] Thirumalai, D. From minimal models to real proteins: time scales for protein folding kinetics. *J. Phys. I (Orsay, Fr.)*, **1995**, *5*, 1457-1469.

[105] Gutin, A.M.; Abkevich, V.I.; Shakhnovich, E.I. Chain length scaling of protein folding time. *Phys. Rev. Lett.*, **1999**, *77*, 5433-5436.

[106] Wolynes, P.G. Folding funnels and energy landscapes of larger proteins within the capillarity approximation. *Proc. Natl. Acad. Sci. USA,* **1997**, *94*(12), 6170-6175.

[107] Li, M.S.; Klimov, D.K.; Thirumalai, D. Thermal denaturation and folding rates of single domain proteins: size matters. *Polymer*, **2004**, *45*, 573-579.

[108] Naganathan, A.N.; Muñoz, V. Scaling of folding times with protein size. *J. Am. Chem. Soc.*, **2005**, *127*(2), 480-481.

[109] Galzitskaya, O.V.; Surin, A.K.; Nakamura, H. Optimal region of average side-chain entropy for fast protein folding. *Protein Sci.*, **2000**, *9*(3), 580-586.

[110] Galzitskaya, O.V.; Garbuzynskiy, S.O. Entropy capacity determines protein folding. *Proteins*, **2006**, *63*(1), 144-154.

[111] Galzitskaya, O.V.; Glyakina, A.V. Nucleation-based prediction of the protein folding rate and its correlation with the folding nucleus size. *Proteins*, **2012**, *80*(12), 2711-2727.

[112] Galzitskaya, O.V. Estimation of protein folding rate from Monte Carlo simulations and entropy capacity. *Curr. Protein Pept. Sci.*, **2010**, *11*(7), 523-537.

[113] Galzitskaya, O.V.; Reifsnyder, D.C.; Bogatyreva, N.S.; Ivankov, D.N.; Garbuzynskiy, S.O. More compact protein globules exhibit slower folding rates. *Proteins*, **2008**, *70*(2), 329-332.

[114] Plaxco, K.V.; Simons, K.T.; Baker, D. Contact order, transition state placement and the refolding rates of single domain proteins. *J. Mol. Biol.*, **1998**, *277*(4), 985-994.

[115] Fersht, A.R. Transition-state structure as a unifying basis in protein-folding mechanisms: contact order, chain topology, stability, and the extended nucleus mechanism. *Proc. Natl. Acad. Sci. USA,* **2000**, *97*(4), 1525-1529.

[116] Gromiha, M.; Selvaraj, S. Comparison between long-range interactions and contact order in determining the folding rate of two-state proteins: application of long-range order to folding rate prediction. *J. Mol. Biol.*, **2001**, *310*(1), 27-32.

[117] Zhou, H.; Zhou, Y. Folding rate prediction using total contact distance. *Biophys. J.*, **2002**, *82*(1 pt. 1), 458-463.

[118] Micheletti, C. Prediction of folding rates and transition-state placement from native-state geometry. *Proteins*, **2003**, *51*(1), 74-84.

[119] Weikl, T.R.; Dill, K.A. Folding rates and low-entropy-loss routes of two-state proteins. *J. Mol. Biol.*, **2003**, *329*(3), 585-598.

[120] Punta, M.; Rost, B. Protein folding rates estimated from contact predictions. *J. Mol. Biol.*, **2005**, *348*(3), 507-512.

[121] Zhang, L.; Li, J.; Jiang, Z.; Xia, A. Folding rate prediction based on neural network model. *Polymer*, **2003**, *44*, 1751-1756.

[122] Eaton, W.A.; Muñoz, V.; Hagen, S.J.; Jas, G.S.; Lapidus, L.J.; Henry, E.R.; Hofrichter, J. Fast kinetics and mechanisms in protein folding. *Annu. Rev. Biophys. Biomol. Struct.*, **2000**, *29*, 327-359.

[123] Ivankov, D.N.; Bogatyreva, N.S.; Lobanov, M.Yu.; Galzitskaya, O.V. Coupling between properties of the protein shape and the rate of protein folding. *PLoS ONE,* **2009**, *4*(8), e6476.

[124] Mirny, L.; Shaknovich, E. Protein folding theory: from lattice to all-atom models. *Annu. Rev. Biophys. Biomol. Struct.*, **2001**, *30*, 361-396.

[125] Ivankov, D.N.; Garbuzynskiy, S.O.; Alm, E.; Plaxco, K.V.; Baker, D.; Finkelstein, A.V. Contact order revisited: Influence of protein size on the folding rate. *Protein Sci.*, **2003**, *12*(9), 2057-2062.

[126] Jones, K.; Wittung-Stafshede, P. The largest protein observed to fold by two-state kinetic mechanism does not obey contact-order correlation. *J. Am. Chem. Soc.*, **2003**, *125*(32), 9606-9607.

[127] Koga, N.; Takada, S. Roles of native topology and chain-length scaling in protein folding: a simulation study with a Go-like model. *J. Mol. Biol.*, **2001**, *313*(1), 171-180.

[128] Jones, D.T. Protein secondary structure prediction based on position-specific scoring matrices. *J. Mol. Biol.*, **1999**, *292*(2), 195-202.

[129] Gong, H.; Isom, D.G.; Srinivasan, R.; Rose, G.D. Local secondary structure content predicts folding rates for simple, two-state proteins. *J. Mol. Biol.*, **2003**, *327*(5), 1149-1154.

[130] Prabhu, N.P.; Bhuyan, A.K. Prediction of folding rates of small proteins: empirical relations based on length, secondary structure content, residue type, and stability. *Biochemistry*, **2006**, *45*(11), 3805-3812.

[131] Shao, H.; Peng, Y.; Zeng, Z-H. A simple parameter relating sequences with folding rates of small alpha helical proteins. *Prot. Pept. Lett.*, **2003**, *10*(3), 277-280.

[132] Shao, H.; Zeng, Z-H. A sequence function reveals new features in beta-protein folding. *Prot. Pept. Lett.*, **2003**, *10*(5), 435-439.

[133] Kuznetsov, I.; Rackovsky, S. Class-specific correlations between protein folding rate, structure-derived, and sequence-derived descriptors. *Proteins*, **2004**, *54*(2), 333-341.

[134] Gromiha, M.M. A statistical model for predicting protein folding rates from amino acid sequence with structural class information. *J. Chem. Int. Model.*, **2005**, *45*(2), 494-501.

[135] Gromiha, M.M.; Thangakani, A.M.; Selvaraj, S. FOLD-RATE: prediction of protein folding rates from amino acid sequence. *Nucl. Acids Res.*, **2006**, *34*(Web Server issue), W70-74.

[136] Ivankov, D.N.; Finkelstein, A.V. Theoretical study of a landscape of protein folding-unfolding pathways. Folding rates at mid-transition. *Biochemistry*, **2001**, *40*(33), 9957-9961.

[137] Galzitskaya, O.V.; Finkelstein, A.V. A theoretical search for folding/unfolding nuclei in three-dimensional protein structures. *Proc. Natl. Acad. Sci. USA,* **1999**, *96*(20), 11299-11304.

[138] Garbuzynskiy, S.O.; Finkelstein, A.V.; Galzitskaya, O.V. Outlining folding nuclei in globular proteins. *J. Mol. Biol.*, **2004**, *336*(2), 509-525.

[139] Garbuzynskiy, S.O.; Finkelstein, A.V.; Galzitskaya, O.V. On prediction of folding nuclei in globular proteins. *Mol. Biol. (Moscow)*, **2005**, *39*(6), 1032-1041.

[140] Galzitskaya, O.V, Garbuzynskiy, S.O.; Finkelstein, A.V. Theoretical study of protein folding: outlining folding nuclei and estimating of protein folding rates. *J. Phys. Cond. Matter*, **2005**, *17*(18), S1539-S1551.

[141] Muñoz, V.; Eaton, W.A. A simple model for calculating the kinetics of protein folding from three-dimensional structures. *Proc. Natl. Acad. Sci. USA,* **1999**, *96*(20), 11311-11316.

[142] Alm, E.; Morozov, A.V.; Kortemme, T.; Baker, D. Simple physical models connect theory and experiment in protein folding kinetics. *J. Mol. Biol.*, **2002**, *322*(2), 463-476.

[143] Finkelstein, A.V. *Proteins: Thermodynamic, structural and kinetic aspects.* In: Slow relaxations and nonequilibrium dynamics in condensed matter. UFJ NATO ASI. Les Houches, Session 77, 2002; Barrat, J.L., Feigelman, M., Kurchan, J., Dalibard, J., Eds.; EDP Sciences: Les Ulis – Paris – Cambridge; Springer-Verlag: Berlin – Heidelberg – New York – Hong Kong – London – Milan – Paris – Tokyo, **2003**; pp. 649-704.

[144] Finkelstein, A.V. Can protein unfolding simulate protein folding? *Protein Eng.*, **1997**, *10*(8), 843-845.

[145] Ivankov, D.N.; Finkelstein, A.V. Protein folding as flow across a network of folding-unfolding pathways. 1. The mid-transition case. *J. Phys. Chem. B*, **2010**, *114*(23), 7920-7929.

[146] Ivankov, D.N.; Finkelstein, A.V. Protein folding as flow across a network of folding-unfolding pathways. 2. The "in-water" case. *J. Phys. Chem. B*, **2010**, *114*(23), 7930-7934.

[147] Taketomi, H.; Ueda, Y.; Gō, N. Studies on protein folding, unfolding and fluctuations by computer simulation. I. The effect of specific amino acid sequence represented by specific inter-unit interactions. *Int. J. Pept. Prot. Res.*, **1975**, *7*(6), 445-459.

[148] Daggett, V.; Li, A.; Itzhaki, L.S.; Otzen, D.E.; Fersht, A.R. Structure of the transition state for folding of a protein derived from experiment and simulation. *J. Mol. Biol.*, **1996**, *257*(2), 430-440.

[149] Shaw, D.E.; Maragakis, P.; Lindorff-Larsen, K.; Piana, S.; Dror, R.O.; Eastwood, M.P.; Bank, J.A.; Jumper, J.M.; Salmon, J.K.; Shah, Y.; Wriggers, W. Atomic-level characterization of the structural dynamics of proteins. *Science*, **2010**, *330*(6002), 341-346.

[150] Voelz V.A.; Bowman G.R.; Beauchamp K.; Pande V.S. Molecular simulation of ab initio protein folding for a millisecond folder NTL9(1-39). *J. Am. Chem. Soc.*, **2010**, *132*(5), 1526-1528.

[151] Lindorff-Larsen, K.; Piana, S.; Dror, R.O.; Shaw, D.E. How fast-folding proteins fold. *Science*, **2011**, *334*(6055), 517-520.

[152] Voelz, V.A.; Jager, M.; Yao, S.; Chen, Y.; Zhu, L.; Waldauer, S.A.; Bowman, G.R.; Friedrichs, M.; Bakajin, O.; Lapidus, J.L.; Weiss, S.; Pande, V.S. Slow unfolded-state structuring in Acyl-CoA binding protein folding revealed by simulation and experiment. *J. Am. Chem. Soc.*, **2012**, *134*(30), 12565-12577.

[153] Best, R.B. Atomistic molecular simulations of protein folding. *Curr. Opin. Struct. Biol.*, **2012**, *22*(1), 52-61.

[154] Shakhnovich, E.; Abkevich, V.; Ptitsyn, O. Conserved residues and the mechanism of protein folding. *Nature*, **1996**, *379*(6560), 96-98.

[155] Sánchez, I.E.; Kiefhaber, T. Origin of unusual phi-values in protein folding: evidence against specific nucleation sites. *J. Mol. Biol.*, **2003**, *334*(5), 1077-1085.

[156] Fersht, A.R.; Sato, S. Phi-value analysis and the nature of protein-folding transition states. *Proc. Natl. Acad. Sci. USA,* **2004**, *101*(21), 7976-7981.

[157] Tseng, Y.Y.; Liang, J. Are residues in a protein folding nucleus evolutionarily conserved? *J. Mol. Biol.,* **2004**, *335*(4), 869-880.

[158] Ptitsyn, O.B. Protein folding and protein evolution: common folding nucleus in different subfamilies of c-type cytochromes? *J. Mol. Biol.*, **1998**, *278*(3), 655-666.

[159] Ptitsyn, O.B.; Ting, K.L. Non-functional conserved residues in globins and their possible role as a folding nucleus. *J. Mol. Biol.*, **1999**, *291*(3), 671-682.

[160] Mirny, L.A.; Shakhnovich, E.I. Universally conserved positions in protein folds: reading evolutionary signals about stability, folding kinetics and function. *J. Mol. Biol.*, **1999**, *291*(1), 177-196.

[161] Plaxco, K.W.; Larson, S.; Ruczinski, I.; Riddle, D.S.; Thayer, E.C.; Buchwitz, B.; Davidson, A.R.; Baker, D. Evolutionary conservation in protein folding kinetics. *J. Mol. Biol.*, **2000**, *298*(2), 303-312.

[162] Nölting, B.; Andret, K. Mechanism of protein folding. *Proteins*, **2000**, *41*(3), 288-298.

[163] Galzitskaya, O.V.; Skoogarev, A.V.; Ivankov, D.N.; Finkelstein, A.V. *Folding nuclei in 3D protein structures*. In: Proceedings of the Pacific Symposium on Biocomputing'2000; Altman, R.B., Dunker, A.K., Hunter, L., Lauderdale, K., Klein, T.E., Eds; World Scientific: Singapore – New Jersey – London – Hong Kong, **1999**, pp. 131-142.

[164] Cota, E.; Steward, A.; Fowler, S.B.; Clarke, J. The folding nucleus of a fibronectin type III domain is composed of core residues of the immunoglobulin-like fold. *J. Mol. Biol.*, **2001**, *305*(5), 1185-1194.

[165] Mirny, L.; Shakhnovich, E. Evolutionary conservation of the folding nucleus. *J. Mol. Biol.*, **2001**, *308*(2), 123-129.

[166] Garbuzynskiy, S.O.; Kondratova, M.S. Structural features of protein folding nuclei. *FEBS Lett.*, **2008**, *582*(5), 768-772.

[167] Efimov, A.V. Structural trees for protein superfamilies. *Proteins*, **1997**, *28*(2), 241-260.

[168] Li, A.; Daggett, V. Identification and characterization of the unfolding transition state of chymotrypsin inhibitor 2 by molecular dynamics simulations. *J. Mol. Biol.*, **1996**, *257*(2), 412-429.

[169] Caflisch, A.; Karplus, M. Acid and thermal denaturation of barnase investigated by molecular dynamics simulations. *J. Mol. Biol.*, **1995**, *252*(5), 672-708.

[170] Brooks, C.L. III; Gruebele, M.; Onuchic, J.N.; Wolynes, P.G. Chemical physics of protein folding. *Proc. Natl. Acad. Sci. USA,* **1998**, *95*(19), 11037-11038.

[171] Lazaridis, T.; Karplus, M. "New view" of protein folding reconciled with the old through multiple unfolding simulations. *Science*, **1997**, *278*(5345), 1928-1931.

[172] Isai, J.; Levitt, M.; Baker, D. Hierarchy of structure loss in MD simulations of src SH3 domain unfolding. *J. Mol. Biol.*, **1999**, *291*(1), 215-225.

[173] Daggett, V.; Fersht, A.R. The present view of the mechanism of protein folding. *Nat. Rev. Mol. Cell. Biol.*, **2003**, *4*(6), 497-502.

[174] Mayor, U.; Johnson, C.M.; Daggett, V.; Fersht, A.R. Protein folding and unfolding in microseconds to nanoseconds by experiment and simulation. *Proc. Natl. Acad. Sci. USA*, **2000**, *97*(25), 13518-13522.

[175] Ferguson, N.; Pires, J.R.; Toepert, F.; Johnson, C. M.; Pan, Y.P.; Volkmer-Engert, R.; Schneider-Mergener, J.; Daggett, V.; Oschkinat, H.; Fersht, A. Using flexible loop mimetics to extend phi-value analysis to secondary structure interactions. *Proc. Natl. Acad. Sci. USA*, **2001**, *98*(23), 13008-13013.

[176] Mayor, U.; Guydosh, N.R.; Johnson, C.M.; Grossman, J.G.; Sato, S.; Jas, G.S.; Freund, S.M.V.; Alonso, D.O.V.; Daggett, V.; Fersht, A.R. The complete folding pathway of a protein from nanoseconds to microseconds. *Nature*, **2003**, *421*(6925), 863-867.

[177] Onuchic, J.N.; Socci, N.D.; Luthey-Schulten, Z.; Wolynes, P.G. Protein folding funnels: the nature of the transition state ensemble. *Fold. Des.*, **1996**, *1*(6), 441-450.

[178] Dill, K.A.; Chan, H.S. From Levinthal to pathways to funnels. *Nat. Struct. Biol.*, **1997**, *4*(1), 10-19.

[179] Shmygelska, A. Search for folding nuclei in native protein structures. *Bioinformatics*, **2005**, *21*(Suppl. 1), i394-i402.

[180] Pande, V.S.; Rocksar, D.S. Folding pathway of a lattice model for proteins. *Proc. Natl. Acad. Sci. USA*, **1999**, *96*(4), 1273-1278.

[181] Shoemaker, B.A.; Wang, J.; Wolynes, P.G. Exploring structures in protein folding funnels with free energy functionals: the transition state ensemble. *J. Mol. Biol.*, **1999**, *287*(3), 675-694.

[182] Shoemaker, B.A.; Wolynes, P.G. Exploring structures in protein folding funnels with free energy functionals: the denatured ensemble. *J. Mol. Biol.*, **1999**, *287*(3), 657-674.

[183] Plotkin, S.S.; Onuchic, J.N. Investigation of routes and funnels in protein folding by free energy functional methods. *Proc. Natl. Acad. Sci. USA*, **2000**, *97*(12), 6509-6514.

[184] Alm, E.; Baker, D. Prediction of protein-folding mechanisms from free-energy landscapes derived from native structures. *Proc. Natl. Acad. Sci. USA*, **1999**, *6*(20), 11305-11310.

[185] Baker, D. A surprising simplicity to protein folding. *Nature*, **2000**, *405*(6782), 39-42.

[186] Takada, S. Go-ing for the prediction of protein folding mechanisms. *Proc. Natl. Acad. Sci. USA*, **1999**, *96*(21), 11698-11700.

[187] Krieger, E.; Darden, T.; Nabuurs, S.B.; Finkelstein, A.; Vriend, G. Making optimal use of empirical energy functions: force field parameterization in crystal space. *Proteins*, **2004**, *57*(4), 678-683.

[188] (a) Baldwin, R.L.; Rose, G.D. Is protein folding hierarchic? I. Local structure and peptide folding. *Trends Biochem. Sci.*, **1999**, *24*(1): 26-33; (b) Baldwin, R.L.; Rose, G.D. Is protein folding hierarchic? II. Folding intermediates and transition states. *Trends Biochem. Sci.*, **1999**, *24*(2), 77-83.

[189] Chan, H.S.; Dill, K.A. Protein folding in the landscape perspective: chevron plots and non-Arrhenius kinetics. *Proteins*, **1998**, *30*(1), 2-33.

[190] Bicout, D.J.; Szabo, A. Entropic barriers, transition states, funnels, and exponential protein folding kinetics: a simple model. *Protein Sci.*, **2000**, *9*(3), 452-465.

[191] Bogatyreva, N.S.; Finkelstein, A.V. Cunning simplicity of protein folding landscapes. *Protein Eng.*, **2001**, *14*(8), 521-523.

[192] Finkelstein, A.V. Cunning simplicity of a hierarchical folding. *J. Biomol. Struct. Dyn.*, **2002**, *20*(3), 311-313.

Send Orders for Reprints to reprints@benthamscience.net
Frontiers in Protein and Peptide Sciences, Vol. 1, 2014, 139-160

CHAPTER 6

Heat Shock Response Agents and the Diseases

Lütfi Tutar[1], Aykut Özgür[2], Esen Tutar[3], Kübra Açıkalın Coşkun[2] and Yusuf Tutar[4,*]

[1]Kahramanmaraş Sütçü İmam University, Faculty of Science and Letters, Department of Biology, Kahramanmaraş, Turkey; [2]Gaziosmanpasa University, Faculty of Natural Sciences and Engineering, Department of Bioengineering, Tokat, Turkey; [3]Kahramanmaraş Sütçü İmam University, ÜSKİM Research Center, Central Laboratory, Kahramanmaraş, Turkey and [4]Cumhuriyet University, Faculty of Pharmacy, Department of Basic Sciences, Division of Biochemistry, Sivas, Turkey

Abstract: Heat shock proteins (Hsps) are highly conserved proteins and have cytoprotective role for maintaining cellular protein conformation. Hsps not only keep proteins in their native state but also involve in several essential biochemical process. This review summarizes structural properties of Hsps (Hsp70, Hsp40, Hsp90, Hsp100, Hsp60, sHsps, and Nucleotide Exchange Factors) and explains their roles in aging, apoptosis, cancer, neurodegeneration, cardio-vascular diseases, obesity and diabetes mellitus, and housekeeping.

Keywords: Apoptosis, cancer, Heat shock proteins, neurodegeneration, protein aggregation, protein folding.

1. INTRODUCTION

Proteins are essential for biochemical functions of an organism. A protein must be properly folded to its native state to perform its function properly. This folding is ciphered in protein primary structure, and weak interactions direct the formation of a protein to the native state. However, several factors influence the folding process such as the milieu, molecular crowding, and cell stress factors. Therefore, a quality control mechanism is required for the cell to survive [1].

***Address correspondence to Yusuf Tutar:** Department of Basic Sciences, Division of Biochemistry, Faculty of Pharmacy, Cumhuriyet University, Sivas, Turkey, 58140; Tel: +90 346-219-1010 ext. 3907; Fax: +90 346-219-1634; E-mails: ytutar@cumhuriyet.edu.tr or ytutar@yahoo.com

Ben M. Dunn (Ed)

Heat shock proteins (Hsps) are highly conserved during evolution, and help non-native proteins to fold to their native state. Hsps can function as molecular chaperones since they facilitate protein folding and prevent protein aggregation. Coordinated activation of Hsps expression is an adaptation mechanism of organisms that helps their survival in their habitat. Hsps also function in protein translocation across membranes, in preservation and restructuring of the cytoskeleton. Hsps can be classified according to their molecular weight; Hsp15-30, Hsp40, Hsp60, Hsp70, Hsp90, and Hsp100. Each Hsp family consists of several isoforms and has analogous functions in different subcellular compartments. And most Hsp families have redundant forms in the same subcellular compartment [2]. Why each Hsp gene family members are constitutively or inducibly regulated, and/or targeted to different compartments in a cell is a question under debate in the literature. However, several lines of evidence indicate that each redundant Hsp member isoform forms a complex to perform specific functions, and induction of Hsps increases Hsp concentration to help constitutive Hsp form under stress [2].

Hsps form different complexes for variety of functions in a cell. If folding is required Hsp70 forms a complex with Hip, Hop and Hsp90. Hip plays a role in substrate protein binding and Hop recruits Hsp90 to form the folding machinery. Similarly Hsp70 forms a complex with Bag-1 and CHIP when a substrate protein sends it for degradation. In this case, Bag-1 and CHIP compete for Hip and Hop, respectively, for binding to Hsp70. Bag-1 recruits proteasome, and CHIP helps in ubiquitination process [3]. Furthermore, Hsp40, Hsp70 and Hsp100 coordinate together to dissolve substrate protein aggregates [1].

Hsps are activated by Heat shock factors (HSF). HSFs bind specifically to heat shock element (HSE). HSEs are stress responsive promoter elements throughout the genome and are located upstream of HSPs. HSEs are positioned upstream of HSP transcription sites in a tandem array of the binding site 5′-nGAAn-3′ and binding of HSFs induces transcription in cells under stress conditions. Under non-stress conditions, HSF exists as unbound monomer; however, upon stress HSF forms trimers and binds to the HSE to activate transcription. Post translational modifications are further involved in regulation and attenuation of HSFs in controlling stress gene expression [4-6].

2. STRUCTURE AND FUNCTION

Chaperones can be classified as "holder" and "folder". Holder chaperones bind to misfolded proteins and prevent their aggregation; however, folder chaperones help misfolded proteins to find their native folded state and prevent aggregation. Disaggregation mechanism is mediated by bi-chaperone system. Structure greatly determines function of Hsps and therefore basic properties of Hsps along with folding and disaggregation mechanisms are discussed briefly below.

2.1. Hsp70

Hsp70 has two major domain; ATPase and substrate binding domain (SBD). ATPase domain has a U shaped structure and binds ATP. Hydrolysis of ATP provides energy for SBD. SBD has a cavity to encapsulate a hydrophobic environment for substrate protein folding. SBD has a lid on top of this cavity to create the hydrophobic environment. The lid opens once the substrate protein is processed and closes when the next substrate is ready. The energy for opening-closing cycle of the lid comes from ATPase domain of the protein. Hsp70 SBD processes approximately seven residues per cycle. During this process substrate protein finds an environment to fold to its native state. After ATP hydrolysis, remaining ADP is taken from the ATPase domain by nucleotide exchange factor (NEF) as will be discussed below [7, 8].

Hsp70 proteins have several redundant forms in a cell. For example, *S. cerevisiae* has six cytosolic Hsp70s. Further, Hsp70 has two functional forms; constitutively expressed and inductively expressed forms. Why organisms have several isoforms is an elusive question in the literature [2].

Hsps cooperate and coordinate with other Hsps to form specific functions. Hsp70 forms complexes with Hsp40 and interacts with Hsp100 to solubilize protein aggregates. Hsp100 interacts with protein aggregates and shows "crowbar" effect and chops off a chunk of the substrate protein and Hsp70-Hsp40 complex folds the unfolded substrate protein to its native state [9, 10].

Beside the protein disaggregation mechanism, Hsp70 is also involved in protein folding machinery. Hsp70 forms a complex with Hsp90 and helps substrate proteins to find their native state as will be discussed in Hsp90 section.

2.2. Hsp40

Hsp40 proteins are called as co-chaperones since they help Hsp70 proteins. Hsp40 submits the substrate protein to Hsp70 and facilitates folding by putting the substrate to an optimum orientation and by increasing Hsp70 ATPase hydrolysis rate. All Hsp40s have J domain and the domain makes contacts with Hsp70 ATPase domain. This interaction stimulates ATP hydrolysis of Hsp70. Hsp70-Hsp40 pair either folds substrate protein or send it for degredation [11-13].

Hsp40 family consists of more than 100 members and some Hsp40s themselves show ATPase activity. These Hsp40s contribute to substrate protein folding as evidenced by luciferase folding experiments [13-18].

Several studies have been made to elucidate the reason for the excess number of the co-chaperons in a cell. The researchers suggested that Hsp70 pairs with Hsp40 to form specific functions and currently details of the mechanism are under investigation [17].

All Hsp40 families contain the J domain, this domain is a 70 kDa N-terminal domain which consists of four helices and a loop between helices II and III. Hsp40 proteins have other conserved domains and based on the differences in these regions, Hsp40 can be classified into three types; I, II, and III. Type I Hsp40s have a J domain, a Gly/Phe rich region, and a zinc finger-like region; however, type II Hsp40s possess a Gly/Phe rich region and a Gly/Met rich region. Type III Hsp40s only have a J domain but not have any other conserved region. Type I and type II Hsp40s may bind to substrate proteins but type II Hsp40s may not bind to substrate proteins. These different domains may determine functional diversity.

2.3. Hsp90

Hsp90 is one of the most abundant proteins expressed in unstressed cells. Like Hsp70, Hsp90 has both inducible and constitutively expressed isoforms in the cytosol in higher eukaryotes. Hsp90 has isoforms in endoplasmic reticulum and mitochondria as well. Hsp90 structure is similar to that of Hsp70; the ATP-binding and substrate protein binding domain. However, Hsp90 structure is different than Hsp70 structure by its dimerizing domain [19, 20]. Hsp90 displays a number of similarities with Hsp70 but unlike Hsp70, Hsp90 does not act on nascent proteins and Hsp90 is more selective than Hsp70 [21-23]. Hsp90 coordinates with Hsp70 and co-chaperones. Hsp70-Hsp40 complex interacts with a co-chaperone Hop (Hsp organizing protein). Then this complex interacts with Hsp90 [24-26]. Hsp90 functions in protein folding, aggregation inhibition, transcriptional regulation, signal transduction, and facilitation in transportation [27-31].

2.4. Hsp100

Hsp100 family, unlike other chaperones, helps to disassemble protein aggregates with Hsp70 and Hp40. Hsp100 protein functions in substrate protein degradation with proteases [32-34]. This highly conserved protein family contains ATP and peptide sites like most folder Hsps but the protein forms six-membered ring complexes. The ring shaped hexamer plays crucial role in protein disaggregation. The hexamer is ATP dependent and substrate proteins are unfolded and threaded through the ring. Similar to Hsp70 and Hsp90 ADP-bound state has low substrate protein binding affinity and ATP-bound state has high affinity [35].

The most accepted theory about Hsp100 is based on its unique structure. The protein shows "crowbar effect" when it interacts with protein aggregates. The hexamer pulls off substrate proteins from the aggregate and Hsp100 may pull down the substrate proteins from aggregates through the hollow spherical tunnel located in the middle of the protein. Hsp70-Hsp40 complex folds the substrate protein. The theory was supported by mutational work [36-38].

2.5. Hsp60

Hsp60 is composed of monomers and the monomers form a two tier heptameric rings. Each subunit consists of three domains: the apical, the equatorial and the intermediate domain. ATP binds to the equatorial domain and intermediate domain connects the equatorial and the apical domain. Hsp60 family works different than other Hsp families. Because Hsp60 barrel shaped structure can accommodate 50 kDa proteins inside and the capacity increases when the dome like Hsp10 closes the barrel [39, 40].

Hsp60 can be found both in the cytoplasm and in mitochondria. Hsp60 functions in protein folding, transport and maintenance of mitochondrial proteins, and transmission and replication of mitochondrial DNA [41]. Hsp70 and Hsp60 both fold substrate proteins but cannot complement each other, probably due to the size of substrate proteins; however, the two chaperone proteins act coordinately [42-44].

2.6. Small Hsps (sHsps)

The sHsps are relatively understudied and plays a crucial role in protein homeostasis and survival under stress conditions. The sHsps family is not conserved among organisms and the number of Hsps can vary among organisms. The protein share a 100 amino acid core, named α-crystallin. The name refers to eye-lens protein since both share similar hollow spherical structure [45].

sHsps are not holder chaperones but they hold substrate proteins. This protects cells from protein losses or toxicity caused by substrate protein aggregation. Then, sHsps release substrate proteins to other folder chaperones. The exact mechanism of holdase function of sHsps is elusive yet. But most sHsps form large oligomeric complexes and under stress oligomer complexes dissociates into smaller subunits. Thus, sHsps-hydrophobic patches of exposed substrate protein prevent protein aggregation [46]. sHsps function without employing ATP energy and it is possible that ATP-dependent Hsps can couple to the reaction and can coordinate sHsps function, *i.e.*, dissociation from substrate protein. sHsps are induced under stress and are involved in several RNA stabilization and cytoskeleton interaction [47-50].

2.7. Nucleotide Exchange Factors (NEF)

Hsp70 processes the substrate proteins in an ATP dependent manner. Energy derived from ATP hydrolysis closes the SBD lid and provide a hydrophobic environment. Once hydrolyzed, ADP must be removed from ATPase domain so that the system can be regenerated. NEFs serve to regenerate acceptor ATPases. NEF act as pliers and help ADP to move out from inside the ATPase domain and Hsp70 starts a new cycle to fold another substrate protein [51, 52]. These ADP-ATP exchange proteins are consisting of evolutionary unrelated proteins, *i.e.*, GrpE, Bag1, Fes1p and Hsp110 [53].

3. HOUSEKEEPING

Stress induces agents in a cell and these agents enhance redox state of the cell, which, in turn, increases misfolded proteins. The cellular response to stress is increased synthesis of Hsps because misfolded proteins often display altered biochemical activities. Hsps play an essential role in protein homeostasis through substrate protein folding and guide misfolded/unfolded proteins to find their native conformation. This also helps unwanted protein-protein interactions and prevents protein aggregation [54, 55]. Heat shock proteins are expressed both inductively and constitutively. And some Hsps are expressed under stress but the rest are expressed under non-stress conditions. Why both types of Hsps are expressed in a cell is still a question to be answered. One plausible explanation is that under stress protein misfolding increases and expression of inducible Hsps increases the concentration of overall Hsp and this helps protein folding activity of the cell. Under non-stress conditions constitutive expression of Hsps monitors the cell protein homeostasis. Hsps also send some proteins to proteasome and help degradation of misfolded or unrepaired substrate proteins. These biochemical activities are part of a cell heat shock response [56-58].

4. AGING

Aging is a commonly used word for senescence. Since it is more convenient, the word aging will be used in this review. Aging can be defined as deteriorative changes that cause mortality. Macromolecular damage and mutations are the primary cause of aging. In organisms several diseases are associated with accumulation of abnormal proteins. The aggregates are typically associated with Hsps. Aging theories define the life span as resistance to intrinsic and extrinsic stress, therefore Hsps are important factors in aging [59, 60]. Heat shock factor (HSF) transcription is regulated by expression of Hsps. Hsp40, Hsp70, and Hsp90 association keeps HSF in an inactive state. Increased protein unfolding causes Hsps to dissociate away from HSF [61, 62]. And this leads a sequence of events; HSF trimerization, activation and translocation to the nucleus. Here, it binds to heat shock elements (HSEs) in the promoters of Hsp genes. This activates Hsps expression and once Hsp levels have raised and/or the stress is removed HSF returns to an inactive monomeric form. Cellular stress causes protein unfolding and this will decrease available Hsps. Insufficient amount of Hsps that will be transduced to their networks will alter biochemical functions of Hsps. The functional alterations in the physiology of the cell can lead to deleterious effects and aging. An example of signaling pathway to aging is the insulin/insulin (IGF)-like signaling (IIS) pathway. The pathway is essential for regulation of life span and aging. IIS acts over transcription factor Forkhead and Hsps are the target [63].

Aging is also associated with altered Hsps gene expression during normal aging in different tissues and in diseases specific patterns under unstressed animal [63].

Details of aging mechanism are elusive yet but several line of evidence indicates the involvement of Hsps in aging process. Several age related human diseases are associated with protein abnormalities; misfolding, modification, and aggregation. The misfolded protein related diseases such as Alzheimer's disease (AD), Huntington's disease (HD), and Parkinson's disease (PD) have age dependence. It is known that misfolded proteins are associated with Hsps and Hsps inhibit protein aggregation; therefore, Hsps involved in aging. Aforementioned IIS-HSF biochemical pathway maybe an example of proteotoxicity since inhibition of IIS and overexpression of HSF activates Hsp expression. The expression of Hsp dissolves the protein aggregates and increases life span in *C. elegans* [64].

Hsps send unprocessed substrate proteins for degradation by using lysosomal pathway. Hsp mediated autophagy declines during aging [65].

Limited number of cell divisions can be defined as cellular aging. The cells can no longer divide and resist the apoptosis pathway. The stress activates p53 and pRB tumor-suppressor proteins and this cause cell-cycle inhibitor accumulation. One way that cells can escape cellular aging is Hsp70 overexpression. The overexpression inhibits p53 pathway and cell-cycle inhibitors [66, 67].

Anti-cancer therapies induce anti-apoptotic sHsp related protein clusterin, Hsp27, and Hsp70. These protein expressions are associated with development and progression of cancers. Both cellular aging and apoptosis are involved in normal tissue maintenance and Hsps play essential roles in regulation of these processes. Abnormal expression of Hsps may disrupt the pathways during aging and Hsp expression play a systemic role in the life span [68, 69].

5. APOPTOSIS

Genes encoding HSPs are regulated at the transcriptional level by cell stress, cell cycle, cell proliferation, and, differentiation [70]. Hsp expression is also mediated by apoptosis, programmed cell death, and is associated with several diseases as will be discussed below [71]. For this reason, it is not surprising that growth factor withdrawal, hypoxia, DNA damage, and cytotoxic factors induce both expression of Hsps and apoptosis [72]. Hsps block cell death and promote cell survival and proliferation. Hsp90, Hsp70, and Hsp27 directly interact with different proteins in apoptotic pathways and thereby block the apoptosis at distinct points.

Apoptotic process can be divided to two pathways; intrinsic and extrinsic. Hsps can block the intrinsic apoptosis pathway upstream of the mitochondria (by modulating signaling pathway), at the mitochondrial level (by controlling cytochrome c), and at the post-mitochondrial level (by inhibiting Hsps interaction with cytochrome c, Apaf-1). Hsps block the extrinsic pathway by interacting with death receptors. Hsp27, Hsp70, or Hsp90 block caspase 8 activation. The extrinsic pathway can also be regulated by caspase independent level through Hsp70 [72]. Therefore, Hsps are key regulators in the balance of apoptosis *versus* cell survival.

6. DISEASES

Before this section, we have discussed the structure and function of Hsps. The expression level of Hsps may be altered by environmental and genetic factors. The level of Hsps is decreased with ageing and it may give rise to many diseases. Furthermore, expression of Hsps is induced under stresses, and thus they can serve as a diagnostic and therapeutic tool in diseases. In this section, we divided the diseases into sub-groups, and focused on protective and therapeutical role of Hsps.

6.1. Cancer

Hsps are involved in all phases of cancer from proliferation to invasion and metastasis. Hsps maintain cellular homeostasis either by refolding denatured proteins or by sending them to lysosome for degradation.

Chemotherapy and radiotherapy have toxic effects such as high temperature and nausea; however, anti-cancer vaccine has low toxicity. There is growing number of evidence suggesting that anti-cancer vaccine stimulation can lead to immune defense to attack cancer cells [73]. Hsps can be overexpressed due to tumor aggressiveness and resistance to therapies [74]. Extracellular Hsps play essential roles in the stimulation of the innate and adaptive immunity [75].

Indeed, production of high levels of heat shock proteins is increased when cells are exposed to high temperatures. A variety of anti-cancer agents, hormones, and tumor related receptors affect expression of Hsps. *Hsp genes* are associated with some oncogenes such as *P53, c-myc, IGF-IR, Stat3 and NF-κB*. Furthermore, they are related with apoptosis-related proteins including caspase 8, caspase 3, Bcl-2, Bcl-X_L and Bax [76-78].

Hsp70 and Hsp90 are key chaperon proteins in oncogenic pathways. Generally, they play significant role in many different types of cancer [79]. These proteins collaborate with steroid receptors, tyrosine kinase receptors, transcription factors, serine-threonine kinases family and the other cell cycle regulator factors in formation and development of tumor [75, 77, 78].

In cancer, Hsp70 works together with Hsp90 coordinately, and it is involved in inhibition of the apoptosis process. Therefore, Hsp70 activates $CD8^+$ and $CD4^+$ lymphocytes, natural killer cells (NKC), dendritic cells and cytokines in the tumor cells. NKC are a type of white blood cells, and they play an important role in apoptosis. To perform this function, NKC attacks membrane bound Hsp70, and this interaction occurs *via* CD94 receptors on the surface of the NKC. At this stage, cells release granzyme B and trigger apoptosis in tumor cells [76, 80].

On the other hand, Hsp90 stabilizes a number of proteins involved in tumor growth. According to the literature, increased estrogen receptor levels have close relationship with the formation of breast cancer. Surprisingly, Hsp90 functions in the stabilization of steroid hormone receptors; therefore, inhibition of Hsp90 is targeted in many anti-cancer drugs, for example, 17-allylamino-17-demethoxygeldanamycin [77, 78]. Furthermore, overexpression of Hsp27 is associated with estrogen receptors levels, and resistance to anti-cancer drugs in tumor cells [77, 81].

Nowadays, the role of all types of Hsps has been considered for all cancer types in cancer studies. According to the results, the mechanisms of cancer will be illuminated, and thus new anticancer agents will be synthesized soon. In this context, Capello and co-workers researched effects of Hsp10, Hsp27, Hsp60, Hsp70, and Hsp90 on urothelial carcinoma of urinary bladder. In this study, tumoral Hsp90 and Hsp70 expression are correlated with cancer pathology. However, Hsp10 is the most important diagnostic and therapeutic component among other Hsps [82]. In other studies, Isomoto and co-workers studied the expression level of Hsp40 and Hsp70 in 81 gastric cancer patients. According to the results, overexpression of Hsp40 and Hsp70 is detected in cancer tissues. Also, these proteins were determined as an effective biomarkers for gastric cancer [83].

6.2. Neurodegenerative Diseases

Certain cellular proteins (prion protein, α-synuclein, β-amyloid, tau protein, *etc.*) are involved in occurrence of different diseases. Protein misfolding and aggregation are proposed to be the main reasons for neurodegenerative diseases. Accumulation of misfolded proteins causes variety of neurodegenerative diseases, such as prion disease, AD, PD, HD, and Hsps appear to play roles in prevention and maintenance of these diseases. To prevent protein misfolding and aggregation, overexpression of Hsps is observed in neuronal cell and mouse models [84-87].

AD is composed of accumulation of neurofibrillary tangles (NFTs) and β-amyloid (Aβ) plaques in brain [88]. NFTs contain microtubule-associated protein tau, which is hyperphosphorylated in AD. Hsp90, Hsp70, Hsp60 and sHsps play a key role in AD, and they are localized in neuritic plaques and NFTs. Hsp90 and Hsp70 increase solubility of tau protein, and prevent aggregation of protein. Dou *et al.* demonstrated that the expression level of Hsp90 and Hsp70is inverse proportionate to accumulated tau protein in transgenic mouse brain. Moreover, Hsp90 and Hsp70 contribute degradation of abnormal folded tau protein, as well as inhibit accumulation of misfolded tau proteins in brain. In AD, Hsp60 is a specific stress protein for tubulin, and it may inhibit tau aggregation [88-90]. Several sHsps, Hsp20 and HspB8, are highly expressed in senile plaques and they are extremely affected on Aβ protein. These sHsps interact with Aβ protein, and inhibits Aβ aggregation. Furthermore, they play a role in correction of misfolded Aβ protein [91].

PD is an age-related neurodegenerative movement disorder and it is characterized by accumulation of Lewy bodies in brain. Lewy bodies contain misfolded α-synuclein protein, neurofilaments, ubiquitinated proteins, and Hsps. Hsp70 is the most important protective protein in PD brain. Outeiro *et al.* indicated that significant effect of sHsps on α-synuclein aggregation and pathogenesis of PD. According to the studies, Hsp27 and αβ-crystallin are potent inhibitors of α-synuclein fibrillization and aggregation. Oxidative stress can lead to mitochondrial dysfunction, which is linked with PD. In a mouse model, electron transport chain inhibitors are lead to PD symptoms. Hsp27 inhibit the apoptosis pathway of mitochondria and also it has a protective role for PD [92-94].

Normal prion (PrP^{c}) is α-helix rich protein, which is highly expressed in brain. However, PrP^{c} can be converted into β-sheet rich scrapie infected prion protein (PrP^{Sc}) with genetic and environmental factors in prion-related diseases such as Creutzfeldt-Jakob disease, mad-cow disease, Kuru, Gerstmann-Sträussler-Scheinker syndrome and fatal familial insomnia (FFI). As might be expected, nearly all Hsps family participates in prion diseases as protective and therapeutic agents. In scrapie infected brain, Hsp70 binds misfolded prions aggregates, and helps their degradation process by the proteasome. Notably, Hsp100 has been studied for its ability in protein disaggregation [95, 96]. In yeast, Hsp100 mediates fragmentation of large prion aggregates into small prion seeds for prion propagation [1]. Apart from Hsp70 and Hsp100, Hsp60 and Hsp10 concomitantly prevent conversation of conversion of PrP^{c} to PrP^{Sc} in *S. cerevisiae* [97].

Finally, the other well-known neurodegenerative diseases such as HD, amyotrophic lateral sclerosis, and multiple sclerosis (MS) are related with expression level of Hsps. All mentioned Hsps provide mechanisms to prevent the neurons from accumulation misfolded protein aggregations, and they enable fragmentation and degradation of protein aggregates.

6.3. Cardiovascular Diseases

Cardiovascular diseases (CVD) are the biggest cause of deaths in the worldwide. Almost all cardiovascular diseases occur due to a number of similar risk factors:

age, high-cholesterol diet, obesity, diabetes, tobacco and alcohol, family history, life-style, hypertension, and hypercholesterolemia [98].

A number of stresses including hypoxia, thermal stress, reactive oxygen species (ROS), and ischemia lead to heart damage and blood vessels disorders. Also, Hsps are prominently expressed in response to these stresses, and they deal with protection of cardiovascular system by inhibition of apoptotic processes. Hence, Hsps are localized in arterial wall, where they have mediator or inducer effects for CVD [99-107].

There is some significant literature on the protector role of Hsp70 in CVD. According to the many studies, high serum levels of Hsp70 reduce the likelihood occurrence of the coronary artery disease, while development of atherosclerosis is associated with low serum levels of Hsp70 [101]. Under hypoxia, hyperthermia and oxidative stress, expression level of Hsp70 is decreased in the cardiac cells. Hsp27 is a critical small heat shock protein which works together with Hsp70 in CVD under various stresses. Hsp27 plays several important roles in oxidative stress, anti-inflammation and apoptosis that all are related with atherosclerosis [103]. Geiger *et al.* indicated that upregulation of Hsp27 and Hsp70 is observed in human pulmonary vasculature, and they are intensely expressed in the pulmonary arteries, and congenital heart disease is associated with abnormal pulmonary hemodynamics [105]. Hsp60 is a mitochondrial cell stress protein and its expression level is an important biomarker in atherosclerosis. Also, variable Hsp60 levels strongly relate with diabetes-related heart diseases [98, 100, 104].

6.4. Obesity and Diabetes Mellitus

Obesity and diabetes have a strong relationship with each other, and also the significant increase of obesity and diabetes can be induced by the occurrence of various different diseases such as cancer, cardiovascular diseases, nephropathy, and vision disorders. Indeed, obesity induces insulin resistance; as a result the frequency of type 2 diabetes is increased in humans [108, 109].

It is now widely accepted that several cellular stresses such as inflammation, oxidative stresses, and lipid oxidation are associated with diabetes mellitus. According to the numerous studies, low Hsps level is an important parameter in diabetes. On the other hand, diabetic nephropathy is due to continual diabetes mellitus, and expression level of Hsps may be increased in different part of the kidney. Remarkably, expression of Hsp25, Hsp60, and Hsp72 is increased in the outer medulla for cytoprotection and counterbalance diabetes-induced cell damage, but Hsp90 levels remains stable [108-112].

Imatoh and co-workers carried out epidemiological study on the effect of Hsp60 on type 2 diabetes. According to the results, Hsp60 levels are down-regulated in type 2 diabetes, and also a low level of expression is determined as 2.03 times lower than in the control group [110].

Finotti and co-worker were focused on Hsp70 in type 1 diabetes. Hsp70 can be fused with α_1-antitrypsin (α_1-AT), and thus composed fusion protein is Hsp70-α_1-AT. This protein has immunogen character and inhibits the occurrence of type 1 diabetes [111].

6.5. Other Diseases

Hsps are also significant therapeutic and diagnostic components in cancer, neurodegenerative diseases, cardiovascular diseases, and diabetes as well as gastrointestinal diseases, cataract, and other autoimmune diseases.

Acute pancreatitis or pancreatic necrosis is an important gastrointestinal disease which is an inflammation of pancreas. In recent years, a number of studies have indicated the effect of the Hsps on the formation, and development of this disease. Hsp70 and Hsp60 are the biggest candidates for providing protection from acute pancreatitis. Tanaka and Mizushima have indicated that Hsp70 and HSF-1 play a protective role in irritant-induced gastric lesions, and inflammatory bowel diseases by using apoptosis mechanisms [113-115].

Cataracts are an age-related disease and its major complication is blindness. According to the literature oxidative stress, protein aggregation, and mutated alpha-crystallin gene lead to formation of cataracts and probably we can describe cataract diseases as a protein aggregation diseases. Alpha-crystallin is a water soluble small heat shock protein which is the main structural component of eye lens. Alpha-crystallin has chaperone-like properties and numerous studies have been suggested that these functions are significant for the maintenance of eye lens and protection from cataract disease [116-119].

7. CLINICAL SIGNIFICANCE

Hsps are related with a wide range of human cancers and are implicated in tumor cell proliferation, differentiation, invasion, metastasis, death, and recognition by the immune system [120]. The levels of stress-inducible Hsps are unusually high in cancer cells. Several authors have proposed the concept that malignant cells are addicted to stress proteins. Hsp70 is an essential factor for tumor cell survival and tumor growth. Most cancer cells highly express Hsp70 at different stages of tumorigenesis and during anti-cancer treatment in contrast to normal cells. Furthermore, the recent clinical trials on Hsp90 inhibitors showed a compensatory increase in Hsp70 levels. Beside the recent work on Hsp70 inhibitors, necessity of Hsp90 inhibitors with simultaneous inhibition of Hsp70 have come forward [121]. Helper and folder function of Hsps in several biochemical process make them potential targets for therapy especially in vaccination and in drug design [1].

8. CONCLUSION

In the recent biotechnological and pharmaceutical developments the importance of Hsps have been understood better for metabolic functions. All members of the Hsp family are involved in protection and maintenance of cells from cellular and environmental stresses. Especially, Hsps are involved in protein folding and maintenance processes, and also elimination of misfolded proteins from the cells. Furthermore, expression level of Hsps is a significant biomarker for almost all diseases and biological events. In anti-cancer drug development, Hsps are discussed as useful and effective biomolecules for cancer vaccine. Now, Hsps are great candidates for treatment and diagnosis of diseases.

ACKNOWLEDGEMENTS

Dr. Tutar acknowledges Turkish National Academy of Science-Young Scientist Program (TUBA-GEBIP/28).

CONFLICT OF INTEREST

The authors confirm that this chapter contents have no conflict of interest.

DISCLOSURE

Declared None.

ABBREVIATIONS

CVD	=	Cardiovascular diseases
HET	=	Heat shock elements
Hop	=	Hsp organizing protein
HSF	=	Heat shock factors
Hsps	=	Heat shock proteins
NEF	=	Nucleotide exchange factors
NKC	=	Natural killer cells
ROS	=	Reactive oxygen species
SBD	=	Substrate Binding Domain
sHsps	=	Small Hsps
α_1-AT	=	α_1-antitrypsin

REFERENCES

[1] Tutar, L.; Tutar, Y. Heat shock proteins; an overview. *Curr. Pharm. Biotechno.,* **2010**, 11(2), 216-222.

[2] Tutar, Y.; Song, Y.; Masison, D.C.; Primate chaperones Hsc70 (constitutive) and Hsp70 (induced) differ functionally in supporting growth and prion propagation in Saccharomyces cerevisiae. *Genetics*, **2006**, 172(2), 851-861.

[3] Lanneau, D.; Wettstein, G.; Bonniaud, P.; Garrido, C. Heat shock proteins: Cell protection through protein triage. *Scientific World Journal,* **2010**, 10, 1543-1552.

[4] Xu, Y.M.; Huang, D.Y.; Chiu, J.F.; Lau, A.T.Y. Post-translational modification of human heat shock factors and their functions: A recent update by proteomic approach. *J. Proteome Res.,* **2012**, 11(5), 2625-2634.

[5] Sorger, P.K. Heat shock factor and the heat shock response. *Cell*, **1991**, 65(3), 363-366.
[6] Morimoto, R.I. Cells in stress: transcriptional activation of heat shock genes. *Science*, **1993**, 259(5100), 1409-1410.
[7] Sharma, D.; Masison, D.C. Hsp70 structure, function, regulation and influence on yeast prions. *Protein Pept. Lett.*, **2009**, 16(6), 571-581.
[8] Young, J.C. Mechanisms of the Hsp70 chaperone system. *Biochem. Cell Biol.*, **2010**, 88(2), 291-300.
[9] Lum, R.; Tkach, J.M.; Vierling, E.; Glover, J.R. Evidence for an unfolding/threading mechanism for protein disaggregation by Saccharomyces cerevisiae Hsp104. *J. Biol. Chem.* **2004**, 279(28), 29139-29146.
[10] Glover, J.R.; Lindquist, S. Hsp104, Hsp70, and Hsp40: a novel chaperone system that rescues previously aggregated proteins. *Cell*, **1998**, 94(1), 73-82.
[11] Li, J.; Qian, X.; Sha, B. Heat shock protein 40: structural studies and their functional implications. *Protein Pept. Lett.*, **2009**, 16(6), 606-612.
[12] Qiu, X.B.; Shao, Y.M.; Miao, S.; Wang, L. The diversity of the DnaJ/Hsp40 family, the crucial partners for Hsp70 chaperones. *Cell Mol. Life Sci.*, **2006**, 63(22), 2560-2570.
[13] Fan, C.Y.; Lee, S.; Cyr, D.M. Mechanisms for regulation of Hsp70 function by Hsp40. *Cell Stress Chaperones*, **2003**, 8(4), 309-316.
[14] Ramos, C.H.; Oliveira, C.L.; Fan, C.Y.; Torriani, I.L.; Cyr, D.M. Conserved central domains control the quaternary structure of type I and type II Hsp40 molecular chaperones. *J. Mol. Biol.*, **2008**, 383(1), 155-166.
[15] Fan, C.Y.; Lee, S.; Ren, H.Y.; Cyr, D.M. Exchangeable chaperone modules contribute to specification of type I and type II Hsp40 cellular function. *Mol. Biol. Cell.*, **2004**, 15(2), 761-773.
[16] Fan, C.Y.; Ren, H.Y.; Lee, P.; Caplan, A.J.; Cyr, D.M.; Lu, Z.J. The type I Hsp40 zinc finger-like region is required for Hsp70 to capture non-native polypeptides from Ydj1. *Biol. Chem.*, **2005**, 280(1), 695-702.
[17] Cyr, D.M. Protein folding activity of Hsp70 is modified differentially by the hsp40 co-chaperones Sis1 and Ydj1. *J. Biol. Chem.*, **1998**, 273(43), 27824-27830.
[18] Cyr, D.M.; Lu, X.; Douglas, M.G. Regulation of Hsp70 function by a eukaryotic DnaJ homolog. *J. Biol. Chem.*, **1992**, 267(29), 20927-20931.
[19] Mayer, M.P.; Bukau B. Molecular chaperones: the busy life of Hsp90. *Curr. Biol.*, **1999**, 9(9), 322-325.
[20] Bukau, B(Ed).; Molecular Chaperones and Folding Catalysis. *Harwood Academic Publishers*, *Singapore*, **1999**.
[21] Frydman, J. Folding of newly translated proteins *in vivo*: the role of molecular chaperones. *Annu. Rev. Biochem.*, **2001**, 70, 603-647.
[22] Pearl, L.H.; Prodromou, C.; Workman, P. The Hsp90 molecular chaperone: an open and shut case for treatment. *Biochem. J.*, **2008**, 410(3), 439-453.
[23] Picard, D. Heat-shock protein 90, a chaperone for folding and regulation. *Cell. Mol. Life Sci.*, **2002**, 59(10), 1640–1648.
[24] Hung, J.J.; Wu, C.Y.; Liao, P.C.; Chang, W.C. Hsp90alpha recruited by Sp1 is important for transcription of 12(S)-lipoxygenase in A431 cells. *J. Biol. Chem.*, **2005**, 280(43), 36283-36292.
[25] Csermely, P.; Schnaider, T.; Soti, C.; Prohászka, Z.; Nardai, G. The 90-kDa molecular chaperone family: structure, function, and clinical applications. A comprehensive review. *Pharmacol. Ther.*, **1998**, 79(2), 129-168.

[26] Roiniotis, J.; Masendycz P.; Ho, S.; Scholz, G.M. Domainmediated dimerization of the Hsp90 cochaperones Harc and Cdc37. *Biochemistry*, **2005**, 44(17), 6662-6669.

[27] Scheufler, C.; Brinker, A.; Bourenkov, G.; Pegoraro, S.; Moroder, L.; Bartunik, H.; Hartl, F.U.; Moarefi, I. Structure of TPR domainpeptide complexes: critical elements in the assembly of the Hsp70-Hsp90 multichaperone machine. *Cell*, **2000**, 101(2), 199-210.

[28] Xu, Y.; Lindquist, S. Heat-shock protein hsp90 governs the activity of pp60v-src kinase. *Proc. Natl. Acad. Sci. USA*, **1993**, 90(15), 7074-7078.

[29] Garnier, C.; Lafitte, D.; Jorgensen, T.J.; Jensen, O.N.; Briand, C.; Peyrot, V. Phosphorylation and oligomerization states of native pig brain HSP90 studied by mass spectrometry. *Eur. J. Biochem.*, **2001**, 268(8), 2402-2407.

[30] Calderwood, S.K.; Khaleque, M.A.; Sawyer, D.B.; Ciocca, D.R. Heat shock proteins in cancer: chaperones of tumorigenesis. *Trends Biochem. Sci.*, **2006**, 31(3), 164-172.

[31] Sasaki, M.; Nie, L.; Maki, C.G. MDM2 binding induces a conformational change in p53 that is opposed by heat-shock protein 90 and precedes p53 proteasomal degradation. *J. Biol. Chem.*, **2007**, 282(19), 14626-14634.

[32] Doyle, S.M.; Shorter, J.; Zolkiewski, M.; Hoskins, J.R.; Lindquist, S.; Wickner, S. Asymmetric deceleration of ClpB or Hsp104 ATPase activity unleashes protein-remodeling activity. *Nat. Struct. Mol. Biol.*, **2007**, 14(2), 114-122.

[33] Parsell, D.A.; Sanchez, Y.; Stitzel, J.D.; Lindquist, S. Hsp104 is a highly conserved protein with two essential nucleotide-binding sites. *Nature*, **1991**, 353(6341), 270-273

[34] Hattendorf, D. A.; Lindquist, S.L. Cooperative kinetics of both Hsp104 ATPase domains and interdomain communication revealed by AAA sensor-1 mutants. *EMBO J.*, **2002**, 21(1-2), 12-21.

[35] Bösl, B.; Grimminger, V.; Walter, S. Substrate binding to the molecular chaperone Hsp104 and its regulation by nucleotides. *J. Biol. Chem.*, **2005**, 280(46), 38170-38176.

[36] Glover, J.R.; Tkach. J.M. Crowbars and ratchets: Hsp100 chaperones as tools in reversing protein aggregation. *Biochem. Cell Biol.*, **2001**, 79(5), 557-568.

[37] Tessarz, P.; Mogk, A.; Bukau, B. Substrate threading through the central pore of the Hsp104 chaperone as a common mechanism for protein disaggregation and prion propagation. *Mol. Microbiol.*, **2008**, 68(1), 87-97.

[38] Wendler, P.; Shorter, J.; Plisson, C.; Cashikar, A.G.; Lindquist, S.; Saibil, H.R. A typical AAA+ subunit packing creates an expanded cavity for disaggregation by the protein-remodeling factor Hsp104. *Cell*, **2007**, 131(7), 1366-1377.

[39] Arsène, F.; Tomoyasu, T.; Bukau, B. The heat shock response of Escherichia coli. *Int. J. Food Microbiol.*, **2000**, 55(1-3), 3-9.

[40] Bukau, B.; Horwich, A.L. The Hsp70 and Hsp60 chaperone machines. *Cell*, **1998**, 92(3), 351-366.

[41] Ranford, J.C.; Coates, A.R.; Henderson, B. Chaperonins are cell-signalling proteins: the unfolding biology of molecular chaperones. *Expert Rev. Mol. Med.*, **2000**, 2(8), 1–17.

[42] Habich, C.; Burkart, V. Heat shock protein 60: regulatory role on innate immune cells. *Cell Mol. Life Sci.*, **2007**, 64(6), 742-751.

[43] Sigler, P.B.; Xu, Z.; Rye, H.S.; Burston, S.G.; Fenton, W.A.; Horwich, A.L. Structure and function in GroEL-mediated protein folding. *Annu. Rev. Biochem.*, **1998**, 67, 581-608.

[44] Richardson, A.; Landry, S.J.; Georgopoulos, C. The ins and outs of a molecular chaperone machine. *Trends Biochem. Sci.*, **1998**, 23(4), 138-143.

[45] De Jong, W.W.; Leunissen, J.A.; Voorter, C.E. Evolution of the alpha-crystallin/small heat-shock protein family. *Mol. Biol. Evol.*, **1993**, 10(1), 103-126.

[46] Kim, K.K.; Kim, R.; Kim, S-H. Crystal structure of a small heatshock protein. *Nature*, **1998**, 394(6693), 595-599.

[47] Lelj-Garolla, B.; Mauk, A.G. Self-association and chaperone activity of Hsp27 are thermally activated. *J. Biol. Chem.*, **2006**, 281(12), 8169-8174.

[48] Pivovarova, A.V.; Chebotareva, N.A.; Chernik, I.S.; Gusev, N.B.; Levitsky, D.I. Small heat shock protein Hsp27 prevents heatinduced aggregation of F-actin by forming soluble complexes with denatured actin. *FEBS J.*, **2007**, 274(22), 5937-5948.

[49] Kocabiyik, S. Essential Structural and Functional Features of Small Heat Shock Proteins in Molecular Chaperoning Process. *Protein Pept. Lett.*, **2009**, 16, 613-622.

[50] Eyles, S.J.; Gierasch, L.M. Nature's molecular sponges: small heat shock proteins grow into their chaperone roles. *PNAS*, **2010**, 107, 7 2727-2728.

[51] Harrison, C.J.; Hayer-Hartl, M.; Di Liberto, M.; Hartl, F.; Kuriyan, J. Crystal structure of the nucleotide exchange factor GrpE bound to the ATPase domain of the molecular chaperone DnaK. *Science*, **1997**, 276(5311), 431-435.

[52] Grimshaw, J.P.; Jelesarov, I.; Schönfeld, H.J.; Christen, P. Reversible thermal transition in GrpE, the nucleotide exchange factor of the DnaK heat-shock system. *J. Biol. Chem.*, **2001**, 276(9), 6098-6104.

[53] Kabani, M. Structural and Functional Diversity Among Eukaryotic Hsp70 Nucleotide Exchange Factors. *Protein Pept. Lett.*, **2009**, 16, 623-630.

[54] Söti, C.; Nagy, E.; Giricz, Z.; Vigh, L.; Csermely, P.,; Ferdinanly, P. Heat shock proteins as emerging therapeutic targets. *Brit. J. Pharmacol.,* **2005**, 146, 769-780.

[55] Pockley, A.G. Heat shock proteins in health and disease: therapeutic targets or therapeutic agents. *Expert Rev. Mol. Med.,* **2001**, 3(23) 1-21.

[56] Brignull, H.R.; Morley, J.F.; Morimoto, R.I. The stress of misfolded proteins: C. elegans models for neurodegenerative disease and aging. *Landes Bioscience and Springer Science* **2007**, 167-189.

[57] Chaudhuri, T.K.; Paul, S. Protein-misfolding diseases and chaperone-based therapeutic approaches. *FEBS J.,* **2006**, 273, 1331-1349.

[58] Adachi, H.; Katsuno, M.; Waza, M.; Minamiyama, M.; Tanaka, F.; Sobue, G. Heat shock proteins in neurodegenerative diseases: Pathogenic roles and therapeutic implications. *Int. J. Hyperther.,* **2009**, 25(8), 647-654.

[59] Morimoto, R.I. Proteotoxic stress and inducible chaperone networks in neurodegenerative disease and aging. *Genes Dev.*, **2008**, 22(11), 1427-1438.

[60] Tower, J. Hsps and aging. *Trends Endocrinol Metab.*, **2009**, 20(5), 216-222.

[61] Voellmy, R. On mechanisms that control heat shock transcription factor activity in metazoan cells. *Cell Stress Chaperon.*, **2004**, 9(2), 122-133.

[62] Soti, C.; Csermely, P. Aging cellular networks: chaperones as major participants. *Exp. Gerontol.* **2007**, 42(1-2), 113-119.

[63] Tower, J. Hsps and aging. *Trends Endocrinol. Metab.*, **2009**, 20(5), 216-222.

[64] Cohen, E.; Bieschke, J.; Perciavalle, R.M.; Kelly, J.W.; Dillin, A. Opposing activities protect against age-onset proteotoxicity. *Science*, **2006**, 313(5793), 1604-1610.

[65] Mizushima, N.; Levine, B.; Cuervo, A.M.; Klionsky, D.J. Autophagy fights disease through cellular self-digestion. *Nature*, **2008**, 451(7182), 1069-1075.

[66] Campisi, J.; Andersen, J.K.; Kapahi, P.; Melov, S. Cellular senescence: a link between cancer and age-related degenerative disease? *Semin. Cancer Biol.*, **2011**, (6), 354-359.

[67] Campisi, J. Senescent cells, tumor suppression, and organismal aging: good citizens, bad neighbors. *Cell*, **2005**, 120(4), 513-522.

[68] Rérole, A.L.; Jego, G.; Garrido, C. Hsp70: anti-apoptotic and tumorigenic protein. *Methods. Mol. Biol.*, **2011**, 787, 205-230.

[69] Garrido, C.; Brunet, M.; Didelot, C.; Zermati, Y.; Schmitt, E.; Kroemer, G. Heat shock proteins 27 and 70: anti-apoptotic proteins with tumorigenic properties. *Cell Cycle*, **2006**, (22), 2592-2601.

[70] Jolly, C.; Morimoto, R.I. Role of the heat shock response and molecular chaperones in oncogenesis and cell death. *J. Natl. Cancer Inst.* **2000**, 92(19), 1564-1572.

[71] Samali, A.; Orrenius, S. Heat shock proteins: regulators of stress response and apoptosis. *Cell Stress Chaperon.*, **1998**, 3(4), 228-236.

[72] Lanneau, D.; Brunet, M.; Frisan, E.; Solary, E.; Fontenay, M.; Garrido, C. Heat shock proteins: essential proteins for apoptosis regulation. *J. Cell Mol. Med.*, **2008**, 12(3), 743-761.

[73] Porter, D.L.; Levine, B.L.; Kalos, M.; Bagg, A.; June, C.H. Chimeric antigen receptor-modified T cells in chronic lymphoid leukemia. *N. Engl. J. Med.*, **2011**, 365(8), 725-733.

[74] Ciocca, D.R.; Fanelli, M.A.; Cuello-Carrion, F.D.; Castro. G.N. Heat shock proteins in prostate cancer: from tumorigenesis to the clinic. *Int. J. Hyperthermia*, **2010**, 26(8), 737-747.

[75] Ciocca, D.R.; Cayado-Gutierrez, N.; MAccioni, M.; Cuello-Carrion, F.D. Heat shock proteins (HSPs) based anti-cancer vaccines. *Curr. Mol. Med.,* **2012**, 12, 1-15.

[76] Tutar, Y. Hsp70 in oncology. *Recent Pat. DNA Gene Seq.,* **2011**, 5(3), 214-218.

[77] Romanucci, M.; Bastow, T.; Salda, L.D. Heat shock proteins in animal neoplasms and human tumours - a comparison. *Cell Stress Chaperon.,* **2008**, 13, 253-262.

[78] Moser, C.; Lang, S.A.; Stoeltzing, O. Heat-shock protein 90 (Hsp90) as a molecular target for theraphy of gastrointestinal cancer. *Anticancer Res.,* **2009**, 29, 2031-2042.

[79] Calderwood, S.K. Heat shock proteins in breast cancer progression – A suitable case for treatment. *Int. J. Hyperther.,* **2010**, 26(7), 681-685.

[80] Multhoff, G.; Pockley A.G.; Streffer, C.; Gaipl, U.S. Dual role of heat shock proteins (HSPs) in anti-tumor immunity. *Curr. Mol. Med.,* **2012**, 1-9.

[81] Sun, Y.; MacRae, T.H. The small heat shock proteins and their role in human disease. *FEBS J.,* **2005**, 272, 2613-2627.

[82] Cappello, F.; David, S.; Ardizzone, N.; Rappa, F.; Marasà, L.; Bucchieri, F., G.; Zummo. Expression of heat shock proteins HSP10, HSP27, HSP60, HSP70, and HSP90 in urothelial carcinoma of urinary bladder. *J. Cancer Mol.*, **2006**, 2(2), 73-77.

[83] Isomoto, H.; Oka, M.; Yano, Y.; Kanazawa, Y.; Soda, H.; Terada, R.; Yasutake, T.; Nakayama, T.; Shikuwa, S.; Takeshima, F.; Udono, H.; Murata, I.; Ohtsuka, K.; Kohno, S. Expression of heat shock protein (Hsp) 70 and Hsp40 in gastric cancer. *Cancer Lett.,* **2003**, 219-228.

[84] Brignull, H.R.; Morley, J.F.; Morimoto, R.I. The stress of misfolded proteins: C. elegans models for neurodegenerative disease and aging. *Adv. Exp. Med. Biol.,* **2007**, 167-189.

[85] Chaudhuri, T.K.; Paul, S. Protein-misfolding diseases and chaperone-based therapeutic approaches. *FEBS J.,* **2006**, 273, 1331-1349.

[86] Adachi, H.; Katsuno, M.; Waza, M.; Minamiyama, M.; Tanaka, F.; Sobue, G. Heat shock proteins in neurodegenerative diseases: Pathogenic roles and therapeutic implications. *Int. J. Hyperther.*, **2009**, 25(8), 647-654.

[87] Noort, J.M.; Stress proteins in CNS inflammation. *J. Pathol.,* **2008**, 214, 267-275.

[88] Wisniewski, T.; Sigurdsson, E.M. Therapeutic approaches for prion and Alzheimer's diseases. *FEBS Lett.*, **2007**, 274, 3784-3798.

[89] Koren, J.; Jinwal, U.K.; Lee, D.C.; Jones, J.R:; Shults, C.L.; Johnson, A.G.; Anderson, L.J.; Dickey, C.A. Chaperone signaling complexes in Alzheimer's disease. *J. Cell. Mol. Med.,* **2009**, 13 (4), 619-630.

[90] Dietz, G.P.H. Cell-penetrating peptide technology to deliver chaperones and associated factors in diseases and basic research. *Curr. Pharm. Biotechno.,* **2010**, 11, 167-174.

[91] Wilhelmus, M.M.M.; Boelens, W.C.; Otte-Holler, I.; Kamps B.; Wall, R.M.W.; Verbeek, M.M. Small heat shock proteins inhibit amyloid-β protein aggregation and cerebrovascular amyloid-β protein toxicity. *Brain Res.,* **2006**, 1089, 67-78.

[92] Licker, V.; Kövari, E.; Hochstrasser, D.F.; Burkhard, P.R. Proteomic in human Parkinson's disease research. *J. Proteomics,* **2009**, 73, 10-29.

[93] Luo, G.R.; Chen, S.; Le, W.D. Are heat shock proteins therapeutic target for Parkinson's disease. *Int. J. Biol. Sci.,* **2007**, 3, 20-26.

[94] Outeiro, T.F.; Klucken, J.; Strathearn, K.E.; Liu, F.; Nguyen, P.; Rochet, J.C:; Hyman, B.T:; McLean, P.J. Small heat shock proteins protect against α-synuclein-induced toxicity and aggregation. *Biochem. Bioph. Res. Co.,* **2006**, 351, 631-638.

[95] Kanapathipillai, M.; Ku, S.H.; Girigoswami, K.; Park, C.B. Small stress molecules inhibit aggregation and neurotoxicity of prion peptide 106-126. *Biochem. Bioph. Res. Co.,* **2008**, 365, 808-813.

[96] Harrison, P.M.; Bamborough, P.; Daggett, V.; Prusiner, S.B.; Cohen, F.E. The prion folding problem. *Curr. Opin. Struc. Biol.,* **1997**, 7(53), 53-59.

[97] Tutar, Y. Yeast prions. *Basic.Clin. Sci.*, **2010**, 1, 45-51.

[98] Scamaei-Tousi, A.; Stepoe, A.; O'Donnell, K.; Palme, J.; Stephens, J.W.; Hurel, S.J.; Marmot, M.; Homer, K.; D'Aiuto, F.; Coates, A.R.M.; Humphries, S.E.; Henderson, B. Plasma heat shock protein 60 and cardiovascular diseases risk: the role of psychosocial, genetic, and biological factors. *Cell Stress Chaperon.,* **2007**, 12(4), 384-392.

[99] Pockley, A.G.; Frostegard, J. Heat shock protein in cardiovascular disease and the prognostic value of heat shock protein related measurement. *Heart,* **2005**, 91, 1124-1126.

[100] Shamaei-Tousi, A.; Stephens, J.W.; Bin, R.; Cooper, J.A.; Steptoe, A.; Coates, A.R.M.; Henderson, B.; Humphries, S.E. Association between plasma levels of heat shock protein 60 and cardiovascular disease in patients with diabetes mellitus. *Eur. Hearth J.,* **2006**, 27, 1565-1570.

[101] Bielecka-Dabrowa, A.; Barylaki, M.; Mikhailidis, D.P.; Rysz, J.; Banach, M. HSP 70 and atherosclerosis- protector or activator. *Expert. Opin. Ther. Tar.,* **2009**, 13(3), 307-317.

[102] Pockley, A.G. Heat shock proteins, inflammations, and cardiovascular diseases. *Am. Heart Ass.,* **2002**, 105, 1012-1017.

[103] Ghayour-Mobarhan, M.; Saber, H.; Ferns, G.A.A. The potential role of heat shock protein 27 in cardiovascular disease. *Clin. Chim. Acta,* **2012**, 413, 15-24.

[104] Mala, J.G.S.; Takeuchi, S. Heat shock proteins in cardiovascular stress. *Clin. Med. Cardiol.,* **2008**, 2, 245-256.

[105] Geiger, R.; Sharma, H.S.; Mooi, W.J.; Berger, R.M.F. Pulmonary vascular remodeling in congenital heart disease: Enhanced expression of heat shock proteins. *Indian J. Biochem. Bio.,* **2009**, 46, 482-490.

[106] Gray, C.C.; Amrani, M.; Yacoub, M.H. Heat stress proteins and myocardial protection: experimental model or potential clinical tool. *Int. J. Biochem. Cell B.,* **1999**, 31, 559-573.

[107] Latchman, D.S. Heat shock proteins and cardiac protection. *Cardiovasc. Res.* **2001**, 51, 637-646.

[108] Gupte, A.A.; Bomhoff, G.L.; Swerdlow, R.H.; Geiger, P.C. Heat treatment improves glucose tolerance and prevents skeletal muscle insulin resistance n rats fed a high-fat diet. *Diabetes* **2009**, 58, 567-577.

[109] Chung, J.; Nguyen, A.K.; Henstridge, D.C.; Holmes, A.G.; Chan, M.H.S.; Mesa, J.L.; Lancaster, G.I.; Southgate, R.J.; Bruce, C.R.; Duffy, S.J.; Hovath, I.; Mestril, R.; Watt, M.; Hooper, P.L.; Kinwell, B.A.; Vigh, L.; Hevener, A.; Febbraio, M.A. HSP72 protects against obesity-induced insulin resistance. *PNAS* **2008**, 105(5), 1739-1744.

[110] Imatoh, T.; Sugie, T.; Miyazaki, m:; Tanihara, S.; Baba, M.; Momose, Y.; Uryu, Y.; Une, H. Is heat shock protein 60 associated with type 2 diabetes mellitus. *Diabetes Res. Clin. Pr.* **2009**, 85, 208-212.

[111] Finotti, P.; Pagetta, A. A heat shock protein 70 fusion protein with α1-antitrypsin in plasma of type 1 diabetic subjects. *Biochem. Bioph. Res. Co.,* **2004**, 315, 297-305.

[112] Hooper, P.L.; Hooper, P.L. Inflammations, heat shock proteins, and type 2 diabetes. Cell *Stress Chaperon.,* **2009**, 14, 113-115.

[113] Tanaka, K.I.; Mizushima, T. Protective role of HSF1 and Hsp70 against gastrointestinal diseases. *Int. J. Hyperther.,* **2009**, 25(8), 668-676.

[114] Feng, J.Y.; Li, Y.Y. Alteration and role of heat shock proteins in acute pancreatitis. *J. Dig. Dis.,* **2010**, 11, 277-283.

[115] Dudeja, V.; Vickers, S.M.; Saluja, A.K. The role of heat shock proteins in gastrointestinal diseases. *Gut,* **2009**, 58(7), 1000-1009.

[116] Marsili, S.; Salganik, R.I.; Albright, C.D.; Freel, C.D.; Johnsen, S.; Peiffer, R.L.; Costello, M.J. Cataract formation in a strain of rats selected for high oxidative stress. *Exp. Eye Res.,* **2004**, 79, 595-612.

[117] Moreau, K.L.; King, J.A. Protein misfolding and aggregation in cataract disease and prospects for prevention. *Cell* **2012**, 15 (5), 273-281.

[118] Graw, J. Genetics of crystallins: Cataract and beyond. *Exp. Eye Res.,* **2009**, 88, 173-189.

[119] Alice, R.C.; Lubsen, N.H.; Slingsby, C. sHSP in the eye lens: Crystallin mutations, cataract and proteostasis. *Int. J. Biochem. Cell B.,* **2012**, 44, 1687-1697.

[120] Ciocca, D.R.; Calderwood, S.K. Heat shock proteins in cancer: diagnostic, prognostic, predictive, and treatment implications. *Cell Stress Chaperon.,* **2005**, 10(2), 86–103.

[121] Goloudinaa, A.R.; Demidova, O.N.; Garrido, C. Inhibition of HSP70: A challenging anti-cancer strategy. *Cancer Lett.,* **2012**, DOI: 10.1016/j.canlet.2012.06.

Send Orders for Reprints to reprints@benthamscience.net
Frontiers in Protein and Peptide Sciences, Vol. 1, **2014, 161-173**

CHAPTER 7

Prediction of Protein Secondary Structure Content by Using the Concept of Chou's Pseudo Amino Acid Composition and Support Vector Machine

Chao Chen[1,*], Lixuan Chen[2], Xiaoyong Zou[3] and Peixiang Cai[3]

[1]School of Traditional Chinese Medicine, Guangdong Pharmaceutical University, Guangzhou 510006, PR China; [2]Guangzhou Institute of Standardization, Guangzhou 510170, PR China and [3]School of Chemistry and Chemical Engineering, Sun Yat-Sen University, Guangzhou 510275, PR China

Abstract: Protein secondary structure carries information about local structural arrangements. Significant majority of successful methods for predicting the secondary structure is based on multiple sequence alignment. However, the multiple alignment fails to achieve accurate results when a protein sequence is characterized by low homology. To this end, we propose a novel method for prediction of secondary structure content through comprehensive sequence representation. The method is featured by employing a support vector machine (SVM) regressing system and adopting a different pseudo amino acid composition (PseAAC), which can partially take into account the sequence-order effects to represent protein samples. It was shown by both the self-consistency test and the independent-dataset test that the trained SVM has remarkable power in grasping the relationship between the PseAAC and the content of protein secondary structural elements, including α-helix, 3_{10}-helix, π-helix, β-strand, β-bridge, turn, bend and the rest random coil. Results prior to or competitive with the popular methods have been obtained, which indicate that the present method may at least serve as an alternative to the existing predictors in this area.

Keywords: Pseudo Amino acid composition, support vector machine, protein secondary structure content, prediction.

1. INTRODUCTION

A priori knowledge of secondary structure content can be of great help for both theoretical and experimental determination of protein structure. Experimental

*****Address correspondence to Chao Chen:** School of Traditional Chinese Medicine, Guangdong Pharmaceutical University, Guangzhou 510006, PR China; Tel: +86 20 39352181; Fax: +86 20 39352174; E-mail: cep02cc@gmail.com

Ben M. Dunn (Ed)

techniques, such as circular dichroism spectroscopy (CD), Fourier transform infrared spectroscopy (FT-IR) and nuclear magnetic resonance spectroscopy (NMR), have been routinely used to probe protein secondary structure content. However, with the avalanche of new protein sequences in the post-genomic era, it is vitally important to develop a computational predictor for fast and accurately determining the protein secondary structure content. In the history of predicting the protein secondary structure content, Krigbaum and Knutton [1] introduced the multiple linear regression (MLR), and only used the conventional amino acid composition (AA) as the input. However, in representing a protein sample solely with its AA, many important features associated with the sequence order are completely missed, undoubtedly reducing the success rate of prediction. In view of this, various descriptors that partly took into account the sequence-order effects were proposed for the purpose of enhancing the prediction quality, including the nonlinear and coupling terms of the AA [2, 3], the 1st-order coupled AA [4], the pair-coupled AA [5], the polypeptide composition [6, 7], the pseudo amino acid (PseAA) composition or PseAAC [8], the functional domain composition [9] and other compositions [10, 11]. The PseAAC was originally introduced by Chou to improve the prediction quality for protein subcellular localization and membrane protein type [8]. It can be used to represent a protein sequence with a discrete model yet without completely losing its sequence-order information. For a brief introduction about Chou's PseAAC, visit the Wikipedia web-page at http://en.wikipedia.org/wiki/Pseudo_amino_acid_composition. Since the concept of Chou's pseudo amino acid composition was introduced, various pseudo amino acid composition approaches have been stimulated to deal with varieties of problems in proteins and protein-related systems [12-27]. Because of its wide usage, recently a very flexible pseudo amino acid composition generator, called "PseAAC" [28], was established at the website http://chou.med.harvard.edu/bioinf/PseAAC/, by which users can generate 63 different kinds of PseAA composition.

As a follow up of our research on protein structure prediction [29-32], in this paper we aim at predicting protein secondary structure content by introducing the PseAAC as the protein sequence representation. Meanwhile, support vector machine (SVM) was adopted as the prediction engine [33]. It demonstrated by both self-consistency test and independent-dataset test that the present SVM system together with the PseAAC exhibited improved performance compared with several published results.

2. MATERIAL AND METHODS

2.1. Dataset

The two datasets constructed by Chou [5] were used here. One dataset contains 244 protein chains and serves as the training dataset, none of which has more than 35% of homology with the others. The other dataset contains 201 protein chains (Note that the Table **3** in ref. [5] at p.479 does not contain 202 chains) and serves as the testing dataset, none of which has more than 35% of homology with the others, either. They are referred to as the Chou244 set and the Chou201 set in this article, respectively.

2.2. Pseudo Amino Acid Composition

Recently, we introduced a different formulation of the PseAAC and successfully applied it to the prediction of protein structural classes [29, 30]. Stimulated by its success, here we adopt it to predict the protein secondary structure content. For readers' convenience, let us give its brief introduction as follows.

The sequence-order effect along a protein chain can be approximately reflected with a set of sequence-order correlation factors as defined below [8]:

$$\theta_m = \frac{1}{L-m}\sum_{i=1}^{L-m}\Theta(R_i, R_{i+m}),\quad (m = 1,2,\ldots,\lambda \ and\ \lambda < L) \tag{1}$$

In Eq.(1), L denotes the length of the protein, θ_m is called the mth rank of coupling factor that harbors the mth sequence-order correlation factor, and the correlation function $\Theta(R_i,R_j)$ is formulated as follows:

$$\Theta(R_i, R_j) = H(R_i)\times H(R_j) \tag{2}$$

where $H(R_i)$, $H(R_j)$ are, respectively, the hydrophobicity values of the amino acid R_i and R_j, taken from ref. [34]. Note that before substituting the hydrophobicity values into Eq.(2), they were all subjected to a standard conversion as described by the following equation:

$$H(R_i) = \frac{H_0(R_i) - \sum_{k=1}^{20} H_0(\Re_k)/20}{\sqrt{\sum_{u=1}^{20} [H_0(\Re_u) - \sum_{k=1}^{20} H_0(\Re_k)/20]^2 / 20}} \quad \textbf{(3)}$$

In Eq.(3), $\Re_i (i = 1, 2, ..., 20)$ denote the 20 native amino acids, and $H_0(R_i)$ is the original hydrophobicity value of amino acid R_i. The optimal value for λ is routinely selected as the one that results in the minimal overall cross-validated average error. For the current study, the optimal value for λ is 17; *i.e.*, the dimension of the PseAAC considered here is 20+17 = 37. Given a protein **X**, its PseAA can thus be defined in a 37-*D* space as given by:

$$\mathrm{X} = \begin{bmatrix} x_1 \\ x_2 \\ \vdots \\ x_i \\ \vdots \\ x_{37} \end{bmatrix} \quad \textbf{(4)}$$

where

$$x_u = \begin{cases} \dfrac{f_u}{\sum_{i=1}^{20} f_i + \omega \sum_{j=1}^{17} \theta_j}, & (1 \le u \le 20) \\ \dfrac{\omega \theta_{u-20}}{\sum_{i=1}^{20} f_i + \omega \sum_{j=1}^{17} \theta_j}, & (20 + 1 \le u \le 20 + 17) \end{cases} \quad \textbf{(5)}$$

In the above equation, f_i (i = 1,2,...,20), the same as in the conventional AA, are the normalized occurrence frequencies of the 20 native amino acids in the protein **X**, and θ_j (j = 1,2,...,17) are the j-tier sequence-order correlation factors computed according to Eqs. (1)-(3). Of the 37-*D* components, the first 20 components reflect the effect of the AA, whereas the components from 20+1 to 20+17 reflect the effect of sequence order. The parameter ω is a weight factor, which adjusts the latter 17 components to be in similar scales with the first 20 components. Here, it is set at 0.1.

2.3. Algorithm

The publicly available LIBSVM software [35] was used to process the SVM regression, and the radial basis function (RBF) was found to have the best performance:

$$K(\vec{u},\vec{v}) = \exp(-\gamma\left\|\vec{u}-\vec{v}\right\|^2) \quad (6)$$

The parameter γ is selected as that which minimized an estimate of the VC-dimension, and it is set at 10. The regularization parameter *C*, controlling the trade-off between low training error and large margin, is set at 1.5. Then, the PseAAC of a protein is taken as input into the SVM regression model, and the output is its predicted secondary structure content.

2.4. Performance Measures

In order to evaluate the performance of prediction, the output from the SVM needs to compare with the expected output, *i.e.* the observed content of secondary structural elements, including H (α-helix), G (3_{10}-helix), I (π-helix), E (β-strand), B (β-bridge), T (turn), S (bend) and - (rest) content, which can be easily derived from the protein DSSP file [36]. The following three criteria [5] were adopted as the performance measures.

The first criterion is the average absolute error, δ^{Θ}, of each secondary structural element, calculated for each protein:

$$\delta^{\Theta} = \frac{\sum_{k=1}^{N}\left|\Theta_k - y_k^{\Theta}\right|}{N} \quad (7)$$

where Θ_k is the predicted content of the secondary structural element Θ for the *k*th protein, and y_k^{Θ} is the content actually observed.

The second criterion is the standard deviation of the average absolute error formulated as follows:

$$\sigma^{\Theta} = sqrt(\frac{\sum_{k=1}^{N}(\delta^{\Theta}-\left|\Theta_k - y_k^{\Theta}\right|)^2}{N-1}) \quad (8)$$

The third criterion is the overall average error given by

$$\langle\delta\rangle = \frac{1}{8}\sum_{\Theta}\delta^{\Theta} \tag{9}$$

3. RESULTS AND DISCUSSION

3.1. Success Rates of Self-Consistency and Independent-Dataset Tests

The results of the self-consistency test on the Chou244 set are listed in Table **1**, where we can see that the prediction of π-helix is the most accurate. Nevertheless, as Table **1** shows, the element of π-helix has the smallest percentage in the dataset. In view of this, the assessment should be focused on the success rates of the prediction for the most abundant elements, including α-helix, β-strand and random coil (denoted as "-, rest" in tables). The average absolute errors for the prediction of them are 0.052, 0.045 and 0.013 with standard deviations of 0.069, 0.062 and 0.027, respectively. Note that the average absolute error of 0.013 is comparatively small, indicating that the prediction for the element of random coil is very successful. The overall average error is 0.018. Such a small error demonstrates that there is definitely a relationship between the PseAAC of a protein and its secondary structure content, and that the SVM can be trained to grasp this relationship and to further predict the secondary structure contents of unknown proteins, provided that their sequences were supplied.

Table 1: Prediction Errors by the Self-consistency Test on the Chou244 Set

Secondary Structure Element	Occurrence Frequency[a] / %	Average Absolute Error δ^{Θ}	Standard Deviation σ^{Θ}
H: α-helix	31.4	0.052	0.069
G: 3_{10}-helix	3.88	0.008	0.014
I: π-helix	0.02	0.000	0.001
E: β-strand	21.1	0.045	0.062
B: β-bridge	1.34	0.002	0.005
T: turn	11.6	0.009	0.017
S: bend	9.56	0.011	0.024
-: rest	21.1	0.013	0.027
Overall average error $<\delta>$	---	0.018	

[a] The values equal to the proportions of the secondary structure elements to the whole residues in the dataset.

As the self-consistency test is based on a limited number of proteins, and more importantly, the rule parameters derived from the training dataset bear the priori information of proteins later plugged back into the testing dataset, this procedure will certainly give a somewhat optimistic error estimate. Hence, it is very necessary to employ cross validation or jackknife tests to evaluate the capability of generalization of our method. In statistical prediction, the sub-sampling test (5-fold or 10-fold cross-validation) and jackknife test are two cross-validation methods often used in literatures for examining the accuracy of a predictor [37]. However, as elucidated by [38] and demonstrated in [39], the sub-sampling (*e.g.*, 5-fold cross-validation) test cannot avoid arbitrariness even for a very simple benchmark dataset. Accordingly, the jackknife test has been increasingly and widely adopted by investigators [13, 18, 21, 22, 24, 26, 40-51]. In order to facilitate comparing with the published methods, however, in this work we further carried a prediction on the Chou201 set as an independent dataset and the detailed results are listed in Table **2**. As expected, the prediction errors become greater when compared with the results of the self-consistency test in Table **1**. However, the overall average error is only 0.038, which is also acceptable.

Table 2: Prediction Errors by the Independent Test on the Chou201 Set

Secondary Structure Element	**Occurrence Frequency / %**	**Average Absolute Error** $\delta^{\ominus}$	**Standard Deviation** $\sigma^{\ominus}$
H: α-helix	31.8	0.091	0.076
G: 3_{10}-helix	3.78	0.022	0.019
I: π-helix	0.04	0.001	0.002
E: β-strand	21.1	0.084	0.074
B: β-bridge	1.32	0.010	0.008
T: turn	12.0	0.029	0.024
S: bend	9.42	0.029	0.029
-: rest	20.6	0.040	0.038
Overall average error $<\delta>$	---	0.038	

3.2. Comparison with Other Experimental and Theoretical Methods

In order to roundly assess our method, we carried a comparison with the published experimental and theoretical methods. CD and FT-IR are two of the routine experimental methods for determining the protein secondary structure. They perform well in predicting the elements of α-helix and β-strand, but perform relatively bad in predicting the element of random coil [52]. From Table **3**, we can observe that the estimation error for random coils is roughly 0.1. Nevertheless, as demonstrated in Table **3** the theoretical methods achieved better accuracies for predicting them, which testifies again that the theoretical methods can play complementary roles to experimental methods. Actually, it was also observed that combining sequence prediction with experimentally determined spectroscopic methods significantly enhanced the accuracy of the overall results obtained [53].

Of the three published theoretical methods the 210-*D* pair-coupled AA was taken as the protein representation, and the prediction engines were MLR, artificial neural network (ANN) and SVM, respectively. By the self-consistency test Chou's method performed best in predicting the contents of α-helices and β-strands, whereas Cai's method performed worst. Nevertheless, it is worth noting that Cai's method performed best by the independent test, not only for the prediction of α-helices and β-strands but for the prediction of random coils. Accordingly, it is not appropriate to simply rank them. As is well known, PHD [54], PSIPRED [55] and PROFsec [56] are of the most successful methods for predicting protein secondary structure. Their improvements are resulted not only from using larger databases, but from using evolutionary information that can be obtained from the multiple sequence alignments [57]. Motivated by this, Lee *et al.* [58] introduced evolutionary information in addition to AA to represent protein samples and achieved an interesting result. According to our recent findings [32] that the multiple protein representations can be fused to increase the prediction accuracy, it can be anticipated that the present PseAAC and the evolutionary information, if complemented, may further improve the success rate.

Last but not least, it is worth to note that user-friendly and publicly accessible web-servers represent the future direction for developing practically more useful models, simulated methods, or predictors [61]. Recently, Chen *et al.* proposed an integrated web-server for prediction of protein secondary structure content using a 3-state classification [62]. Thus, the method presented in this paper may be also implemented in a web-server for accurate prediction of protein secondary structure content.

Table 3: Comparison with Other Experimental and Theoretical Methods

Statistical Significance	Input Information	Average Absolute Error δ^{Θ}			Overall Average Error <δ>	Reference
		H: α-Helix	E: β-Strand	-: Rest		
Self-consistency	CD Spectroscopy	0.054	0.087	0.101	n/a[a]	[52]
	Pair-coupled AA	0.036	0.031	0.012	0.018	[5]
	Pair-coupled AA	0.060	0.057	0.031	0.028	[59]
	Pair-coupled AA	0.049	0.037	0.008	0.013	[31]
	PseAAC	0.052	0.045	0.013	0.018	this paper
Independent test	FT-IR Spectroscopy	0.017	0.023	n/a	n/a	[60]
	Pair-coupled AA	0.077	0.076	0.088	0.061	[5]
	Pair-coupled AA	0.071	0.069	0.037	0.034	[59]
	Pair-coupled AA	0.104	0.097	0.048	0.044	[31]
	PseAAC	0.091	0.084	0.040	0.038	this paper

[a] n/a, not available.

ACKNOWLEDGEMENTS

C. Chen acknowledges the financial support from the Scientific Research Foundation for Doctoral Program of Guangdong Pharmaceutical University (2007ZYX05) and the Natural Science Foundation of Guangdong Province (No. S2012010009166). X.-Y. Zou acknowledges the financial support from the National Natural Science Foundation of China (No. 20475068, 20575082), the Natural Science Foundation of Guangdong Province (No. 7003714) and the Scientific Technology Project of Guangdong Province (No. 2005B30101003).

CONFLICT OF INTEREST

The authors confirm that this chapter contents have no conflict of interest.

DISCLOSURE

The chapter submitted for eBook series titled "Frontiers in Protein and Peptide Sciences, Vol. 1" is an update of our article published in Protein and Peptide Letters, Volume 16, Number 1 (January), Pages 27-31, 2009 (PMID: 19149669), with additional text and references.

REFERENCES

[1] Krigbaum, W.R.; Knutton, S.P. Prediction of the amount of secondary structure in a globular protein from its aminoacid composition. *Proc. Natl. Acad. Sci. USA*, **1973**, *70*(10), 2809-2813.

[2] Zhang, C.T.; Zhang, Z.; He, Z. Prediction of the secondary structure content of globular proteins based on structural classes. *J. Protein Chem.*, **1996**, *15*(8), 775-786.

[3] Zhang, C.T.; Zhang, Z.; He, Z. Prediction of the secondary structure contents of globular proteins based on three structural classes. *J. Protein Chem.*, **1998**, *17*(3), 261-272.

[4] Liu, W.; Chou, K.C. Prediction of protein secondary structure content. *Protein Eng.*, **1999**, *12*(12), 1041-1050.

[5] Chou, K.C. Using pair-coupled amino acid composition to predict protein secondary structure content. *J. Protein Chem.*, **1999**, *18*(4), 473-480.

[6] Luo, R.Y.; Feng, Z.P.; Liu, J.K. Prediction of protein structural class by amino acid and polypeptide composition. *Eur. J. Biochem.*, **2002**, *269*(17), 4219-4225.

[7] Sun, X.D.; Huang, R.B. Prediction of protein structural classes using support vector machines. *Amino Acids*, **2006**, *30*(4), 469-475.

[8] Chou, K.C. Prediction of protein cellular attributes using pseudo-amino acid composition. *Proteins*, **2001**, *43*(3), 246-255.

[9] Chou, K.C.; Cai, Y.D. Predicting protein structural class by functional domain composition. *Biochem. Biophys. Res. Commun.*, **2004**, *321*(4), 1007-1009.

[10] Du, Q.S.; Wei, D.; Chou, K.C. Correlations of amino acids in proteins. *Peptides*, **2003**, *24*(12), 1863-1869.

[11] Du, Q.S.; Jiang, Z.Q.; He, W.Z.; Li, D.P.; Chou, K.C. Amino acid principal component analysis (aapca) and its applications in protein structural class prediction. *J. Biomol. Struct. Dyn.*, **2006**, *23*(6), 635-640.

[12] Chou, K.C.; Shen, H.B. Predicting eukaryotic protein subcellular location by fusing optimized evidence-theoretic k-nearest neighbor classifiers. *J. Proteome Res.*, **2006**, *5*(8), 1888-1897.

[13] Ding, Y.S.; Zhang, T.L.; Chou, K.C. Prediction of protein structure classes with pseudo amino acid composition and fuzzy support vector machine network. *Protein Pept. Lett.*, **2007**, *14*(8), 811-815.

[14] Gonzalez-Diaz, H.; Vilar, S.; Santana, L.; Uriarte, E. Medicinal chemistry and bioinformatics--current trends in drugs discovery with networks topological indices. *Curr. Top Med. Chem.*, **2007**, *7*(10), 1015-1029.

[15] Shen, H.B.; Chou, K.C. Hum-mploc: an ensemble classifier for large-scale human protein subcellular location prediction by incorporating samples with multiple sites. *Biochem. Biophys. Res. Commun.*, **2007**, *355*(4), 1006-1011.

[16] Shen, H.B.; Chou, K.C. Virus-ploc: a fusion classifier for predicting the subcellular localization of viral proteins within host and virus-infected cells. *Biopolymers*, **2007**, *85*(3), 233-240.

[17] Shen, H.B.; Chou, K.C. Nuc-ploc: a new web-server for predicting protein subnuclear localization by fusing pseaa composition and psepssm. *Protein Eng. Des. Sel.*, **2007**, *20*(11), 561-567.

[18] Xiao, X.; Chou, K.C. Digital coding of amino acids based on hydrophobic index. *Protein Pept. Lett.*, **2007**, *14*(9), 871-875.

[19] Fang, Y.; Guo, Y.; Feng, Y.; Li, M. Predicting dna-binding proteins: approached from chou's pseudo amino acid composition and other specific sequence features. *Amino Acids*, **2008**, *34*(1), 103-109.

[20] Gonzalez-Diaz, H.; Gonzalez-Diaz, Y.; Santana, L.; Ubeira, F.M.; Uriarte, E. Proteomics, networks and connectivity indices. *Proteomics*, **2008**, *8*(4), 750-778.

[21] Jiang, X.; Wei, R.; Zhang, T.; Gu, Q. Using the concept of chou's pseudo amino acid composition to predict apoptosis proteins subcellular location: an approach by approximate entropy. *Protein Pept. Lett.*, **2008**, *15*(4), 392-396.

[22] Li, F.M.; Li, Q.Z. Predicting protein subcellular location using chou's pseudo amino acid composition and improved hybrid approach. *Protein Pept. Lett.*, **2008**, *15*(6), 612-616.

[23] Lin, H. The modified mahalanobis discriminant for predicting outer membrane proteins by using chou's pseudo amino acid composition. *J. Theor. Biol.*, **2008**, *252*(2), 350-356.

[24] Lin, H.; Ding, H.; Guo, F.B.; Zhang, A.Y.; Huang, J. Predicting subcellular localization of mycobacterial proteins by using chou's pseudo amino acid composition. *Protein Pept. Lett.*, **2008**, *15*(7), 739-744.

[25] Nanni, L.; Lumini, A. Genetic programming for creating chou's pseudo amino acid based features for submitochondria localization. *Amino Acids*, **2008**, *34*(4), 653-660.

[26] Zhang, G.Y.; Li, H.C.; Gao, J.Q.; Fang, B.S. Predicting lipase types by improved chou's pseudo-amino acid composition. *Protein Pept. Lett.*, **2008**, *15*(10), 1132-1137.

[27] Zhang, S.W.; Zhang, Y.L.; Yang, H.F.; Zhao, C.H.; Pan, Q. Using the concept of chou's pseudo amino acid composition to predict protein subcellular localization: an approach by incorporating evolutionary information and von neumann entropies. *Amino Acids*, **2008**, *34*(4), 565-572.

[28] Shen, H.B.; Chou, K.C. PseAAC: a flexible web server for generating various kinds of protein pseudo amino acid composition. *Anal. Biochem.*, **2008**, *373*(2), 386-388.

[29] Chen, C.; Tian, Y.X.; Zou, X.Y.; Cai, P.X.; Mo, J.Y. Using pseudo-amino acid composition and support vector machine to predict protein structural class. *J. Theor. Biol.*, **2006**, *243*(3), 444-448.

[30] Chen, C.; Zhou, X.; Tian, Y.; Zou, X.; Cai, P. Predicting protein structural class with pseudo-amino acid composition and support vector machine fusion network. *Anal. Biochem.*, **2006**, *357*(1), 116-121.

[31] Chen, C.; Tian, Y.; Zou, X.; Cai, P.; Mo, J. Prediction of protein secondary structure content using support vector machine. *Talanta*, **2007**, *71*(5), 2069-2073.

[32] Chen, C.; Chen, L.X.; Zou, X.Y.; Cai, P.X. Predicting protein structural class based on multi-features fusion. *J. Theor. Biol.*, **2008**, *253*(2), 388-392.

[33] Cortes, C.; Vapnik, V. Support-vector networks. *Mach. Learn.*, **1995**, *3*(20), 273-297.

[34] Kyte, J.; Doolittle, R.F. A simple method for displaying the hydropathic character of a protein. *J. Mol. Biol.*, **1982**, *157*(1), 105-132.

[35] Chang, C.C.; Lin, C.J., Software available at http://www.csie.ntu.edu.tw/~cjlin/libsvm. In 2001.

[36] Kabsch, W.; Sander, C. Dictionary of protein secondary structure: pattern recognition of hydrogen-bonded and geometrical features. *Biopolymers*, **1983**, *22*(12), 2577-2637.

[37] Chou, K.C.; Zhang, C.T. Prediction of protein structural classes. *Crit. Rev. Biochem. Mol. Biol.*, **1995**, *30*(4), 275-349.

[38] Chou, K.C.; Shen, H.B. Cell-ploc: a package of web servers for predicting subcellular localization of proteins in various organisms. *Nat. Protoc.*, **2008**, *3*(2), 153-162.

[39] Chou, K.C.; Shen, H.B. Recent progress in protein subcellular location prediction. *Anal. Biochem.*, **2007**, *370*(1), 1-16.

[40] Chou, K.C.; Shen, H.B. Hum-ploc: a novel ensemble classifier for predicting human protein subcellular localization. *Biochem. Biophys. Res. Commun.*, **2006**, *347*(1), 150-157.

[41] Niu, B.; Cai, Y.D.; Lu, W.C.; Li, G.Z.; Chou, K.C. Predicting protein structural class with adaboost learner. *Protein Pept. Lett.*, **2006**, *13*(5), 489-492.

[42] Chou, K.C.; Shen, H.B. Memtype-2L: a web server for predicting membrane proteins and their types by incorporating evolution information through Pse-PSSM. *Biochem. Biophys. Res. Commun.*, **2007**, *360*(2), 339-345.

[43] Chou, K.C.; Shen, H.B. Signal-CF: a subsite-coupled and window-fusing approach for predicting signal peptides. *Biochem. Biophys. Res. Commun.*, **2007**, *357*(3), 633-640.

[44] Chou, K.C.; Shen, H.B. Euk-mPLoc: a fusion classifier for large-scale eukaryotic protein subcellular location prediction by incorporating multiple sites. *J. Proteome Res.*, **2007**, *6*(5), 1728-1734.

[45] Chou, K.C.; Shen, H.B. Large-scale plant protein subcellular location prediction. *J. Cell. Biochem.*, **2007**, *100*(3), 665-678.

[46] Shen, H.B.; Chou, K.C. Ezypred: a top-down approach for predicting enzyme functional classes and subclasses. *Biochem. Biophys. Res. Commun.*, **2007**, *364*(1), 53-59.

[47] Shen, H.B.; Chou, K.C. Signal-3L: a 3-layer approach for predicting signal peptides. *Biochem. Biophys. Res. Commun.*, **2007**, *363*(2), 297-303.

[48] Chen, W.J.; Huang, P.T.; Cheng, Y.C.; Liao, T.H. Putative secondary active site of bovine pancreatic deoxyribonuclease i. *Protein Pept. Lett.*, **2008**, *15*(6), 640-646.

[49] Jin, Y.H.; Niu, B.; Feng, K.Y.; Lu, W.C.; Cai, Y.D.; Li, G.Z. Predicting subcellular localization with adaboost learner. *Protein Pept. Lett.*, **2008**, *15*(3), 286-289.

[50] Niu, B.; Jin, Y.H.; Feng, K.Y.; Liu, L.; Lu, W.C.; Cai, Y.D.; Li, G.Z. Predicting membrane protein types with bagging learner. *Protein Pept. Lett.*, **2008**, *15*(6), 590-594.

[51] Wang, T.; Yang, J.; Shen, H.B.; Chou, K.C. Predicting membrane protein types by the llda algorithm. *Protein Pept. Lett.*, **2008**, *15*(9), 915-921.

[52] Sreerama, N.; Woody, R.W. Estimation of protein secondary structure from circular dichroism spectra: comparison of contin, selcon, and cdsstr methods with an expanded reference set. *Anal. Biochem.*, **2000**, *287*(2), 252-260.

[53] Lees, J.G.; Janes, R.W. Combining sequence-based prediction methods and circular dichroism and infrared spectroscopic data to improve protein secondary structure determinations. *BMC Bioinform.*, **2008**, *9*, 24-29.

[54] Rost, B. PHD: predicting one-dimensional protein structure by profile-based neural networks. *Methods Enzymol.*, **1996**, *266*, 525-539.

[55] Jones, D.T. Protein secondary structure prediction based on position-specific scoring matrices. *J. Mol. Biol.*, **1999**, *292*(2), 195-202.

[56] Rost, B. Review: protein secondary structure prediction continues to rise. *J. Struct. Biol.*, **2001**, *134*(2-3), 204-218.

[57] Homaeian, L.; Kurgan, L.A.; Ruan, J.; Cios, K.J.; Chen, K. Prediction of protein secondary structure content for the twilight zone sequences. *Proteins*, **2007**, *69*(3), 486-498.

[58] Lee, S.; Lee, B.C.; Kim, D. Prediction of protein secondary structure content using amino acid composition and evolutionary information. *Proteins*, **2006**, *62*(4), 1107-1114.

[59] Cai, Y.D.; Liu, X.J.; Chou, K.C. Prediction of protein secondary structure content by artificial neural network. *J. Comput. Chem.*, **2003**, *24*(6), 727-731.

[60] Smith, B.M.; Oswald, L.; Franzen, S. Single-pass attenuated total reflection Fourier transform infrared spectroscopy for the prediction of protein secondary structure. *Anal. Chem.*, **2002**, *74*(14), 3386-3391.

[61] Chou, K.C.; Shen, H.B. Review: recent advances in developing web-servers for predicting protein attributes. *Nat. Sci.*, **2009**, *2*(1), 63-92 (openly accessible at http://www.scrip.org/journal/NS/).

[62] Chen, K.; Stach, W.; Homaeian, L.; Kurgan, L. iFC(2): an integrated web-server for improved prediction of protein structural class, fold type, and secondary structure content. *Amino Acids*, **2011**, *40*(3), 963-973.

Send Orders for Reprints to reprints@benthamscience.net

174 *Frontiers in Protein and Peptide Sciences, Vol. 1,* **2014, 174-193**

CHAPTER 8

Prediction of Golgi-Resident Protein Types Using Computational Method

Hao Lin[1,*], Hui Ding[1,*] and Wei Chen[2,*]

[1]Key Laboratory for NeuroInformation of Ministry of Education, Center of Bioinformatics, School of Life Science and Technology, University of Electronic Science and Technology of China, Chengdu, 610054, China and [2]Department of Physics, School of Sciences, Center for Genomics and Computational Biology, Hebei United University, Tangshan, China

Abstract: The Golgi apparatus plays a key role in sorting, modifying, packaging and distributing proteins. Thus, correct identification of the type of a Golgi-resident protein will provide in-depth insights into its function involved in various biological processes. In this article, we will briefly outline the prediction of protein subGolgi localizations using bioinformatics method. We introduced the available databases for training predicted models. Moreover, three kinds of methods were presented for discriminating between *cis*-Golgi and *trans*-Golgi proteins.

Keywords: *Cis*-Golgi proteins, *trans*-Golgi proteins, benchmark dataset, information parameter, feature selection, increment of diversity, analysis of variance, binomial distribution, support vector machine, cross-validation.

1. INTRODUCTION

In eukaryotic cell, the Golgi apparatus is one of the most important organelle, which consists of a stack of membrane-bounded cisternae located between the endoplasmic reticulum and the cell surface [1]. The stack of membrane-bounded cisternae contains two specialized cisternae: *cis*-Golgi and *trans*-Golgi apparatus [2]. The *cis*-Golgi apparatus receive and process proteins output from endoplasmic reticulum. Subsequently, these proteins are transported to *trans*-Golgi apparatus.

***Address correspondence to Hao Lin and Hui Ding:** Key Laboratory for NeuroInformation of Ministry of Education, Center of Bioinformatics, School of Life Science and Technology, University of Electronic Science and Technology of China, Chengdu, 610054, China; Tel/Fax: +86-28-83202351; E-mails: hlin@uestc.edu.cn (H. Lin), hding@uestc.edu.cn (H. Ding);

Wei Chen: Department of Physics, School of Sciences, Center for Genomics and Computational Biology, Hebei United University, Tangshan, China; E-mail: greatchen@heuu.edu.cn

The *trans*-Golgi apparatus packages and sends these proteins to the required destination. Many important enzymes and proteins retained in *cis*- and *trans*-Golgi apparatus play key roles in the above mentioned two biological processes. The defects in the Golgi apparatus can result in many neurodegenerative diseases [3], such as Parkinson's disease [4] and Alzheimer's disease [5]. Therefore, correctly identifying the types of Golgi-resident proteins is helpful for analysis of the function of Golgi-resident proteins in the process of transport and modification of transited proteins.

Experimentally identifying the types of Golgi-resident proteins is a good starting point for in-depth study of their functions. However, it is time-consuming and costly for biochemical experiments to systematically investigate the types of Golgi-resident proteins because large amounts of potential proteins are required to interrogate. Therefore, it is necessary to develop theoretical method to recognize the types of Golgi-resident proteins. Phylogenetic tree is an accepted method for this task. However, the defect of this method is time-consuming. Moreover, because the sequence similarity is fundamental of phylogenetic tree, if no homologous sequences can be found in benchmark dataset, the prediction results will be unauthentic.

Machine learning techniques are powerful tools which can overcome these limitations. Although some works have focused on subnuclear, submitochondria and subchloroplast localization of proteins [6], few bioinformatics methods have been specifically designed for the study of Golgi-resident proteins. Based on transmembrane domains of proteins, Yuan and Teasdale predicted the Golgi type II membrane proteins [7]. Chou *et al.* [8] have designed another tool called GolgiP to identify Golgi-resident proteins in plants. However, these works can not provide any information for the recognition of *cis*-Golgi and *trans*-Golgi proteins.

The discrimination between *cis*-Golgi and *trans*-Golgi proteins based on their primary sequence is an important problem toward the advancement of functional disease proteomics. In this review, we focused on three classification algorithms developed for the prediction of *cis*-Golgi and *trans*-Golgi proteins based on their sequence information. At first, we introduced a pseudo amino acid composition based method for predicting *cis*-Golgi and *trans*-Golgi proteins. Secondly, the analysis of variance (ANOVA) was presented to optimize dipeptides for improving accuracy. Finally, a powerful feature optimized method, the binomial distribution, was developed to select the best tripeptide feature set.

2. INCREMENT OF DIVERSITY BASED METHOD

2.1. Materials

This dataset was constructed in 2010 [9]. All sequence and annotation of the Golgi proteins were extracted from the Universal Protein Resource (Uniprot). The following procedures were performed to obtain the high quality dataset. At first, only those sequences which have subcellular location annotation containing 'Golgi Apparatus' with confidence 'experimental' were selected because the current study was focused on Golgi Apparatus proteins only. Secondly, the sequences which are fragment of other proteins were excluded. Thirdly, the protein sequences containing ambiguous residues (such as "X", "B" and "Z") were excluded. Fourthly, only those sequences annotated by one location were selected because the number of multiplex proteins in the existing Golgi protein database is not large enough to allow us to construct a statistically meaningful benchmark dataset for studying multiplex Golgi proteins. Finally, the CD-HIT program [10] was used to remove the homologous sequences from the benchmark dataset. A stringent criterion that the cutoff threshold is set to 25% will make benchmark dataset more objective and strict. However, if we use such criterion, the numbers of proteins for some subsets would be too few to have statistical significance. Thus, we set the cutoff threshold value at 40%. By doing so, 95 Golgi resident proteins including 46 *cis*-Golgi proteins and 49 *trans*-Golgi proteins were obtained. The sequences can be downloaded from http://immunet.cn/SubGolgi/.

2.2. The Information Parameter

It is one of the most important parts for pattern recognition to generate a set of informative parameters. To avoid losing the important information hidden in protein sequences, the pseudo amino acid composition (PseAAC) was proposed [11] to replace the simple amino acid composition (AAC) for representing the sample of a protein. For a summary about its development and applications, see a recent comprehensive review [12], in which 16 different modes of PseAAC are summarized including dipeptide component mode. Ever since the concept of Chou's pseudo amino acid composition (PseAAC) was proposed, it has been widely used to study various problems in proteins and protein-related systems. However, so far no report whatsoever has been seen that the PseAAC was used for predicting the Golgi protein types. Here, we propose a different mode of PseAAC, the so-called g-gap dipeptide components, to address this problem. The *g*-gap dipeptides and AACs can be formulated as:

$$X_{400}^{g} = [x_1^g \cdots x_i^g \cdots x_{400}^g]^T \text{ and } Y_{20} = [y_1 \cdots y_j \cdots y_{20}]^T \quad \textbf{(1)}$$

where x_i^g (i=1,2,...,400) and y_j (j=1,2,...,20) denote the absolute frequencies of 400 *g*-gap dipeptides and 20 AACs, respectively. Symbol **T** denotes the transposition of vector.

2.3. Increment of Diversity

Based on the information theory, in the state space of *d* dimension, the Shannon entropy (or called information entropy) of data source X:{n_1, n_2, ... n_d} is defined by

$$H(p_X) = -\sum_i^d p(n_i)\log_b p(n_i) = -\sum_i^d \frac{n_i}{N}\log_b \frac{n_i}{N} \; (N = \sum_i^d n_i) \quad \textbf{(2)}$$

According to the definition of diversity by Laxton [13], the diversity measure of data source X:{n_1, n_2, ... n_d} equals N fold of Shannon entropy and can be formulated as [14]:

$$D(X) = -\sum_i^d n_i \log_b p(n_i) = -\sum_i^d n_i \log_b \frac{n_i}{N} = N \cdot H(p_X) \; (N = \sum_i^d n_i) \quad \textbf{(3)}$$

For measuring the similarity between X:{n_1, n_2, ... n_d} and Y:{m_1,m_2,...m_d}, the increment of diversity (*ID*) can be defined by the following formula:

$$ID(X,Y) = D(X+Y) - D(X) - D(Y) \quad \textbf{(4)}$$

where $D(X+Y)$, $D(X)$ and $D(Y)$ denote the diversity measure of data source: $X+Y$, X and Y calculated by Eq.(3), respectively.

It is easily deduced that the *ID* satisfies nonnegative and symmetry. Therefore, *ID* is a quantitative measure of the similarity of two data sources. It can be proved that the higher the similarity of two data sources, the smaller the *ID*.

2.4. Modified Mahalanobis Discriminant

For the prediction of sub-Golgi proteins, the modified Mahalanobis Discriminant (*MD*) [15] between the test sequence x and training set x_ξ can be calculated by:

$$MD_\xi = (x - \overline{x}_\xi)^T \sum\nolimits_\xi^{-1} (x - \overline{x}_\xi) + \log\left|\Sigma_\xi\right| \qquad \textbf{(5)}$$

where $\overline{x}_\xi$ and Σ_ξ are the group mean and covariance matrix for ξ-th training set, respectively. Symbol **T** denotes the transposition of vector $(x - \overline{x}_\xi)$. Σ_ξ^{-1} and $\left|\Sigma_\xi\right|$ are the inverse matrix and determinant, respectively. Thus, $(x - \overline{x}_\xi)^T \sum\nolimits_\xi^{-1} (x - \overline{x}_\xi)$ denotes the square of Mahalanobis distance between test sequence x and training set x_ξ.

According to the principle of similarity, the smaller the *MD* between the test sequence x and training set x_ξ, the higher the probability that the sequence x belongs to class ξ.

2.5. Performance Evaluation

In statistical prediction, the following three cross-validation methods are often used to examine a predictor for its effectiveness in practical application: independent dataset test, sub sampling test, and jackknife test [16]. Among the three cross-validation methods, the jackknife test can always yield a unique result for a given benchmark dataset. Therefore, jackknife test was used to evaluate the proposed method. For evaluating the performance of the algorithm, we calculated the sensitivity (*Sn*), false positive rate (*FPR*) and overall accuracy (*Ac*) as the following formulas:

$$Sn = TP / (TP + FN) \qquad \textbf{(6)}$$

$$FPR = FP / (FP + TN) \qquad \textbf{(7)}$$

$$Ac = (TP + TN) / (TP + TN + FP + FN) \qquad \textbf{(8)}$$

here *TP* denotes true positives, *FN* denotes false negatives, *FP* denotes false positives, and *TN* denotes true negatives.

To describe the performance of models across the entire range of *MD* decision values, the receiver operating characteristic (ROC) curves were performed. A ROC space is defined by *FPR* and *Sn* as x and y axes, respectively, which depicts relative trade-offs between true positive and false positive. Since *FPR* is equal to 1−specificity (1−*Sp*), the ROC graph is sometimes called the *Sn vs* (1−*Sp*) plot. Here, the area under ROC curve (auROC) is used to evaluate models.

3. RESULTS AND DISCUSSION

At first, *ID* combined with *MD* (called *IDMD*) was used to discriminate between *cis*-Golgi proteins and *trans*-Golgi proteins based on AAC and *g*-gap dipeptides. The parameter *g* represents the distance between two considered residues in one sequence. The minimal g=0 denotes two nearest neighbor residues. The maximal *g* can not be larger than the length of the shortest sequence in the dataset. A small *g* represents short-range correlation of residues, whereas a large *g* represents long-range of residues. Usually, the more the *g*-gap dipeptides select, the more information the representation bears. However, if information parameter contains too many components, it would reduce the cluster-tolerant capacity so as to lower down the jackknife success rate.

Therefore, we must investigate which g-gap could achieve the highest accuracy. After examining the accuracies of *IDMD* with different *g*-gap (g=0, 1, 2, 3, 4) dipeptides using jackknife cross-validation (Fig. **1**), we found that the 0-gap dipeptides can achieve the maximum area under ROC curve (auROC) of 0.709 with the overall accuracy of 69.5%. In order to improve predictive performance, we have done a great deal of experiments to estimate the accuracy of different combined parameters of *g*-gap dipeptides and AAC using jackknife test. Results in Fig. **2** show that the auROC of 0.772 is achieved when using the combined parameters AAC, 2-gap and 3-gap dipeptides. The sensitivities of *cis*-Golgi proteins and *trans*-Golgi proteins are 69.6% and 79.6% and 73.6%, respectively. The overall accuracy is 74.7%.

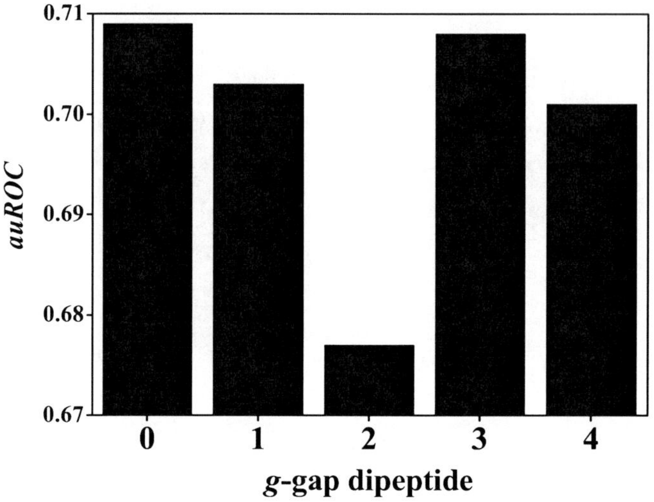

Figure 1: The area under ROC curve for different *g*-gap dipeptides.

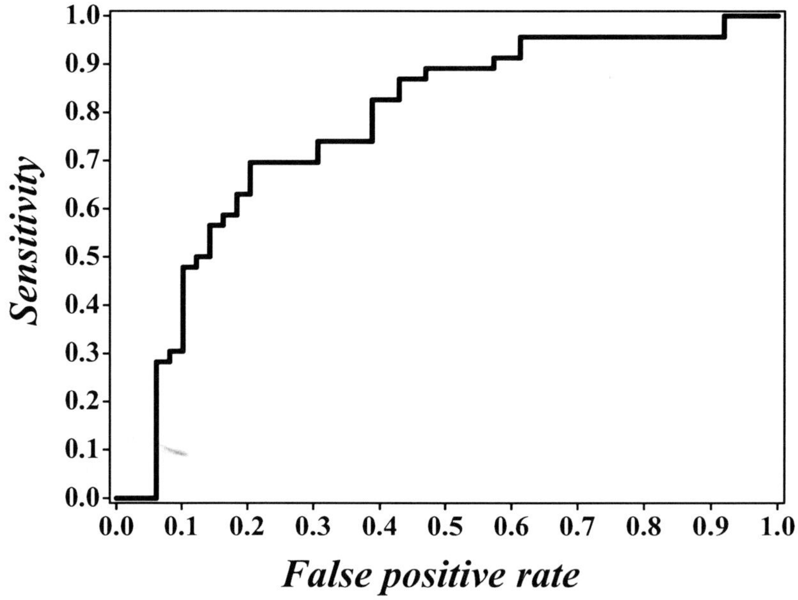

Figure 2: The ROC curve for predicting *cis*-Golgi and *trans*-Golgi proteins.

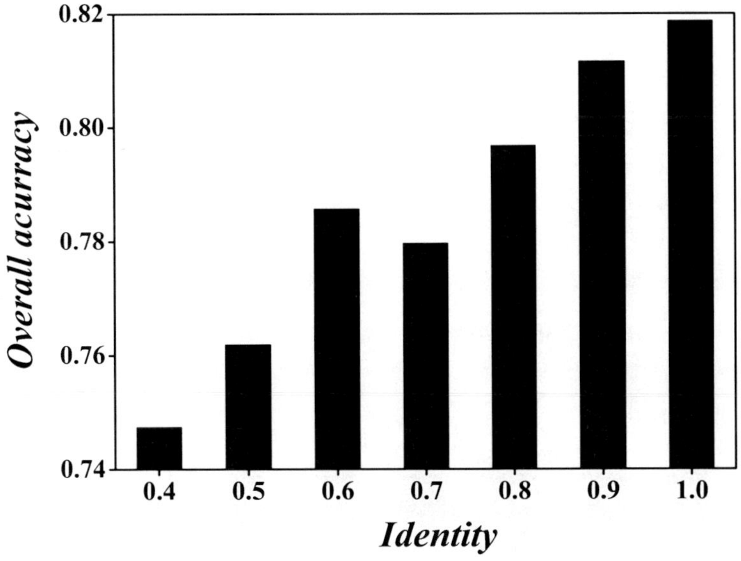

Figure 3: The overall accuracies for the datasets with different sequence identities.

It is well known that the sequence identity strongly affects the predictive accuracy. Many methods can achieve good performances on highly similar data sets. But, the performance decreases dramatically when the sequences with high similarity are removed from data sets. Highly similar data will surely lead to overestimate the performance of proposed methods. Therefore, we evaluated the predictive accuracy of *IDMD* for the datasets with different sequence identity using jackknife test. Fig. **3** shows that, with the sequence identity decreasing from 100% to 40%, the overall accuracies decrease from 81.9% to 74.7%. This demonstrates that the accuracy has a close relationship with sequence identity in protein prediction, and using redundant data can surely lead to overestimate the performance of the methods.

4. ANALYSIS OF VARIANCE BASED METHOD

4.1. Materials

This dataset was constructed in 2012 [17]. Both amino acid sequences and annotation information of Golgi-resident proteins were also extracted from

Uniprot. We still applied several selected strategies to obtain reliable and objective benchmark dataset. First of all, only those proteins annotated by one subGolgi location were selected, because the number of proteins with two subGolgi locations is too small to have statistical significance. Subsequently, proteins with ambiguous protein existence annotations, such as 'uncertain', 'predicted' and 'inferred from homology' were excluded because they lack confidence. Furthermore, only those proteins with experimental confirmed subGolgi location were included because they can provide correct and validated information. Moreover, the sequences which are fragments of other proteins were excluded because their information is redundant and not integrity. Finally, sequences containing nonstandard letters, such as 'B', 'X' or 'Z', were excluded because their meanings are ambiguous. Because sequence similarity in dataset correlates with estimated accuracy, it is necessary to get rid of redundancy and homology bias. Thus, the PISCES program [18] was used to remove the highly similar sequences by setting the cutoff of sequence identity to 25%. As a result, total of 42 *cis*-Golgi and 95 *trans*-Golgi proteins were obtained.

4.2. The Information Parameter

In the above section, we have demonstrated that *g*-gap dipeptide composition is powerful parameter for prediction. Thus, we still used *g*-gap dipeptide compositions to represent the proteins. A Golgi-resident protein with length of L can be characterized by a 400 dimensional feature vector and described as follows:

$$F_{400}^{g} = \left[f_1^g, f_2^g, \cdots, f_i^g, \cdots, f_{400}^g \right]^T \quad \textbf{(9)}$$

here symbol **T** denotes the transposition of the vector. f_i^g is the frequency of the *i*-th *g*-gap dipeptide and expressed as:

$$f_i^g = n_i^g / \sum_{i=1}^{400} n_i^g = n_i^g / (L - g - 1) \quad \textbf{(10)}$$

here n_i^g denotes the number of the *i*-th *g*-gap dipeptide. 0-gap dipeptides denote proximate dipeptides.

4.3. ANOVA

The original feature set generally contains redundant information or noise which will reduce the predictive accuracy. Thus, it is necessary to pick out informative parameters. Here, we proposed the ANOVA to perform feature selection. The ANOVA method can rank the features by measuring the ratio between their variances between groups and within groups [19]. The ratio reveals how strong the λ-th feature is related to the group variables. The ratio F value ($F(\lambda)$) of the λ-th g-gap dipeptide in two benchmark datasets is defined as the following equation:

$$F(\lambda)=\frac{\sum_{i=1}^{K} n_i\left(\frac{\sum_{j=1}^{n_i} f_{ij}(\lambda)}{n_i}-\frac{\sum_{i=1}^{K}\sum_{j=1}^{n_i} f_{ij}(\lambda)}{\sum_{i=1}^{K} n_i}\right)^2/(K-1)}{\sum_{i=1}^{K}\sum_{j=1}^{n_i}\left(f_{ij}(\lambda)-\frac{\sum_{i=1}^{K}\sum_{j=1}^{n_i} f_{ij}(\lambda)}{\sum_{i=1}^{K} n_i}\right)^2/(N-K)} \quad \textbf{(11)}$$

here K and N represent the number of groups and total number of samples, respectively. $f_{ij}(\lambda)$ denotes the frequency of the λ-th feature of the j-th sample in the i-th group. n_i denotes the number of sample in the i-th group.

The $F(\lambda)$ reveals how strong the λ-th feature is related to the group variables. If there are m g-gap dipeptides whose $F(\lambda)$ is larger than a given cutoff F_{cutoff}, the frequencies of these g-gap dipeptides are selected as optimized feature set and expressed as:

$$F_m=[f_1, f_2, \cdots, f_i, \cdots, f_m]^T \quad \textbf{(12)}$$

If F_{cutoff} is set to zero, 400 g-gap dipeptides are all selected. The larger the F_{cutoff} is, the less the number of features is. By setting an appropriate F_{cutoff}, high-dimensional data can be projected into a low-dimensional space and the best accuracy can be achieved. The parameter m or F_{cutoff} was chosen by using cross-validation.

4.4. Support Vector Machine

Support vector machine (SVM) is a powerful machine learning method and has been successfully applied in protein structure and function prediction. The SVM can find a decision boundary that separates two training data. The decision boundary is a hyperplane which maximizes the margin between the two sets in the feature vector space. In this study, our implementation utilized the software LibSVM [20]. The RBF is used to perform the prediction. The grid search program was applied to tune the regularization parameter C and kernel width parameter γ by using cross-validation.

To estimate the robustness of the prediction system, the jackknife cross-validation was carried out. In the jackknife cross-validation, each sequence in the training dataset is in turn singled out as an independent test sample and all the rule-parameters are calculated without including the one being identified. As a unique outcome can be yielded, jackknife cross-validation was employed to evaluate the performance of all models.

5. RESULTS AND DISCUSSION

We used ANOVA to optimize features and used SVM to perform prediction. Let us use as an example the 2-gap dipeptides, whose frequencies can be achieved by Eq. (9). We ranked the importance of the 400 2-gap dipeptides according to their F values as defined by Eq. (11). A larger F value of a 2-gap dipeptide suggests that it is more important for the prediction of the types of Golgi proteins. Furthermore, by adding the ranked 2-gap dipeptides one by one, 400 individual predictors can be built for the 400 sub-feature sets. If large F values are used, the selected features will give more reliable information for classification. However, the number of the selected features is too small to afford enough information, which deduces the poor predictive accuracy. For example, when setting F values larger than 7, the optimized 2-gap dipeptides set contains 10 features. The overall accuracy is only 77.4% in jackknife cross-validation. On the contrary, if the F values are set small, the number of the selected features is so large that the cluster-tolerant capacity is reduced which also deduces a bad prediction in cross-validations. An example is that 400 2-gap dipeptides can only produce the overall accuracy of 70.8% in jackknife cross-validation. Therefore, using appropriate 2-gap dipeptides can yield a prediction with higher accuracy.

By testing the predictive performances of each of 400 predictors, we plotted a 3-dimension graph in Fig. **4** for F value, feature dimension and overall accuracy. As shown in the Fig. **4**, the overall accuracy reaches its maximum when 83 features (F_{cutoff}=2.17) are used. Such 83 features were regarded as the optimal sub-feature set of our classifier. Based on these features, the ROC curves of jackknife cross-validation was drawn in Fig. **5** for investigating the change of the sensitivity of *cis*-Golgi proteins *vs.* false *cis*-Golgi proteins rate (*trans*-Golgi proteins are predicted as *cis*-Golgi proteins) by varying decision values of SVM. It shows that the auROC achieves 0.878 with the overall accuracy of 85.4%.

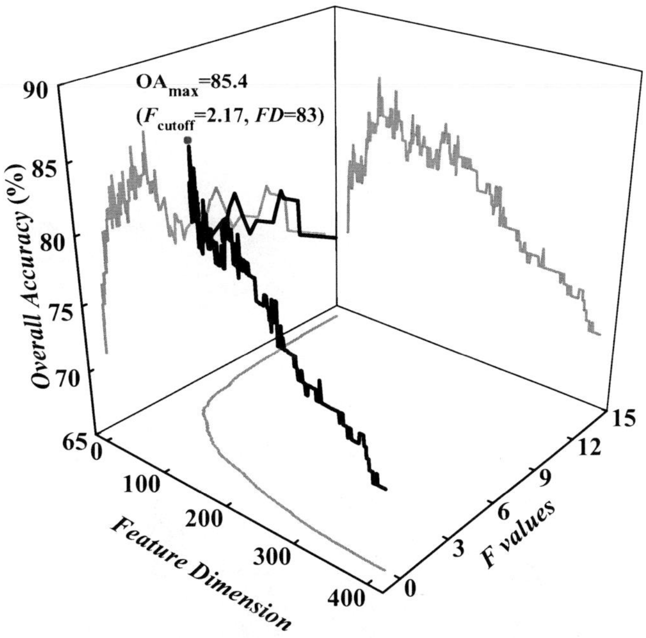

Figure 4: The graph for predicting the types of Golgi-resident proteins. Dark line denotes 3-D curve. Three gray lines are projections on three planes (overall accuracy/feature dimension plane, overall accuracy/*F* value, *F* value/feature dimension plane).

We should also investigate whether other *g*-gap sub-feature sets can obtain higher accuracies or not. We varied the *g* from 0 to 6 and repeated the feature selection process for finding the maximum accuracy of each *g*-gap dipeptides. Results in Fig. **6** show that the best optimized feature set is derived from 2-gap dipeptides. 73.8% of *cis*-Golgi and 90.5% *trans*-Golgi proteins can be correctly identified by this optimized feature set.

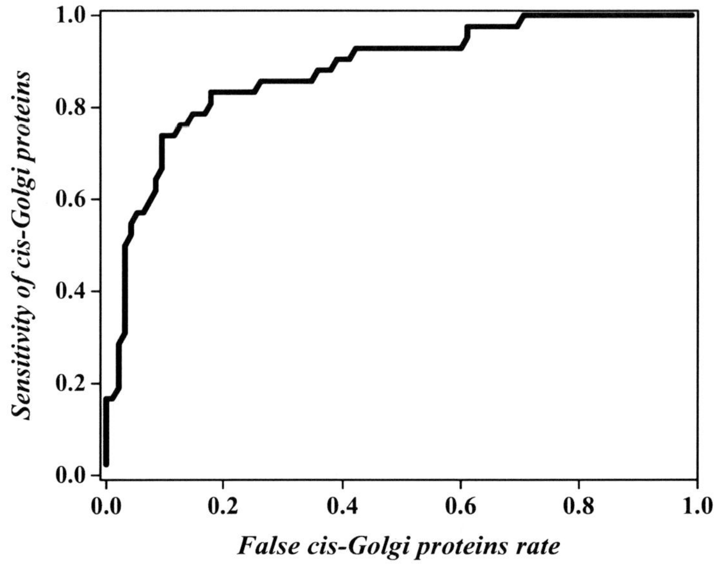

Figure 5: The ROC curves of the ANOVA-based method.

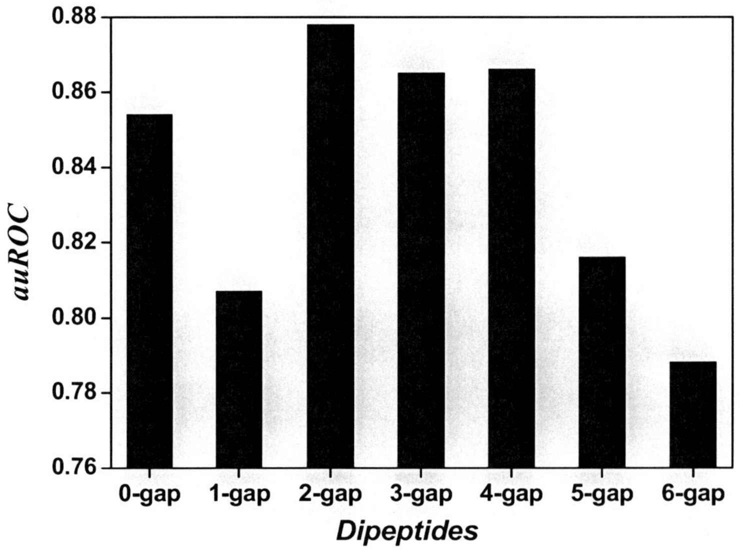

Figure 6: The maximum auROC of 7 optimized dipeptide sets by ANOVA.

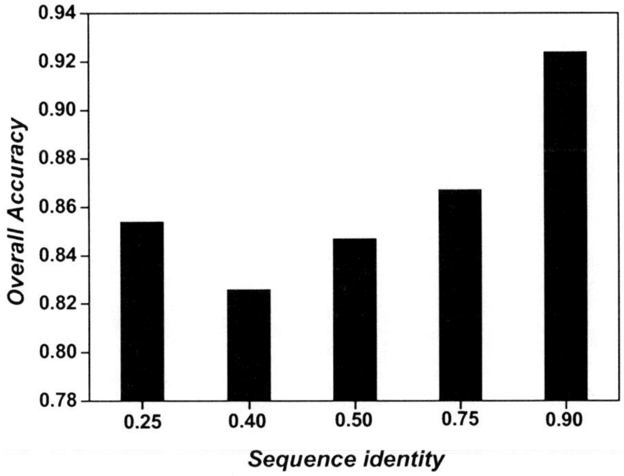

Figure 7: The predictive accuracies for the datasets with different sequence identities.

For the further display of relationship between sequence identity and predicted accuracy, we evaluated the predictive accuracies of proposed model on the datasets with different sequence identity using jackknife cross-validation. Results in Fig. 7 show that, with the sequence identity varying from 100% to 25%, the overall accuracies decrease and stabilize at about 85%, demonstrating that our proposed method is robust.

We compared the results obtained by ANOVA-based method and the results obtained by *IDMD*-based method. It is found an interesting phenomenon that the best result is obtained by 0-gap dipeptides with IDQD method, while the highest accuracy is obtained by 2-gap dipeptides in this section. We think that both are reasonable. The difference is derived from different algorithms. The proximate dipeptides can reflect the short range correlation which is based on nucleotide correlation in DNA sequence [21]. The 2-gap dipeptides can reflect the interaction between two residues in space.

6. BIONOMIAL DISTRIBUTION BASED MEHOD

6.1. Materials

The benchmark dataset including 42 *cis*-Golgi and 95 *trans*-Golgi proteins was described in the Section of **ANALYSIS OF VARIANCE BASED METHOD**

6.2. The Information Parameter

By scanning one sequence using a sliding window of three residues with one residue step, the frequency of each tripeptide appeared in the protein sequence can be calculated according to the following equation:

$$f_i = n_i \Big/ \sum_{i=1}^{8000} n_i = n_i \,/\, (L-2) \tag{13}$$

where n_i and L denote the number of the i-th tripeptide and the length of the protein sequence, respectively. Then the protein can be expressed as an 8000-dimensional vector described as:

$$F_{8000} = \left[f_1, f_2, \cdots, f_i, \cdots f_{8000} \right]^T \tag{14}$$

where symbol T denotes the transposition of vector and f_i is the frequency of the i-th tripeptide.

6.3. Binomial Distribution

Three contiguous amino acids constitute a useful and minimal biological recognition signal for discovering peptides and small organic molecule mimics that are useful modulators of biological function. Tripeptide may also be used to predict plausible structures for oligopeptides as well as denovo protein design. Therefore, tripeptide compositions were utilized to represent the sample of membrane proteins. There are 8000 kinds of tripeptides in benchmark dataset, and each kind of tripeptide occurring in one class may be a stochastic event, therefore, we must judge whether each kind of tripeptide occurring in one class is a stochastic event or not. The binomial distribution was used to determine the probabilities or confidence levels (1-probability) of 8000 kinds of tripeptides occurred in *cis*-Golgi and *trans*-Golgi protein datasets.

Suppose the total occurrence frequency of all tripeptides in the dataset is M, the probability of tripeptides occurred in class i is denoted as p_i and calculated as:

$$p_i = m_i \,/\, M \tag{15}$$

where the m_i is the number of tripeptides that appear in the class i.

Let N_j represent the total occurrence number of a given tripeptide j in dataset, the probability of tripeptide j randomly occurring n_{ij} or more times in the class i can be defined by:

$$P(n_{ij}) = \sum_{m=n_{ij}}^{N_j} \frac{N_j!}{m!(N_j - m)!} p_i^m (1-p_i)^{N_j - m} \quad \textbf{(16)}$$

If $P(n_{ij})$ is a small value, it means the tripeptide j appearing in class i should not be random. The confidence level of this case is defined by CL_{ij}:

$$CL_{ij} = 1 - P(n_{ij}) \quad \textbf{(17)}$$

In the present work, we had two types of Golgi-resident proteins. Thus, an arbitrary tripeptide j has two confidence levels ($CL_{j\ cis}$ and $CL_{j\ trans}$) which describe the probabilities belonging to non-stochastic event of the j-th tripeptide occurring two types. Then we defined the confidence level of tripeptide j in benchmark dataset as follows:

$$CL_j = \max\{CL_{j\ cis},\ CL_{j\ trans}\} \quad \textbf{(18)}$$

If there are k tripeptides whose CL is larger than a given cutoff CL_0, the frequencies of these tripeptides are selected as optimized features that can be described as:

$$F_k = [f_1, f_2, \cdots, f_i, \cdots, f_k]^T \quad \textbf{(19)}$$

Then, high-dimensional feature parameters can be projected into low-dimensional space. The parameter k or CL_0 can be chosen by the use of five-fold cross-validation.

7. RESULTS AND DISCUSSION

According to Eqs. (16)-(19), the optimized tripeptides can be obtained. For economizing time and improving efficiency, the five-fold cross-validation was used to optimize the regularization parameter C and kernel parameter γ. In general, the tripeptide sets with high confidence level can provide more reliable information for classification. However, the number of these words is too small to afford enough information, which deduces the poor predictive accuracy. For example, using >99.9% as confidence level, we achieved 12 over-represented tripeptides. But the overall accuracy is only 79.6% in five-fold cross-validation. In contrast, the tripeptide sets with low confidence contains too many components. But it would reduce the cluster-tolerant capacity so as to lower down the cross-validation accuracy. For instance, 4956 tripeptides with >50% of confidence level only produced the overall accuracy of 75.9% in five-fold cross-validation.

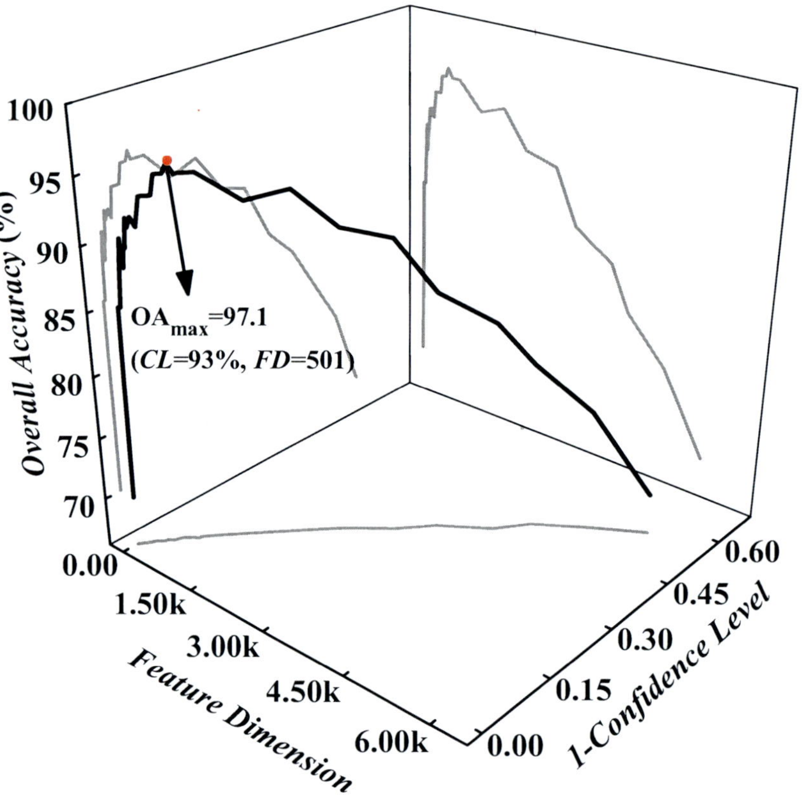

Figure 8: The graph for predicting the types of Golgi-resident proteins. Dark line denotes 3-D curve. Three gray lines are projections on three planes (overall accuracy/feature dimension plane, overall accuracy/confidence level, confidence level /feature dimension plane).

Therefore, using appropriate tripeptide sets would yield a prediction with higher accuracy. By varying the cutoff of confidence level, we obtained a series of tripeptide sets and then used SVM to predict *cis*-Golgi and tran-Golgi proteins. The 3-dimension graph for feature dimension, confidence level and overall accuracy was drawn in Fig. **8**. It exhibits that the five-cross validated accuracy increases to 97.1% when using 93% as cutoff of confidence level. The optimized tripeptide set contains 501 parameters. The regularization parameter C and kernel parameter γ are 6 and 0.03125, respectively. In the jackknife cross-validation, the maximum overall accuracy is 97.8%.

We also investigated the performances of the PseAAC [12]. We selected hydrophobicity, hydrophilicity and Mass as amino acid characters. The correlation parameter lambda varied from 0 to 7 with the step of 1. The weight factor varied from 0.05 to 0.7 with the step of 0.05. By examining 98 sets of PseAAC parameters, the highest overall accuracy of 75.2% was obtained when $\lambda=1$ with $w=0.05$ in jackknife cross-validation. Comparison showed that the optimized tripeptide set was the best feature set.

8. CONCLUDING REMARKS

This review focused on the development of prediction methods for subGolgi localization. The datasets developed for subGolgi resident proteins have been outlined. Three kinds of methods to predict subGolgi location of proteins have been described. In addition, we also draw another two conclusions. At first, high similarity data can make the sequence data lacking enough representativeness and the results being overestimated. Secondly, high dimension parameters can produce over-fitting, information redundancy or noise and dimension disaster which results in low generalization ability of prediction model, bad prediction accuracy and a handicap for the computation. Using feature selection techniques to optimize features can not only economize the time for computation, but also build robust prediction model. In essence, we hope this comprehensive review will provide insights into understanding of the subGolgi localization of proteins.

ACKNOWLEDGEMENT

This work was supported by Grants from The National Nature Scientific Foundation of China (No. 61202256, 61301260, 61100092), the Fundamental Research Funds for the Central Universities (ZYGX2012J113, ZYGX2013J102).

CONFLICT OF INTEREST

The authors confirm that this chapter contents have no conflict of interest.

DISCLOSURE

The chapter submitted for eBook series titled "Frontiers in Protein and Peptide Science" Volume 1 is an update of our article "Identify Golgi protein types with modified Mahalanobis discriminant algorithm and pseudo amino acid composition" published in Protein and Peptide Letter, Volume 18, Number 1, Pages 58-63, 2011 (PMID: 20955168)

REFERENCES

[1] Rhee, S.W.; Starr, T.; Forsten-Williams, K.; Storrie, B. The steady state distribution of glycosyltransferases between the Golgi apparatus and the endoplasmic reticulum is approximately 90:10. *Traffic*, 2005, 6, 978-990.

[2] Pfeffer, S.R. Constructing a Golgi complex. *J. Cell Biol.*, 2001, 155, 873-875.

[3] Hu, Z.; Zeng, L.; Xie, L.; Lu, W.; Zhang, J.; Li, T.; Wang, X. Morphological alteration of Golgi apparatus and subcellular compartmentalization of TGF-beta1 in Golgi apparatus in gerbils following transient forebrain ischemia. *Neurochem. Res.*, 2007, 32, 1927-1931.

[4] Fujita, Y.; Ohama, E.; Takatama, M.; Al-Sarraj, S.; Okamoto, K. Fragmentation of Golgi apparatus of nigral neurons with alphasynuclein-positive inclusions in patients with Parkinson's disease. *Acta Neuropathol.*, 2006, 112, 261-265.

[5] Gonatas, N.K.; Gonatas, J.O.; Stieber, A. The involvement of the Golgi apparatus in the pathogenesis of amyotrophic lateral sclerosis, Alzheimer's disease, and ricin intoxication. *Histochem. Cell Biol.*, 1998, 109, 591-600.

[6] Du, P.; Xu, C. Predicting multisite protein subcellular locations: progress and challenges. *Expert. Rev. Proteomics*. 2013, 10, 227-237.

[7] Yuan, Z.; Teasdale, R.D. Prediction of Golgi Type II membrane proteins based on their transmembrane domains. *Bioinformatics*, 2002, 18, 1109-1115.

[8] Chou, W.C.; Yin, Y.; Xu, Y. GolgiP: prediction of Golgi-resident proteins in plants. *Bioinformatics*, 2010, 26, 2464-2465.

[9] Ding, H.; Liu, L.; Guo, F.B.; Huang, J.; Lin, H. Identify Golgi protein types with modified Mahalanobis discriminant algorithm and pseudo amino acid composition. *Protein Pept. Lett.*, 2011, 18, 58-63.

[10] Li, W.; Godzik, A. Cd-hit: a fast program for clustering and comparing large sets of protein or nucleotide sequences. *Bioinformatics*, 2006, 22, 1658-1659.

[11] Chou, K.C. Using amphiphilic pseudo amino acid composition to predict enzyme subfamily classes. *Bioinformatics*, 2005, 21, 10-19.

[12] Chou, K.C. Pseudo amino acid composition and its applications in bioinformatics, proteomics and system biology. *Curr. Proteomics*, 2009, 6, 262-274.

[13] Laxton, R.R. The measure of diversity. *J. Theor. Biol.*, 1978, 71, 51-67.

[14] Lin, H.; Ding, H.; Guo, F.B.; Zhang, A.Y.; Huang, J. Predicting subcellular localization of mycobacterial proteins by using Chou's pseudo amino acid composition. *Protein Pept. Lett.*, 2008, 15, 739-744.

[15] Lin, H. The modified Mahalanobis Discriminant for predicting outer membrane proteins by using chou's pseudo amino acid composition. *J. Theor. Biol.*, 2008, 252, 350-356.

[16] Chou, K.C.; Zhang, C.T. Review: Prediction of protein structural classes. *Crit. Rev. Biochem. Mol. Biol.*, 1995, 30, 275-349.

[17] Ding, H.; Guo, S.H.; Deng, E.Z.; Yuan, L.F.; Guo, F.B.; Huang, J.; Rao, N.; Chen, W.; Lin, H. Prediction of Golgi-resident protein types by using feature selection technique. *Chem. Intel. Lab. Syst.*, 2013, 124, 9-13.

[18] Wang, G.; Dunbrack Jr., R.L. PISCES: recent improvements to a PDB sequence culling server. *Nucleic Acids Res.*, 2005, 33, W94–W98.

[19] Lin, H.; Ding, H. Predicting ion channels and their types by the dipeptide mode of pseudo amino acid composition. *J. Theor. Biol.*, 2011, 269, 64–69.

[20] Fan, R.E.; Chen, P.H.; Lin, C.J. Working set selection using the second order information for training SVM. *J. Mach. Learn. Res.*, 2005, 6, 1889-1918.

[21] Luo, L.F., Li, H. The statistical correlation of nucleotides in protein-coding DNA sequences. *Bull. Math. Biol.*, 1991, 53, 345-53.

CHAPTER 9

Predicting Membrane Protein Types with Bagging Learner

Bing Niu[1,2], Yu-Huan Jin[4], Kai-Yan Feng[6], Liang Liu[4], Wen-Cong Lu[4,*], Yu-Dong Cai[3,5,*] and Guo-Zheng Li[7]

[1]College of Life Science, Shanghai University, 99 Shang-Da Road, Shanghai, China, 200444, [2]School of Materials Science and Engineering, Shanghai University, 149 Yan-Chang Road, Shanghai 200072, People's Republic of China, [3]Department of Combinatorics and Geometry, CAS-MPG Partner Institute for Computational Biology, Shanghai Institutes for Biological Sciences, Chinese Academy of Sciences, Shanghai 200031, People's Republic of China, [4]Department of Chemistry, College of Sciences, Shanghai University, 99 Shang-Da Road, Shanghai 200444, People's Republic of China, [5]Department of Mathematics, University of Manchester, Manchester M60, 1QD, UK, [6]Division of Imaging Science & Biomedical Engineering, Room G424 Stopford Building, The University of Manchester, M13 9PT, UK and [7]School of Computer Science & Engineering, Shanghai University, Shanghai, 200072, China

Abstract: The membrane protein type is an important feature in characterizing the overall topological folding type of a protein or its domains therein. Many investigators have put their efforts to the prediction of membrane protein type. Here, we propose a new approach, the bootstrap aggregating method or bragging learner, to address this problem based on the protein amino acid composition. As a demonstration, the benchmark dataset constructed by K.C. Chou and D.W. Elrod (Proteins, 1999, 34, 137-153) was used to test the new method. The overall success rate thus obtained by jackknife cross-validation was over 84%, indicating that the bragging learner as presented in this paper holds a quite high potential in predicting the attributes of proteins, or at least can play a complementary role to many existing algorithms in this area. It is anticipated that the prediction quality can be further enhanced if the pseudo amino acid composition (K.C. Chou, Proteins, 2001, 43, 246-255) can be effectively incorporated into the current predictor. An online membrane protein type prediction web server developed in our lab is available at http://chemdata.shu.edu.cn/protein/protein.jsp.

Keywords: Bagging, neural network, SVM, membrane protein types, jackknife cross-validation.

***Address correspondence to Yu-Dong Cai:** Department of Combinatorics and Geometry, CAS-MPG Partner Institute for Computational Biology, Shanghai Institutes for Biological Sciences, Chinese Academy of Sciences, Shanghai 200031, People's Republic of China; E-mail: cai_yud@yahoo.com.cn

Ben M. Dunn (Ed)

1. INTRODUCTION

Due to the rapid development of genome and protein science, the information of bioinformatics has expanded dramatically. Therefore, it is highly desirable for computers to manage, organize and interpret the information. As part of biochemistry, study of membrane protein has become a hot topic for both experimental and theoretical purposes. A biological cell is highly organized with many functional units or organelles, and various suborganelle existing in superior organism. Most of these units are limited or bounded by one or more membranes. Subcellular membrane, subcellular and cell membrane are generally called biomembrane. Biomembrane is a structure bounding a cell or cell organelle as a barrier; it contains lipid bilayers, proteins, glycolipids, steroids, *etc.* Biomembrane is closely related to the exchange of energy, recognition and transmission of information [1,2]. The membrane contains specific proteins and lipid components that enable it to perform its unique roles in the functioning of the cell or organelle. Moreover, several studies show that many membrane proteins are also the key targets of drug discovery (see, *e.g.*, [3]), particularly membrane channel proteins [4-7].

Generally, membrane proteins can be classified to two classes – the transmembrane proteins and anchored membrane proteins. Transmembrane proteins contain one or more transmembrane segments with one or more hydrophobic segments to ensure stable association with the hydrophobic interior of the membrane, and hence can be easily discriminated from non-membrane proteins [8]. Unlike the transmembrane proteins, the anchored membrane proteins do not have the hydrophobic membrane spanning portions. Instead they have a consensus sequence motif at either the N- or C-terminus, which can be used to discriminate the anchored membrane proteins from nonmembrane proteins [9,10]. Membrane proteins can be further classified into the following five types [11,12]:

(1) type-I membrane protein: This single-pass transmembrane protein (the polypeptide crosses the lipid bilayer only once) has an extracellular (or luminal) N-terminus and cytoplasmic C-terminus for a cell (or organelle) membrane; (2) type-II membrane protein: This single-pass transmembrane protein has an extracellular (or luminal) C-terminus and cytoplasmic N-terminus for a cell (or

organelle) membrane; (3) multipass transmembrane proteins: the polypeptide crosses the lipid bilayer multiple times; (4) lipidchain-anchored membrane proteins: they are bound to the membrane by one or more covalently attached fatty acid chains or other types of lipid chains called prenyl groups; (5) GPI-anchored membrane proteins: they are bound to the membrane by a glycosylphosphatidylinositol (GPI) anchor. It can be tedious and costly to classify each new-found protein through biological experiments. An accurate and effective prediction method is needed to associate a membrane protein with its functions. Instead of separating the membrane proteins from non-membrane proteins, which is relatively straightforward, we will only focus on classifying the membrane proteins into the aforementioned five types.

Chou and Elrod firstly introduced the covariant discriminant algorithm to predict the types of membrane proteins based on their amino acid composition [12-14]. Later, several other efforts have been made to predict membrane protein types: Chou and Cai used incorporating amphipathic effects to study the above types of membrane protein [15]; Cai used artificial neural network and support vector machine to predict the types of membrane protein [16-18]; Wang used weighted-support vector machines to predict membrane protein types based on the pseudo-amino acid composition [19]; Cai and Chou predict membrane protein types based on functional domain composition and pseudo-amino acid composition [20]; Shen and Chou used optimized evidence-theoretic K-nearest neighbor classifier and ensemble learning to predict membrane protein types [21,22]. The PseAA composition [23] was originally proposed for predicting protein subcellular localization and membrane protein type [23]; while the amphiphilic PseAA composition [24] proposed for predicting the enzyme functional classification. The essence of PseAA composition is to use a discrete model to represent a protein sample yet without complete losing its sequence-order information. Ever since the concept of PseAA composition was introduced, various PseAA composition approaches have been stimulated to deal with different problems in proteins and protein-related systems [19-22,25-56]. Owing to its wide usage, recently a very flexible PseAA composition generator, called "PseAAC" [57], was established at the website http://chou.med.harvard.edu/bioinf/PseAA/, by which users can generate 63 different kinds of PseAA composition. Recently,

Chou and Shen [58] proposed a novel idea in predicting the type of membrane proteins; in that paper, membrane proteins are classified into 8 types, and a web-server called "MemType-2L" at http://chou.med.harvard.edu/bioinf/MemType/ has been provided.

In this paper, Bagging method is applied in membrane protein types prediction based on the database derived from [12].

2. MATERIALS AND METHODS

2.1. Materials

The test database taken from [12] contains 2059 prokaryotic proteins(Type I membrane protein: 435; Type II membrane protein: 152; Multipass transmembrane proteins: 1311; Lipid chain-anchored membrane proteins: 51; GPI-anchored membrane proteins: 110).

According to its amino acid composition, a protein domain can be represented as a point in a 20-D space [14]. The amino acid composition is taken as the input of the Bagging algorithm.

2.2. Methods

Bagging is one of the most popular and important ensemble learning algorithms which are designed to combine many weak classifiers to gain a strong classifier. The Bagging algorithm [59] is similar to another popular ensemble learning algorithm – the boosting classifiers [60, 61]. Bagging creates some independent training data sets to generate a number of independent classifiers and then combine them to create the final classifier by majority voting. In boosting each classifier depends on the previous one by focusing on the misclassification data. Bagging and Boosting have their own advantages and disadvantages: bagging is very suitable for parallel computation while boosting is more straightforward at dealing with misclassification data. Nevertheless, both classifiers can improve the classification results. Considering that bioinformatics data are usually very large, bagging can be chosen as an appropriate classifier to deal with those data if it can produce promising results. Bagging has been applied to solve various classification and prediction problems in biology and social sciences [62-66].

In bagging the training data set is bootstrapped [67] to create another training data set with the same size as the original training set by randomly choosing data from the original training data. Notice that any individual data from the original training data set may be excluded or can be repeatedly chosen through chance. In this way, the training data set is bootstrapped many times to create many bootstrapped training sets. Each training set will generate a classifier by training its data. Thus many classifiers are generated, which can then be combined to build the final classifier through majority vote. For example, if 10 classifiers are built and a new data needs to be classified into one class or another, 6 classifiers predict that the data belong to class 1 and 4 classifiers predict that the data belong to class 2, then this new data shall be classified into the first class as it has the majority vote. A mathematical description of the algorithm is given below.

The input dataset can be expressed as

$$D = (\mathbf{x}_1, y_1), \ldots, (\mathbf{x}_m, y_m), \ldots, y \in \{\ 1,2,3,\ldots, j\ \},$$

where $\mathbf{x}_m$ is a vector representing the amino acid composition, and the label y_m is the target class. The original dataset D is randomly bootstrapped *k* times with replacement to produce *k* training sets. Choose a base machine learning method to train all these k training sets to get k classifiers, *i.e.* a classifier h_i is obtained by training i^{th} dataset D^i. Given a new data $\vec{x}$, let $N_j(\vec{x}) = \sum c(i)$, where $c(i) = 1$ if $h_i(\vec{x}) = j$, otherwise $c(i) = 0$. Here, N_j represents the number of each class's vote, and $c(i)$ represents the account of each class. The final classifier $H(\vec{x})$ can be defined as $H(\vec{x}) = arc\max(N_j(\vec{x}))$.

The KNN (K Nearest Neighbor) method was selected as the base learning machine (weak classifier). The parameter K, the number of the nearest neighbors, was set as 1. After being trained, the thumb of rules output by the Bagging method was obtained, indicating that the trained model could be used for performing the prediction concerned.

2.3. Implementation

The software package Weka that includes the Bagging algorithm ran under the Java environment on a 1400 MHz Pentium IV computer. All the learning input data were range-scaled to [0, 1] in this work.

3. RESULTS AND DISCUSSION

In statistical prediction the independent dataset test, subsampling test (self-consistency test), and jackknife test are the three cross-validation methods usually utilized to examine the quality of a predictor [68]. However, as elucidated by [32] and demonstrated in [31], among these three methods, the jackknife test is deemed the most objective and can always yield a unique result for a given benchmark dataset, and hence has been widely and increasingly adopted by investigators to examine the accuracy of various predictors [22,27,28,33,34,38,39,42,48,50,52,54,56,69-97].

Tables are suggested for illustrating the following rates.

As a result, the rates of correct prediction of self-consistency for the five possible membrane protein types are 99.03% (type-I membrane protein: 98.85%; type-II membrane protein: 96.71%; multipass transmembrane proteins: 99.62%; lipidchain- anchored membrane proteins: 98.04%; GPI-anchored membrane proteins: 97.27%).

And the rates of correct prediction of Jackknife Cross-Validation for the five possible membrane protein types are 84.18 % (type-I membrane protein: 79.80%; type-II membrane protein: 48.68%; multipass transmembrane proteins: 93.21%; lipidchain-anchored membrane proteins: 49.02%; GPI-anchored membrane proteins: 60.91%).

We compare the Bagging Learning method with other pattern classification methods — the covariant discriminant algorithm [12], neural networks [98] and support vector machines (SVM) [99,100] as is shown in Tables **1** and **2**. The correct classification rates of Covariant discriminant algorithm and neural networks were taken from references [12,18]. As we can see from the tables, the self-consistency test has reached a very high correct classification rate indicating that the membrane protein type can potentially be predicted very accurately. It also means that Bagging has learned very well through the training process. The correct classification rate of the Jackknife test also outperformed other algorithms though it decreased about 15% from the self-consistency test. We predict that if more data samples are used in the training, Jackknife test shall increase its correct rate a lot. Otherwise, bagging classifier may suffer from the over-fitting problem. We suggest that this shall be investigated in a future research.

Table 1: Self-consistency success rates by Covariant Discriminant, Neural Network, SVM and Bagging

Algorithm	**Rate of Correct Prediction for Each Class (%)**					**Overall Rate of Correct Prediction**
	Type I Transmembrane	**Type II Transmembrane**	**Multipass Transmembrane**	**Lipid-Chain Anchored Membrane**	**GPI-Anchored Membrane**	
Covariant discriminant	79.8	63.2	85.1	84.3	61.8	81.1
Neural Network	99.31	91.45	96.57	90.20	62.72	94.8
SVM	77.24	66.45	90.85	80.39	79.09	85.24
Bagging	98.85	96.71	99.62	98.04	97.27	99.03

Table 2: Jackknife Cross-Validation success rates by Covariant Discriminant, Neural Network, SVM and Bagging

Algorithm	**Rate of Correct Prediction for Each Class (%)**					**Overall Rate of Correct Prediction**
	Type I Transmembrane	**Type II Transmembrane**	**Multipass Transmembrane**	**Lipid-Chain Anchored Membrane**	**GPI-Anchored Membrane**	
Covariant discriminant	74.0	52.0	83.7	49	45.4	76.4
Neural Network	75.63	30.92	88.86	50.98	30.91	77.76
SVM	77.7	28.3	92.5	52.9	35.5	80.9
Bagging	79.80	48.68	93.21	49.02	60.91	84.18

4. CONCLUSION

The Bagging Learner has remarkably outperformed the Covariant discriminant algorithm, neural network and SVM classifiers in predicting the membrane protein types for the same dataset constructed by previous investigators. It is thus anticipated that the Bagging classifier can also be used to predict other protein attributes, such as subcellular localization [11,17,44,62,101], protein structural class, enzyme family and subfamily classes [24,102,103], enzyme active sites, G-protein coupled receptor classification [104,105], and protein quaternary structure types [106], among many others; an effective combination of the current bagging learner with PseAA composition will further enhance the prediction quality. An online membrane protein type prediction web server developed in our lab is available at http://chemdata.shu.edu.cn/protein/protein.jsp.

ACKNOWLEDGEMENTS

Financial support to this work from the National Natural Science Foundation of China under Grant 20503015, from Shanghai Leading Academic Discipline Project (Project Number: J50101) and from Systems Biology Research Foundation of Shanghai University are gratefully acknowledged. The author wishes to thank Prof. Dr. Kuo-Chen for providing the data set.

We appreciate Mr. Kun Tu for editing the manuscript of this paper.

CONFLICT OF INTEREST

The authors confirm that this chapter contents have no conflict of interest.

DISCLOSURE

The chapter submitted for eBook series titled "Frontiers in Protein and Peptide Science" Volume 1" is an update of our article "Predicting Membrane Protein Types with Bagging Learner" published in Protein and Peptide Letter, 2008,15(6): 590-594.

REFERENCES

[1] Alberts, B., Bray, D., Lewis, J., Raff, M., Raff, M., Roberts, K.andWatson, J.D. *Molecular Biology of the Cel*; (**1994**).Garland Publishing, New York

[2] Lodish, H., Baltimore, D., A.Berk, S.L.Zipursky, P.MatsudairaandJ.Darnell *Molecular Cell Biology*; (**1995**).Scientific American Books, New York

[3] Chou, K.C. (**2004**) *Current Medicinal Chemistry*, *11*, 2105-2134.

[4] Chou, K.C. (**2004**) *Journal of Proteome Research*, *3*, 856-861.
[5] Doyle, D.A., Cabral, J.M., Pfuetzner, R.A., Kuo, A.L., Gulbis, J.M., Cohen, S.L., Chait, B.T. and Mackinnon, R. (**1998**) *Science*, *280*, 69-77.
[6] Oxenoid, K. and Chou, J.J. (**2005**) *Proceedings of the National Academy of Sciences of the United States of America*, *102*, 10870-10875.
[7] Schnell, J.R. and Chou, J.J. (**2008**) *Nature*, *451*, 591-U12.
[8] Rost, B., Casadio, R., Fariselli, P. and Sander, C. (**1995**) *Protein Science*, *4*,, 521–533.
[9] Casey, P. (**1995**) *Science*, *268*, 221–225.
[10] Resh, M.D. (**1994**) *Cell*, *76* 411–413.
[11] Chou, K.C. and Elrod, D.W. (**1999**) *Protein Engineering*, *12*, 107-118.
[12] Chou, K.C. and Elrod, D.W. (**1999**) *Proteins-Structure Function and Genetics*, *34*, 137-153.
[13] Mahalanobis, P.C. (**1936**) *Proceedings of the National Institute of Sciences of India*, *2*, 49–55.
[14] Chou, K.C. (**1995**) *Proteins-Structure Function and Bioinformatics*, *21*, 319-344.
[15] Chou, K.C. and Cai, Y.D. (**2005**) *Journal of Chemical Information and Modeling*, *45*, 407-413.
[16] Cai, Y.D., Liu, X.J. and Chou, K.C. (**2001**) *Journal of Biomolecular Structure & Dynamics*, *18*, 607-610.
[17] Cai, Y.D., Zhou, G.P. and Chou, K.C. (**2003**) *Biophysical Journal*, *84*, 3257-3263.
[18] Cai, Y.D., Ricardo, P.W., Jen, C.H. and Chou, K.C. (**2004**) *Journal of Theoretical Biology*, *226*, 373-376.
[19] Wang, M., Yang, J., Liu, G.P., Xu, Z.J. and Chou, K.C. (**2004**) *Protein Engineering Design & Selection*, *17*, 509-516.
[20] Cai, Y.D. and Chou, K.C. (**2006**) *Journal of Theoretical Biology*, *238*, 395-400.
[21] Shen, H.B. and Chou, K.C. (**2005**) *Biochemical and Biophysical Research Communications*, *334*, 288-292.
[22] Shen, H.B. and Chou, K.C. (**2007**) *Amino Acids*, *32*, 483-488.
[23] Chou, K.C. (**2001**) *Proteins-Structure Function and Genetics*, *43*, 246-255.
[24] Chou, K.C. (**2005**) *Bioinformatics*, *21*, 10-19.
[25] Aguero-Chapin, G., Gonzalez-Diaz, H., Molina, R., Varona-Santos, J., Uriarte, E. and Gonzalez-Diaz, Y. (**2006**) *FEBS Lett*, *580*, 723-30.
[26] Caballero, J., Fernandez, L., Garriga, M., Abreu, J.I., Collina, S. and Fernandez, M. (**2007**) *Journal of Molecular Graphics & Modelling*, *26*, 166-178.
[27] Chen, C., Tian, Y.X., Zou, X.Y., Cai, P.X. and Mo, J.Y. (**2006**) *Journal of Theoretical Biology*, *243*, 444-448.
[28] Chen, C., Zhou, X.B., Tian, Y.X., Zou, X.Y. and Cai, P.X. (**2006**) *Analytical Biochemistry*, *357*, 116-121.
[29] Chen, Y.L. and Li, Q.Z. (**2007**) *Journal of Theoretical Biology*, *248*, 377-381.
[30] Chen, Y.L. and Li, Q.Z. (**2007**) *Journal of Theoretical Biology*, *245*, 775-783.
[31] Chou, K.C. and Shen, H.B. (**2007**) *Analytical Biochemistry*, *370*, 1-16.
[32] Chou, K.-C. and Shen, H.-B. (**2008**) *Nat Protoc*, *3*, 153-62.
[33] Du, P.F. and Li, Y.D. (**2006**) *Bmc Bioinformatics*, *7*, 518.
[34] Gao, Y., Shao, S., Xiao, X., Ding, Y., Huang, Y., Huang, Z. and Chou, K.C. (**2005**) *Amino Acids*, *28*, 373-376.

[35] Gonzalez-Diaz, H., Perez-Bello, A., Uriarte, E. and Gonzalez-Diaz, Y. (**2006**) *Bioorganic & Medicinal Chemistry Letters*, *16*, 547-553.
[36] Gonzalez-Diaz, H., Vilar, S., Santana, L. and Uriarte, E. (**2007**) *Current Topics in Medicinal Chemistry*, *7*, 1015-1029.
[37] Kurgan, L.A., Stach, W. and Ruan, J. (**2007**) *Journal of Theoretical Biology*, *248*, 354-366.
[38] Lin, H. and Li, Q.Z. (**2007**) *Biochemical and Biophysical Research Communications*, *354*, 548-551.
[39] Lin, H. and Li, Q.Z. (**2007**) *Journal of Computational Chemistry*, *28*, 1463-1466.
[40] Liu, H., Wang, M. and Chou, K.C. (**2005**) *Biochemical and Biophysical Research Communications*, *336*, 737-739.
[41] Liu, H., Yang, J., Wang, M., Xue, L. and Chou, K.C. (**2005**) *Protein Journal*, *24*, 385-389.
[42] Mondal, S., Bhavna, R., Babu, R.M. and Ramakumar, S. (**2006**) *Journal of Theoretical Biology*, *243*, 252-260.
[43] Mundra, P., Kumar, M., Kumar, K.K., Jayaraman, V.K. and Kulkarni, B.D. (**2007**) *Pattern Recognition Letters*, *28*, 1610-1615.
[44] Pan, Y.X., Zhang, Z.Z., Guo, Z.M., Feng, G.Y., Huang, Z.D. and He, L. (**2003**) *Journal of Protein Chemistry*, *22*, 395-402.
[45] Pua, X., Guo, H., Leung, H. and Lin, Y.L. (**2007**) *Journal of Theoretical Biology*, *247*, 259-265.
[46] Shen, H.B. and Chou, K.C. (**2005**) *Biochemical and Biophysical Research Communications*, *337*, 752-756.
[47] Shen, H.B. and Chou, K.C. (**2006**) *Bioinformatics*, *22*, 1717-1722.
[48] Shen, H.B. and Chou, K.C. (**2007**) *Biochemical and Biophysical Research Communications*, *355*, 1006-1011.
[49] Shen, H.B., Yang, J. and Chou, K.C. (**2006**) *Journal of Theoretical Biology*, *240*, 9-13.
[50] Shen, H.B., Yang, J. and Chou, K.C. (**2007**) *Amino Acids*, *33*, 57-67.
[51] Wang, S.Q., Yang, J. and Chou, K.C. (**2006**) *Journal of Theoretical Biology*, *242*, 941-946.
[52] Xiao, X., Shao, S., Ding, Y., Huang, Z. and Chou, K.C. (**2006**) *Amino Acids*, *30*, 49-54.
[53] Xiao, X., Shao, S.H., Huang, Z.D. and Chou, K.C. (**2006**) *Journal of Computational Chemistry*, *27*, 478-482.
[54] Zhang, S.W., Pan, Q., Zhang, H.C., Shao, Z.C. and Shi, J.Y. (**2006**) *Amino Acids*, *30*, 461-468.
[55] Zhang, T.L., Ding, Y.S. and Chou, K.C. (**2006**) *Computational Biology and Chemistry*, *30*, 367-371.
[56] Zhang, T.L. and Ding, Y.S. (**2007**) *Amino Acids*, *33*, 623-629.
[57] Shen, H.B. and Chou, K.C. (**2008**) *Analytical Biochemistry*, *373*, 386-388.
[58] Chou, K.C. and Shen, H.B. (**2007**) *Analytical Biochemistry*, *370*, 1-16.
[59] Breiman, L. (**1996**) *Machine Learning*, *24*, 123-140.
[60] Freund, Y. and Schapire, R.E. (**1997**) *J. Comput. System Sci.*, *1*, 119-139.
[61] Schapire, R.E. and Y., S. (**1999**) *Mach. Learning.*, *37*, 297–336.
[62] Dudoit, S. and Fridlyand, J. (**2003**) *Bioinformatics*, *19*, 1090-1099.
[63] Mardin, C.Y., Hothorn, T., Peters, A., Junemann, A.G., Nguyen, N.X. and Lausen, B. (**2003**) *Journal of Glaucoma*, *12*, 340-346.
[64] Huang, Y., Cai, J., Ji, L. and Li, Y.D. (**2004**) *Computational Biology and Chemistry*, *28*, 275-280.

[65] Dong, L.H., Yuan, Y. and Cai, Y.D. (**2006**) *Journal of Biomolecular Structure & Dynamics*, *24*, 239-242.
[66] Prasad, A.M., Iverson, L.R. and Liaw, A. (**2006**) *Ecosystems*, *9*, 181-199.
[67] Efron, B. and R, T. *An Introduction to the Bootstrap*; (**1993**).Chapman & Hall, NY.
[68] Chou, K. and Zhang, C. (**1995**) *Critical Reviews in Biochemistry and Molecular Biology*, *30*, 275-349.
[69] Chen, J., Liu, H., Yang, J. and Chou, K.C. (**2007**) *Amino Acids*, *33*, 423-428.
[70] Chou, K.C. and Shen, H.B. (**2006**) *Biochemical and Biophysical Research Communications*, *347*, 150-157.
[71] Chou, K.C. and Shen, H.B. (**2006**) *Journal of Proteome Research*, *5*, 1888-1897.
[72] Chou, K.C. and Shen, H.B. (**2007**) *Journal of Proteome Research*, *6*, 1728-1734.
[73] Chou, K.C. and Shen, H.B. (**2007**) *Journal of Cellular Biochemistry*, *100*, 665-678.
[74] Chou, K.C. and Shen, H.B. (**2007**) *Biochemical and Biophysical Research Communications*, *360*, 339-345.
[75] Chou, K.C. and Shen, H.B. (**2007**) *Biochemical and Biophysical Research Communications*, *357*, 633-640.
[76] Diao, Y., Ma, D., Wen, Z., Yin, J., Xiang, J. and Li, M. (**2008**) *Amino Acids*, *34*, 111-117.
[77] Ding, Y.S., Zhang, T.L. and Chou, K.C. (**2007**) *Protein and Peptide Letters*, *14*, 811-815.
[78] Fang, Y., Guo, Y., Feng, Y. and Li, M. (**2008**) *Amino Acids*, *34*, 103-109.
[79] Guo, Y.Z., Li, M., Lu, M., Wen, Z., Wang, K., Li, G. and Wu, J. (**2006**) *Amino Acids*, *30*, 397-402.
[80] Kedarisetti, K.D., Kurgan, L. and Dick, S. (**2006**) *Biochemical and Biophysical Research Communications*, *348*, 981-988.
[81] Li, F.M. and Li, Q.Z. (**2008**) *Amino Acids*, *34*, 119-125.
[82] Liu, D.Q., Liu, H., Shen, H.B., Yang, J. and Chou, K.C. (**2007**) *Amino Acids*, *32*, 493-496.
[83] Niu, B., Cai, Y.D., Lu, W.C., Li, G.Z. and Chou, K.C. (**2006**) *Protein and Peptide Letters*, *13*, 489-492.
[84] Shen, H.B. and Chou, K.C. (**2007**) *Protein Engineering Design & Selection*, *20*, 39-46.
[85] Shen, H.B. and Chou, K.C. (**2007**) *Biopolymers*, *85*, 233-240.
[86] Shi, J.Y., Zhang, S.W., Pan, Q., Cheng, Y.M. and Xie, J. (**2007**) *Amino Acids*, *33*, 69-74.
[87] Sun, X.D. and Huang, R.B. (**2006**) *Amino Acids*, *30*, 469-475.
[88] Tan, F., Feng, X., Fang, Z., Li, M., Guo, Y. and Jiang, L. (**2007**) *Amino Acids*, *33*, 669-675.
[89] Wang, M., Yang, J. and Chou, K.C. (**2005**) *Amino Acids*, *28*, 395-402.
[90] Wen, Z., Li, M., Li, Y., Guo, Y. and Wang, K. (**2007**) *Amino Acids*, *32*, 277-283.
[91] Xiao, X. and Chou, K.-C. (**2007**) *Protein Pept Lett*, *14*, 871-5.
[92] Xiao, X., Shao, S., Ding, Y., Huang, Z., Chen, X. and Chou, K.C. (**2005**) *Amino Acids*, *28*, 29-35.
[93] Xiao, X., Shao, S., Ding, Y., Huang, Z., Huang, Y. and Chou, K.C. (**2005**) *Amino Acids*, *28*, 57-61.
[94] Zhou, G.P. (**1998**) *Journal of Protein Chemistry*, *17*, 729-738.
[95] Zhou, G.P. and Assa-Munt, N. (**2001**) *Proteins-Structure Function and Genetics*, *44*, 57-59.
[96] Zhou, G.P. and Doctor, K. (**2003**) *Proteins-Structure Function and Genetics*, *50*, 44-48.
[97] Zhou, X.B., Chen, C., Li, Z.C. and Zou, X.Y. (**2007**) *Journal of Theoretical Biology*, *248*, 546-551.
[98] Bishop, C.M. *Neural networks for pattern recognition*; (**1995**).Oxford Press, London

[99] Cristianini, N. and Shawe-Taylor, J. *An introduction to support vector machines*; (**2000**).Cambridge University Press, Cambridge

[100] Vapnik, V. *Statistical learning theory*; (**1998**).Wiley-Interscience, New York

[101] Chou, K.C. and Cai, Y.D. (**2003**) *Journal of Cellular Biochemistry*, *90*, 1250-1260.

[102] Chou, K.C. and Cai, Y.D. (**2004**) *Protein Science*, *13*, 2857-2863.

[103] Chou, K.C. and Elrod, D.W. (**2003**) *Journal of Proteome Research*, *2*, 183-190.

[104] Chou, K.C. and Elrod, D.W. (**2002**) *Journal of Proteome Research*, *1*, 429-433.

[105] Elrod, D.W. and Chou, K.C. (**2002**) *Protein Engineering*, *15*, 713-715.

[106] Chou, K.C. and Cai, Y.D. (**2003**) *Proteins-Structure Function and Genetics*, *53*, 282-289.

Send Orders for Reprints to reprints@benthamscience.net

CHAPTER 10

Prediction of Human Protein Subcellular Locations with Feature Selection and Analysis

Bi-Qing Li[2], Tao Huang[3], Lei Chen[4], Kai-Yan Feng[5] and Yu-Dong Cai[1*]

[1]Institute of Systems Biology, Shanghai University, Shanghai, People's Republic of China, [2]Key Laboratory of Systems biology, Shanghai Institutes for Biological Sciences, Chinese Academy of Sciences, Shanghai, People's Republic of China, [3]Department of Genetics and Genomic Sciences, Mount Sinai School of Medicine, New York, USA, [4]College of Information Engineering, Shanghai Maritime University, Shanghai, People's Republic of China and [5]Beijing Genomics Institute, Shenzhen, People's Republic of China

Abstract: In this paper, we propose a strategy to predict subcellular locations of human proteins using multi-step feature selection. Each protein is firstly coded by features derived from KEGG and GO enrichment scores. After an initial feature reduction, 9958 features remain and they are sorted by the Minimum Redundancy Maximum Relevance (mRMR) method. The sorted features are then filtered by an incremental feature selection (IFS) procedure and a compact set of features are obtained. Random forest (RF) is used as the prediction model and achieved an overall prediction accuracy of 67.72%, evaluated by ten-fold cross-validation. The corresponding KEGG pathways and GO terms of the resultant features are analyzed in-depth, and are deemed as the most important terms relating to human protein subcellular location.

Keywords: Subcellular location, minimum redundancy maximum relevance, incremental feature selection, random forest algorithm, ten-fold cross-validation.

1. INTRODUCTION

A cell consists of different compartments. On one hand, each compartment contains its own characteristic set of proteins exerting specific functions. On the other hand, proper functioning of a protein requires appropriate subcellular location

***Address correspondence to Yu-Dong Cai:** Institute of Systems Biology, Shanghai University, 99 ShangDa Road, Shanghai, 200444, People's Republic of China; Tel: 86-21-13764837251; Email: cai_yud@126.com of the protein, *i.e.*, proteins may lose their functions if not correctly located, *e.g.*, autoimmune polyendocrinopathy-candidiasis-ectodermal dystrophy (APECED) is caused by a defective gene AIRE, the mutation of which dramatically alters the subcellular location of its protein products [1]. Proteins are transferred to the compartments of a cell by different distribution systems, *e.g.*, a cell may transport proteins related to energy production to mitochondria by the utilization of signal sequences and a protein translocator [2]. Therefore, the determination of protein subcellular location could be helpful in elucidating the mechanisms by which the protein trafficking occurs. In addition, knowing the specific location of a protein is beneficial for the design of effective new drugs [3]. Therefore, it is important to annotate the subcellular location of a protein.

Although determining the subcellular location of proteins is important, experimental methods are usually time-consuming and costly. With the rapid increase of protein sequences in databanks in the post-genomics era, there is a desire to develop computational methods to predict protein subcellular location. Much effort was already put into predicting protein subcellular location using computational methods [4-6] about 10 years ago. More recent progresses have also been made [7-9]. Many of these efforts were reviewed in a recent paper written by Chou and Shen [10].

This study attempts to develop a novel method for predicting human protein subcellular localization based on the KEGG pathway and GO enrichment scores, in hopes that it may become a useful complementary tool to the existing methods in the related areas. In this research, we incorporate multiple feature selection procedures including feature reduction, mRMR and incremental feature selection, which not only allow us to select an optimized feature set, but also enable us to analyze which KEGG pathway and GO terms are most important in the determination of human subcellular locations. As a result, 142 features are selected including 5 KEGG pathways and 137 GO terms, achieving a prediction accuracy of 67.72%. Feature analysis shows that cell (GO: 0005623), organelle (GO: 0043226), binding (GO: 0005488), metabolic process (GO: 0008152) and hsa01100 (Metabolic pathways) are critical for the prediction of human protein subcellular location. Lastly, the biological meanings of these most important properties are also discussed.

2. MATERIALS AND METHODS

2.1. Data Preparation

Data analyzed in this research are obtained from Swiss-Prot (http://cn.expasy.org/, release 54.0), by searching the database for proteins annotated with "subcellular location". 53427 protein sequences are initially downloaded. Proteins with length shorter than 50 amino acids (*e.g.*, protein fragments) and those with length longer than 5000 (*e.g.*, protein complexes) are excluded. Proteins containing unknown amino acid abbreviation such as X are also excluded. We also remove protein sequences with high degree of similarity using program CD-HIT [11], with a cutoff value of 0.7. And finally only human proteins are reserved for the study. As a result, 5960 protein sequences remain after the exclusion (please refer to Additional file S1 for the Uniprot ID for these proteins). These proteins are classified into 16 categories. Table **1** lists the 16 categories, and the number of proteins in each category.

Table 1: Number of proteins in each category.

Category	**Number of Proteins**	**Category**	**Number of Proteins**
Biological membrane	1864	Golgi apparatus	114
Cell periphery	42	Microtubule cytoskeleton	57
Cytoplasm	605	Mitochondrion	391
Cytoplasmic vesicle	83	Nuclear periphery	41
Endoplasmic reticulum	215	Nucleolus	139
Endosome	31	Nucleus	1525
Extracellular space or cell surface	737	Peroxisome	51
Flagellum or cilium	3	Vacuole	62

2.2. STRING Protein-Protein Interaction (PPI) Data

The initial PPI network was retrieved from STRING (version 9.0) [12] (http://string.embl.de/), which is a large database containing of known and predicted protein interaction. The interaction network takes proteins as its nodes, with an edge between two proteins if they interact with each other. These interactions contain direct (physical) and indirect (functional) interactions, derived from numerous sources such as experimental repositories and computational prediction methods.

2.3. KEGG Enrichment Scores of Proteins

The immediate neighborhood method is the simplest and most direct option for predicting the function of a query protein based on other proteins of known function [13]. A query protein interacts with many neighboring proteins in the STRING network [12]. The KEGG (Kyoto Encyclopedia of Genes and Genomes, http://www.genome.jp/kegg/) enrichment score of the protein and its neighbors was defined as the -$\log_{10}$ of the p-value generated by hypergeometric test [14] on all KEGG pathways, which can be computed by.

$$\mathrm{Score}_{\mathrm{KEGG}}(i,j) = -\log_{10}\left(\sum_{k=m}^{n} \frac{\binom{M}{k}\binom{N-M}{n-k}}{\binom{N}{n}}\right) \quad \textbf{(1)}$$

where N is the number of overall protein in human, M is the number of proteins that are in the j-th KEGG pathway, n is the number of proteins in the protein set Pi, m is the number of proteins in Pi that are annotated to the j-th KEGG pathway. The larger the enrichment score, the more representation the term is. There were 229 KEGG enrichment scores for each of the proteins investigated. (See Additional file S2).

2.4. Gene Ontology (GO) Enrichment Scores of Proteins

The network context of a protein can tell much about its functions. In this study, the network context we used was from STRING [12]. The GO (Gene Ontology: http://www.geneontology.org/) enrichment score of a protein and its neighbors was defined as the -$\log_{10}$ of the p-value generated by hypergeometric test on all GO terms [15, 16], which can be formulated as

$$\mathrm{Score}_{\mathrm{GO}}(i,j) = -\log_{10}\left(\sum_{k=m}^{n} \frac{\binom{M}{k}\binom{N-M}{n-k}}{\binom{N}{n}}\right) \quad \textbf{(2)}$$

where N is the number of overall protein in human, M is the number of proteins that are annotated to the j-th Gene Ontology term, n is the number of proteins in the protein set Pi consisting of protein *i* and its direct neighbors in STRING network, m is the number of proteins in Pi that are annotated to the *j*-th Gene Ontology term. The larger the enrichment score of a GO term, the more representation the term is. There were 15942 GO enrichment score features. (See Additional file S2).

2.5. Feature Reduction

We calculated the Cramer's V coefficient [17, 18] between features and classes of samples. Cramer's V coefficient is a statistical measurement derived from the Pearson Chi-square test [19]. It ranges from 0 to 1. A smaller Cramer's V coefficient indicates weaker association. The features with Cramer's V coefficient smaller than 0.1 were removed. After feature reduction, a total of 9958 features remained, including 192 KEGG pathway features and 9766 GO terms features. (See Additional file S3).

2.6. mRMR METHOD

We used Minimum Redundancy Maximal Relevance (mRMR) method to rank the importance of the features [20]. The mRMR method ranks features based on both their relevance to the target and the redundancy between features. A smaller index of a feature denotes that it has a better trade-off between maximum relevance to the target and minimum redundancy.

Both relevance and redundancy were quantified by mutual information (MI), which estimates the extent to which one vector is related to another. The MI equation is defined as:

$$I(x,y)=\iint p(x,y)\log\frac{p(x,y)}{p(x)p(y)}dxdy \quad \textbf{(3)}$$

Where x and y are vectors, $p(x,y)$ is their joint probabilistic density, and $p(x)$ and $p(y)$ are the marginal probabilistic densities.

Ω denotes the whole feature set. Ω_s denotes the already-selected feature set containing m features and Ω_t was used to denote the to-be-selected feature set containing n features. The relevance D between the feature f in Ω_t and the target c can be calculated by:

$$D = I(f, c) \tag{4}$$

The redundancy R between the feature f in Ω_t and all the features in Ω_s can be calculated by:

$$R = \frac{1}{m} \sum_{f_i \in \Omega_s} I(f, f_i) \tag{5}$$

To obtain the feature f_j in Ω_t with maximum relevance and minimum redundancy, the mRMR function combined equation (4) and equation (5) and are defined as below:

$$\max_{f_j \in \Omega_t} \left[I(f_j, c) - \frac{1}{m} \sum_{f_i \in \Omega_s} I(f_j, f_i) \right] (j = 1, 2, \cdots, n) \tag{6}$$

The mRMR evaluation would continue N rounds when given a feature set with N (N=m+n) features. After the mRMR evaluation, we get a feature set S :

$$S = \left\{ f_1^{'}, f_2^{'}, \ldots, f_h^{'}, \ldots, f_N^{'} \right\} \tag{7}$$

In this feature set S , the index h of each feature indicates at which round that the feature is selected. The smaller the index h is, the earlier the feature satisfied equation (6) and the better the feature is.

3. RANDOM FOREST

Random Forest (RF) approach is a popular machine-learning algorithm that has been successfully used in handling various biological prediction problems [21-25]. Developed by Loe Breiman [26], RF is an ensemble predictor that consists of multiple decision trees. To classify a new queried sample with an input vector, the input vector is predicted by each decision tree in the forest. Each tree provides a predicted class. And the class with the most votes will be output as the predicted class of the random forest. Each tree is constructed using the following procedure:

1) Suppose the number of training cases is N, take N samples at random, but with replacement from the original data, which will be the training set for growing the tree.

2) If there are M input variables, choose a number *m* which ought to be much less than M. At each node, *m* variables are selected randomly out of the M variables and the most optimized split on these *m* variables is employed to split the node. The value of *m* does not change during the growth of the forest.

3) Each tree is fully grown and not pruned.

In this study, we employed Random Forest developed in Weka 3.6.4 [27], which implements the algorithm described above. Notably, it was run with default parameters.

4. TEN-FOLD CROSS-VALIDATION METHOD

Ten-fold cross-validation was often used to evaluate the performance of a classifier [28]. In ten-fold cross-validation the data are first divided equally into ten folds. Subsequently, each fold of data is in turn used as the test data and the remaining 9-folds of data as the training data. Thus, each data is tested exactly once. To evaluate the performance of the predictor, the total accuracy of prediction is calculated below:

$$AC = \frac{\sum_{i=1}^{16} M_i}{N} \qquad (i = 1, 2, ..., 16) \tag{8}$$

where M_i stands for the number of correctly classified proteins for the i-th subcellular location, and N represents the total number of proteins in the benchmark dataset.

5. INCREMENTAL FEATURE SELECTION (IFS)

Based on the features ranked by mRMR, we used Incremental Feature Selection (IFS) [23, 29, 30] to determine the optimal number of features. During IFS procedure, features in the ranked feature set are added one by one from higher to lower rank. A new feature set is composed when one feature is added. Thus N feature sets will be composed when N ranked features are given. The i-th feature set is:

$$S_i = \{f_1, f_2, \ldots, f_i\}\ (1 \leq i \leq N) \tag{9}$$

For each of the N feature sets, a random forest is constructed and tested using ten-fold cross-validation test. With N prediction accuracies calculated, an IFS table is obtained with one column being the index i and the other columns being the prediction accuracy for each type of subcellular locations and total prediction accuracy for all subcellular locations. We then can get the optimal feature set (S_{optimal}), using the predictor that achieves the best total overall prediction accuracy.

6. RESULTS AND DISCUSSIONS

6.1. Feature Reduction Result

Each protein was originally encoded by 229 KEGG pathway features and 15942 GO features. (See Additional file S2). After feature reduction by the method described in Materials and Methods, there were 192 KEGG features and 9766 GO features left. (See Additional file S3). The following analysis was based on these 9958 features.

6.2. The mRMR Result

After feature reduction, a total of 9958 features remained including 192 KEGG pathway features and 9766 GO terms features, which is too many to calculate by mRMR. Therefore, by setting the parameter "-n" of mRMR software to be 300, it only lists the first 300 features. Listed in the Additional file S4 are two tables obtained by running the mRMR software: one is a MaxRel feature table that ranks the 300 features according to their relevance to the class of samples; the other is called the mRMR feature table that lists the ranked 300 features according to mRMR criteria. In the mRMR feature table, a feature with a smaller index implies that it is more important for subcellular location prediction. Such a list of ranked features has been used in the following IFS procedure for optimal feature set selection.

6.3. IFS Results

Referring to IFS in Materials and Methods, by setting 300 for N and 1 for the feature-increasing step, 300 individual predictors corresponding to 300 feature subsets were constructed to predict the subcellular location of human proteins. Listed in the Additional file S5 are the rates of prediction accuracy for each subcellular location and overall prediction accuracy for all subcellular location by each of the 300 predictors. Shown in Fig. **1** is the IFS curve plotted based on the data in Additional file S5. From Fig. **1** we can see that the predictor achieved a maximum of overall accuracy equaling to 67.72% when 142 features were included. These 142 features were deemed as the optimal feature set of our classifier. The optimal 142 features are given in the Additional file S6. Hereafter, all the analyses are based on these 142 optimal features.

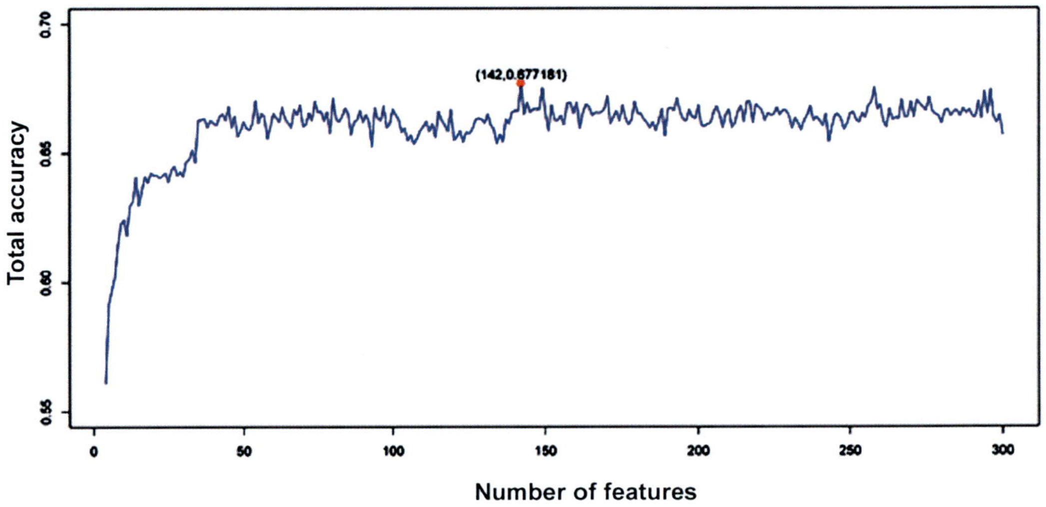

Figure 1: A plot showing the change of overall prediction accuracies *versus* feature numbers. The IFS curves were drawn based on the data in Additional file S5. The overall prediction accuracy reached the peak when the number of features was 142. The 142 features thus obtained were used to compose the optimal feature set for the subcellular location predictor.

7. FEATURE ANALYSIS

The distribution of each type of features in the final optimal feature set was investigated and shown in Table **2**. Of the 142 optimal features, 5 were obtained from KEGG pathway enrichment scores and 137 were obtained from GO enrichment scores including 58 Cellular Component (CC) terms, 29 Molecular Function (MF) terms and 50 Biological Process (BP) terms.

Table 2: Distribution of 142 optimal features

Feature Type	Feature Subtype	Number of Features
KEGG pathway	KEGG pathway	5
GO	CC	58
	MF	29
	BP	50

7.1. GO Number and Percentage

To illustrate the biological meanings of the selected optimal feature subset, we analyzed the distribution of GO features in the biological process, cellular component and molecular function GO terms. The 137 GO features in optimal feature set were mapped to the children of the three root GO terms. The frequency of each GO term in the optimal feature set and the ratio of the number of each GO term to the scale of the number of its children terms are shown in Fig. **2** to Fig. **4**.

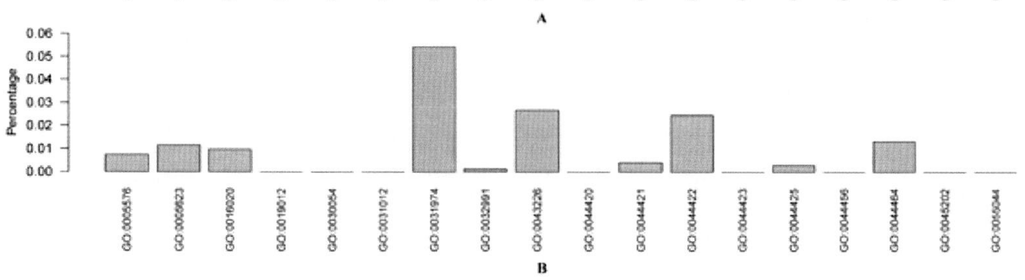

Figure 2: Illustrating the distribution of GO terms of Cellular Component in the optimal feature set. (A) The frequency of GO terms of Cellular Component. (B) The percentage of GO terms of Cellular Component.

7.2. GO Terms of Cellular Component

Shown in Fig. **2A** is the frequency of GO terms mapped to cellular component related terms. The top six GO cellular component terms are GO: 0005623: cell (47), GO: 0044464: cell part (47), GO: 0043226: organelle (43), GO: 0044422: organelle part (30), GO: 0031974: membrane-enclosed lumen (17) and GO: 0016020: membrane (8) (also See the table of “CC” in Additional file S7). In Fig. **2B**, the top six GO cellular component terms with the highest percentage are GO: 0031974: membrane-enclosed lumen (5.43%), GO: 0043226: organelle(2.68%), GO: 0044422: organelle part (2.45%), GO: 0044464: cell part (1.33%), GO: 0005623: cell (1.15%) and GO: 0016020: membrane (0.98%) (also See the table of “CC” in Additional file S7).

According to both the number of GO cellular component terms and percentage distribution analysis, the GO terms corresponding to the cell (GO: 0005623) and cell part (GO: 0044464) were highlighted within the top six GO terms, which was consistent with the finding in the study of subcellular locations for proteins in budding yeast [31]. This suggests that the cell or cell part GO term could contribute to the prediction of protein subcellular location. This may be attributed to the reason that proteins located to proper cellular compartments should be guaranteed.

The inclusion of both organelle (GO: 0043226) and organelle part (GO: 0044422) suggest that organelles, which have specific structural and functional attributes, may require more definite protein localization to exert their specific functions. Since organelles are the location for specialized functions in the cell, such as transcription of ribosomal RNA in nucleoli, oxidative metabolism in mitochondria and maturation of newly synthesized proteins in the endoplasmic reticulum; therefore, the determination of which organelle a protein is located can help to reveal in which metabolism pathway it is involved and the proteins with which it interacts [32]. Furthermore, this also implicated that these GO terms play crucial roles in localizing proteins to organelles and could contribute relatively more to the overall protein subcellular localization.

The inclusion of membrane-enclosed lumen (GO: 0031974) suggests that this cellular component could contribute to protein subcellular localization.

The membrane (GO: 0016020) was also selected as an important feature for protein subcellular location prediction. Membrane proteins are a key group of proteins, which form part of the boundary of the organelles and execute many important functions such as transporters, receptors, and trafficking [33]. In eukaryotes, numerous complex subcellular structures exist, most of which are delineated by membranes. Many proteins are trafficked to membranes in order to be able to exert their correct physiological function [33].

7.3. GO Terms of Molecular Function

Shown in Fig. **3A** is the frequency of GO terms mapped to molecular function related terms. The top six GO molecular function terms are GO: 0005488: binding (17), GO: 0003824: catalytic activity (6), GO: 0004872: receptor activity (5), GO: 0001071: nucleic acid binding transcription factor activity (4), GO: 0000988: protein binding transcription factor activity (3) and GO: 0060089: molecular transducer activity (1) (also See the table of "MF" in Additional file S7). Proteins assigned to these six GO terms required to be localized to specific subcellular locations to carry out their structural or functional activities. This suggests that these six GO terms require more definite localization and contribute profoundly to the protein subcellular localization. In Fig. **3B**, the top six GO molecular function terms with highest percentage are GO: 0005488: binding (0.98%), GO: 0001071: nucleic acid binding transcription factor activity (0.90%), GO: 0004872: receptor activity (0.70%), GO: 0000988: protein binding transcription factor activity (0.29%), GO: 0060089: molecular transducer activity (0.14%) and GO: 0003824: catalytic activity (0.09%) (also See the table of "MF" in Additional file S7).

According to both the number of GO molecular function terms and percentage distribution analysis, the GO terms corresponding to the binding (GO: 0005488) were highlighted within the top six GO terms, which suggested it is a pivotal feature for the prediction of protein subcellular location. It is known that protein functions are closely related to their location and different subcellular locations are usually corresponding to distinct molecular functions. For example, DNA binding (GO: 0003677), transcription factor binding (GO: 0000989), chromatin binding (GO: 0003682) and pre-mRNA binding (GO: 0036002) are usually corresponding to nucleus location. rRNA binding (GO: 0019843) is usually corresponding to ribosome location.

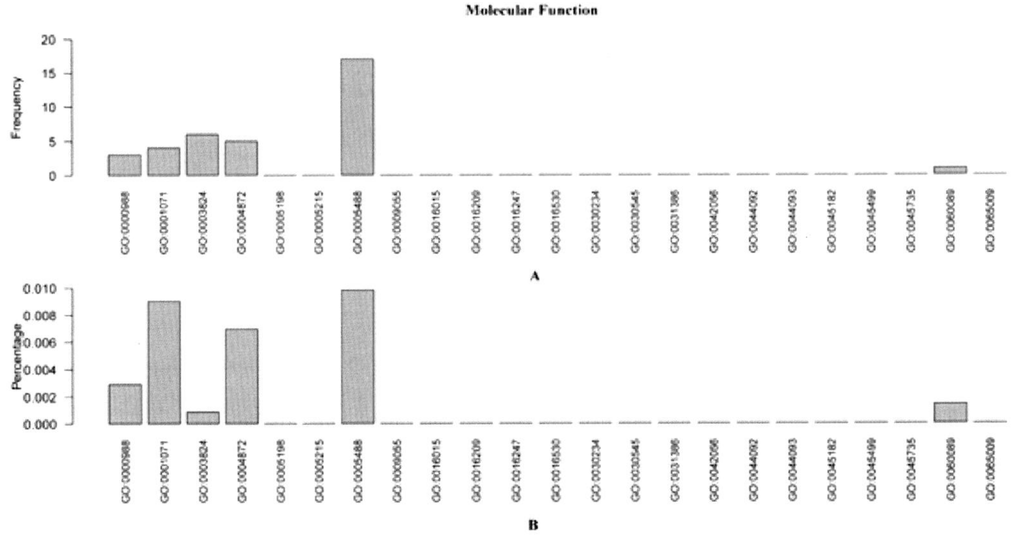

Figure 3: Illustrating the distribution of GO terms of Molecular Function in the optimal feature set. (A) The frequency of GO terms of Molecular Function. (B) The percentage of GO terms of Molecular Function.

The highlight of receptor activity (GO: 0004872) may be attributed to the relatively more definite localization of proteins on cell membrane or nuclear receptor located in the nucleus. The highlight of nucleic acid binding transcription factor activity (GO: 0001071) and protein binding transcription factor activity (GO: 0000988) may be due to the reason that transcriptional regulation mostly occurred in the nucleus.

7.4. GO Terms of Biological Process

Shown in Fig. **4A** is the frequency of GO terms mapped to biological process related terms. The top seven GO biological process terms are GO: 0009987: cellular process (53), GO: 0008152: metabolic process (46), GO: 0019740: nitrogen utilization (22), GO: 0050789: regulation of biological process (16), GO: 0065007: biological regulation (16), GO: 0050896: response to stimulus (11) and GO: 0071840: cellular component organization or biogenesis (9) (also See the table of "BP" in Additional file S7). In Fig. **4B**, the top seven GO biological processes with highest ratio are GO:

0001906: cell killing (0.73%), GO: 0019740: nitrogen utilization (0.43%), GO: 0002376: immune system process (0.32%), GO: 0071840: cellular component organization or biogenesis (0.31%), GO: 0008152: metabolic process (0.28%), GO: 0009987: cellular process (0.27%) and GO: 0016265: death (0.24%) (also See the table of "BP" in Additional file S7).

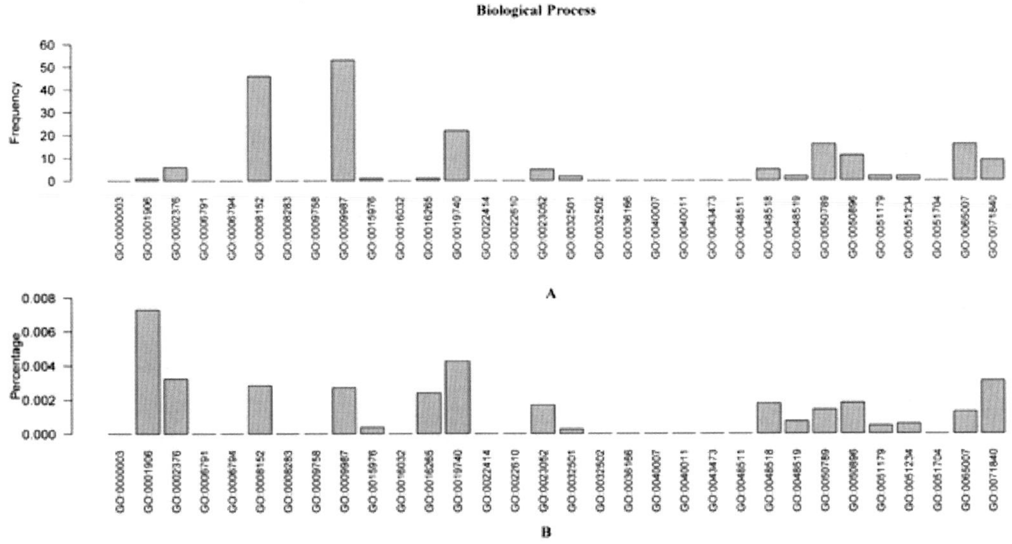

Figure 4: Illustrating the distribution of GO terms of Biological Process in the optimal feature set. (A) The frequency of GO terms of Biological Process. (B) The percentage of GO terms of Biological Process.

According to both the number of GO biological process terms and percentage distribution analysis, the GO term corresponding to the cellular component organization or biogenesis (GO: 0071840) was highlighted within the top seven GO terms indicating that this GO term influences more on protein subcellular location determination.

The inclusion of cellular process (GO: 0009987) within the top seven GO terms suggests that the biological processes at the cellular level performed by certain proteins may influence greatly on the subcellular localization of those proteins, and this is consistent with the common knowledge that specific cellular biological activities need the proteins to be specifically localized. This finding is also in agreement with Wan *et al.*'s study [31].

The highlight of metabolic process (GO: 0008152) suggests that the metabolic processes, which are essential for cell survival and development, may affect protein localization, implicating that proteins involved in a specific metabolic process require more explicit localizations.

7.5. KEGG Pathway Analysis

The five KEGG pathways in the optimal feature set are hsa01100 (Metabolic pathways), hsa00020 (Citrate cycle (TCA cycle)), hsa00630 (Glyoxylate and dicarboxylate metabolism), hsa05016 (Huntington's disease) and hsa03008 (Ribosome biogenesis in eukaryotes). Among these KEGG pathways, metabolic pathways (hsa01100) has an index of 14 in the optimal feature set, indicating it is an important determinant for protein subcellular location prediction, which is consistent with that the GO terms of metabolic process (GO: 0008152) is also selected as the critical feature for the determination of protein subcellular location.

To investigate the relationship between these KEGG pathways and protein subcellular location. We carried out the GO enrichment analysis of genes in these 5 pathways using DAVID, a functional annotation tool [34]. The result showed that the genes in metabolic pathways (hsa01100) are significantly related to organelle membrane (GO: 0031090), mitochondrion (GO: 0005739) and mitochondrial part (GO: 0044429). In eukaryotes, cells are subdivided into membrane-bound subcellular organelles [35]. Subcellular localization of metabolic enzymes can control precisely over where these are synthesized and operated [36]. Such compartmentalization of metabolic enzymes and metabolites of pathways provides a regulatory mechanism to control metabolic pathways.

The genes in Citrate cycle (TCA cycle) (hsa00020) are closely related to mitochondrial matrix (GO: 0005759), mitochondrion (GO: 0005739) and mitochondrial part (GO: 0044429). In eukaryotes, the enzymes of the TCA cycle are located in mitochondria, which is bound to the inner mitochondrial membrane as part of Complex II, indicating subcellular location precisely [37].

The genes in Glyoxylate and dicarboxylate metabolism (hsa00630) are also tightly associated with mitochondrial matrix (GO: 0005759), mitochondrion (GO: 0005739) and mitochondrial part (GO: 0044429). That may due to the process being carried out in mitochondria. The genes in Huntington's disease (hsa05016) are significantly enriched in mitochondrial inner membrane (GO: 0005743), organelle inner membrane (GO: 0019866), respiratory chain (GO: 0070469) and envelope (GO: 0031975). It has been reported that mitochondrial loss, dysfunction and altered dynamics can cause Huntington's disease [38]. Furthermore, it has been shown that the mitochondrial DNA (mtDNA) may play an essential role in the pathogenesis of the respiratory chain complex activities in neurodegenerative disorders such as Huntington's disease [39].

The genes in Ribosome biogenesis in eukaryotes (hsa03008) are closely related to nucleolus (GO: 0005730) and nuclear lumen (GO: 0031981), which is supported by the report that most proteins involved in ribosome biogenesis are not stationary in the nucleolus, but they rapidly exchange between the nucleolus and the nucleoplasm [40].

8. CONCLUSION

In this study, we developed a new method for the prediction of subcellular location of human proteins by a multi-step feature selection procedure. Our method utilized KEGG and GO enrichment scores of proteins and its neighbors in STRING PPI network as the source of the prediction features. On the basis of the feature selection algorithm, an optimal set of features were selected, which are regarded as the features that contribute most significantly to the prediction of subcellular location. As a result, the predictor based on Random Forest achieved an overall prediction accuracy of 67.72% with 142 features selected, including 5 KEGG pathway features and 137 GO term features. The biological significances of these important features are also discussed in depth. It is anticipated that our method may become a useful high throughput tool for predicting the subcellular locations of proteins.

ACKNOWLEDGEMENTS

This work was supported by grants from the National Basic Research Program of China (2011CB510102, 2011CB510101), the National Natural Science Foundation of China (31371335), the Innovation Program of Shanghai Municipal

Education Commission (12ZZ087) and the grant of "The First-class Discipline of Universities in Shanghai".

CONFLICT OF INTEREST

The authors confirm that this chapter contents have no conflict of interest.

DISCLOSURE

This is an updated report of the earlier review article by Cai YD "Predicting protein subcellular locations with feature selection and analysis". Protein Pept. Lett. 2010 Apr;17(4):464-72.

REFERENCES

[1] Bjorses P, Halonen M, Palvimo JJ, Kolmer M, Aaltonen J, Ellonen P, *et al.* Mutations in the AIRE gene: effects on subcellular location and transactivation function of the autoimmune polyendocrinopathy-candidiasis-ectodermal dystrophy protein. Am J Hum Genet. 2000 Feb;66(2):378-92.

[2] Alberts B, Johnson A, Lewis J, Raff M, Roberts K, Walter P. Internal organization of the cell: Intracellular compartments and protein sorting In: Molecular Biology of the Cell, 4th ed. Garland Science. 2002:659-710.

[3] Veenman L, Gavish M. The peripheral-type benzodiazepine receptor and the cardiovascular system. Implications for drug development. Pharmacol Ther. 2006 Jun;110(3):503-24.

[4] Chou KC, Cai YD. Using functional domain composition and support vector machines for prediction of protein subcellular location. J Biol Chem. 2002 Nov 29;277(48):45765-9.

[5] Chou KC, Elrod DW. Protein subcellular location prediction. Protein Eng. 1999 Feb;12(2):107-18.

[6] Zhou GP, Doctor K. Subcellular location prediction of apoptosis proteins. Proteins. 2003 Jan 1;50(1):44-8.

[7] Chou KC, Shen HB. Hum-PLoc: a novel ensemble classifier for predicting human protein subcellular localization. Biochem Biophys Res Commun. 2006 Aug 18;347(1):150-7.

[8] Chou KC, Shen HB. Large-scale plant protein subcellular location prediction. J Cell Biochem. 2007 Feb 15;100(3):665-78.

[9] Shen HB, Yang J, Chou KC. Euk-PLoc: an ensemble classifier for large-scale eukaryotic protein subcellular location prediction. Amino Acids. 2007 Jul;33(1):57-67.

[10] Chou KC, Shen HB. Recent progress in protein subcellular location prediction. Anal Biochem. 2007 Nov 1;370(1):1-16.

[11] Li W, Godzik A. Cd-hit: a fast program for clustering and comparing large sets of protein or nucleotide sequences. Bioinformatics. 2006 Jul 1;22(13):1658-9.

[12] Szklarczyk D, Franceschini A, Kuhn M, Simonovic M, Roth A, Minguez P, *et al.* The STRING database in 2011: functional interaction networks of proteins, globally integrated and scored. Nucleic Acids Res. 2011 Jan;39(Database issue):D561-8.

[13] Sharan R, Ulitsky I, Shamir R. Network-based prediction of protein function. Mol Syst Biol. 2007;3:88.

[14] Carmona-Saez P, Chagoyen M, Tirado F, Carazo JM, Pascual-Montano A. GENECODIS: a web-based tool for finding significant concurrent annotations in gene lists. Genome Biol. 2007;8(1):R3.

[15] Li B-Q, Huang T, Liu L, Cai Y-D, Chou K-C. Identification of Colorectal Cancer Related Genes with mRMR and Shortest Path in Protein-Protein Interaction Network. PLoS ONE. 2012;7(4):e33393.

[16] Li B-Q, Zhang J, Huang T, Zhang L, Cai Y-D. Identification of retinoblastoma related genes with shortest path in a protein–protein interaction network. Biochimie. 2012;94(9):1910-7.

[17] Cramér H. Mathematical Methods of Statistics. Princeton: Princeton University Press; 1946.

[18] Kendall M, Stuart A. The advanced theory of statistics: vol.2 - inference and relationship. New York: Macmillan; 1979.

[19] Harrison KM, Kajese T, Hall HI, Song R. Risk factor redistribution of the national HIV/AIDS surveillance data: an alternative approach. Public Health Rep. 2008 Sep-Oct;123(5):618-27.

[20] Peng H, Long F, Ding C. Feature selection based on mutual information: criteria of max-dependency, max-relevance, and min-redundancy. IEEE Trans Pattern Anal Mach Intell. 2005 Aug;27(8):1226-38.

[21] Li B-Q, Cai Y-D, Feng K-Y, Zhao G-J. Prediction of Protein Cleavage Site with Feature Selection by Random Forest. PLoS One. 2012;7(9):e45854.

[22] Zhang N, Li B-Q, Gao S, Ruan J-S, Cai Y-D. Computational Prediction and Analysis of Protein [gamma]-carboxylation Sites Based on a Random Forest Method. Molecular BioSystems. 2012.

[23] Li B-Q, Hu L-L, Chen L, Feng K-Y, Cai Y-D, Chou K-C. Prediction of Protein Domain with mRMR Feature Selection and Analysis. PLoS One. 2012;7(6):e39308.

[24] Li B-Q, Feng K-Y, Chen L, Huang T, Cai Y-D. Prediction of Protein-Protein Interaction Sites by Random Forest Algorithm with mRMR and IFS. PLoS ONE. 2012;7(8):e43927.

[25] Gao Y-F, Li B-Q, Cai Y-D, Feng K-Y, Li Z-D, Jiang Y. Prediction of active sites of enzymes by maximum relevance minimum redundancy (mRMR) feature selection. Molecular BioSystems. 2013.

[26] Breiman L. Random forests. Machine learning. 2001;45(1):5-32.

[27] Witten IH, Frank E. Data Mining: Practical machine learning tools and techniques: Morgan Kaufmann Pub; 2005.

[28] Kohavi R, editor. A Study of Cross-Validation and Bootstrap for Accuracy Estimation and Model Selection. Proceedings of the Fourteenth International Joint Conference on Artificial Intelligence; 1995; San Mateo.

[29] Li BQ, Hu LL, Niu S, Cai YD, Chou KC. Predict and analyze S-nitrosylation modification sites with the mRMR and IFS approaches. J Proteomics. 2012 Feb 16;75(5):1654-65.

[30] Li B-Q, Huang T, Zhang J, Zhang N, Huang G-H, Liu L, *et al.* An Ensemble Prognostic Model for Colorectal Cancer. PLoS ONE. 2013;8(5):e63494.

[31] Wan S-B, Hu L-L, Niu S, Wang K, Cai Y-D, Lu W-C, *et al.* Identification of Multiple Subcellular Locations for Proteins in Budding Yeast. Current Bioinformatics. 2011;6(1):71-80.

[32] Glory E, Murphy RF. Automated Subcellular Location Determination and High-Throughput Microscopy. Developmental Cell. 2007;12(1):7-16.

[33] Sadowski PG, Groen AJ, Dupree P, Lilley KS. Sub-cellular localization of membrane proteins. Proteomics. 2008 Oct;8(19):3991-4011.

[34] Huang da W, Sherman BT, Lempicki RA. Systematic and integrative analysis of large gene lists using DAVID bioinformatics resources. Nat Protoc. 2009;4(1):44-57.

[35] Lehninger A, Nelson D, Cox M. Lehninger Principles of Biochemistry: W. H. Freeman; 2008.

[36] Ovadi J, Saks V. On the origin of intracellular compartmentation and organized metabolic systems. Mol Cell Biochem. 2004 Jan-Feb;256-257(1-2):5-12.

[37] King A, Selak MA, Gottlieb E. Succinate dehydrogenase and fumarate hydratase: linking mitochondrial dysfunction and cancer. Oncogene. 2006 Aug 7;25(34):4675-82.

[38] Kim J, Moody JP, Edgerly CK, Bordiuk OL, Cormier K, Smith K, *et al.* Mitochondrial loss, dysfunction and altered dynamics in Huntington's disease. Hum Mol Genet. 2010 Oct 15;19(20):3919-35.

[39] Banoei MM, Houshmand M, Panahi MS, Shariati P, Rostami M, Manshadi MD, *et al.* Huntington's disease and mitochondrial DNA deletions: event or regular mechanism for mutant huntingtin protein and CAG repeats expansion?! Cell Mol Neurobiol. 2007 Nov;27(7):867-75.

[40] Chen D, Huang S. Nucleolar components involved in ribosome biogenesis cycle between the nucleolus and nucleoplasm in interphase cells. J Cell Biol. 2001 Apr 2;153(1):169-76.

Supplementary Files

Additional file S1: The 16 categories of human protein subcellular locations, and the number of proteins in each category.

Additional file S2: The original features used to encode each protein including 229 KEGG pathways and 15942 GO terms.

Additional file S3: The features remained after feature reduction including 192 KEGG pathways and 9766 GO terms.

Additional file S4: This file contains two sheets. The first one shows the MaxRel feature table, which ranked the features according to the relevance between features and class of the samples. The second one shows the mRMR feature table, which ranked the features according to both the redundancy and relevance criteria.

Additional file S5: The prediction accuracy for each subcellular location and overall prediction accuracy for all subcellular location.

Additional file S6: The 142 features selected by the IFS procedure.

Additional file S7: This file contains three sheets. They show the frequency of each GO term in the optimal feature set and the ratio of the number of each GO term to the number of its children terms for Cellular Component (CC), Molecular Function (MF) and Biological Process (BP), respectively.

Additional file S8: The GO enrichment results for genes on the five KEGG pathways in the optimal feature set.

Send Orders for Reprints to reprints@benthamscience.net
226 *Frontiers in Protein and Peptide Sciences, Vol. 1*, 2014, 226-272

CHAPTER 11

Analysis of Sequence Divergence in Metabolic Proteins of *Plasmodium falciparum:* Implications for Remote Homology Detection

Smita Mohanty[1], Gayatri Ramakrishnan[1,2], Pratik Dave[3] and Narayanaswamy Srinivasan[1,*]

[1]Molecular Biophysics Unit, [2]Indian Institute of Science Mathematics Initiative, [3]Department of Microbiology and Cell Biology, Indian Institute of Science, Bangalore-560012, India

Abstract: *Plasmodium falciparum*, an obligate intracellular parasite and causative agent of malaria, exhibits an atypical metabolic organization of proteins which are diverged extensively from their homologues in other organisms. This divergence makes homology-based on structural and functional annotation of the proteins difficult. With an emphasis on determinant aspects such as domain duplication, divergence and domain splitting, we provide a comprehensive basis to understand the significance of sequence divergence observed in metabolic proteins of *P. falciparum.* Such a pronounced sequence divergence in proteins renders the homology-based annotation transfer non-trivial. Additionally the unavailability of function annotation for about 40% of the parasitic proteome poses a limitation to understand the physiological basis of metabolic flexibility exhibited by the parasite to facilitate its multi-host survival. Thus, detection and exploration of remotely related homologues with the help of highly sensitive sequence search techniques become indispensable. Employment of a combination of highly sensitive profile-based approaches has enabled enhanced structural and functional characterization of metabolic proteins along with identification of proteins potentially involved in metabolic pathways. An enriched function annotation of the parasitic protein repertoire not only facilitates recognition of putative pathway proteins but also paves way for prioritizing targets for chemotherapeutic interventions.

Keywords: Metabolic enzymes, metabolic pathways, *Plasmodium falciparum*, protein evolution, remote homology detection.

1. INTRODUCTION

Malaria, a potentially life threatening disease responsible for the mortality levels

***Address correspondence to Narayanaswamy Srinivasan:** Molecular Biophysics Unit, Indian Institute of Science, Bangalore-560012, India; Tel: +91-80-2293-2837; Fax: + 91-80-2360-0535; Email: ns@mbu.iisc.ernet.in

Ben M. Dunn (Ed)

that remain as high as more than a million each year, is reported to inflict nearly half of the world's population at risk [1]. Malaria is mostly endemic to sub-Saharan Africa and other regions such as Asia and Latin America show high incidence. This disease is caused by organisms belonging to the genus Plasmodium. Amongst the known species belonging to this genus, four are known to be pathogenic to humans, namely *Plasmodium falciparum, Plasmodium vivax, Plasmodium malariae* and *Plasmodium ovale.* Of the four, *Plasmodium falciparum* is most virulent manifesting the most lethal form of malaria [2]. *P. falciparum*, an obligate intracellular protozoan parasite [3], completes its entire life cycle within two hosts, namely humans, where asexual stages occur and the female Anopheles mosquitoes, where the sexual stages are observed [4]. The life cycle of the parasite is quite complex, with the parasite residing in 5 different tissue types and exhibiting 10 different morphological stages [5]. The aetiology of intracellular pathogenesis typically comprises of 4 distinct phases:

- Invading the host.
- Survival within the host.
- Replication.
- Exit.

All these phases have been studied quite extensively in *P. falciparum* and other apicomplexans (organisms with an apicoplast) and many of the features involved in these stages are found to be common across these organisms [4]. In each of the stages within the host, the parasite faces a series of challenges in terms of cellular invasion, environmental condition depicted by different tissues and host immunity [4,6]. Vast re-organization is therefore brought about in the cellular processes facilitating the parasite to overcome these challenges. These changes are short lived transient alterations in gene expression and regulation [4]. However, many adaptive changes are fairly long-term such as the presence or absence of pathways based on essentiality.

Being an obligate intracellular parasite, *P. falciparum* resides in nutrient enriched medium [4] and thus many of its biosynthetic pathways seem not to be essential.

Most of the nutritional requirements can be met through the import of host metabolites thereby making the parasite auxotrophic for many of the nutrients [4,7].

Fig. **1** shows the distribution of various enzymatic steps characterized in different unicellular eukaryotic organisms, which includes organisms belonging to the genus Plasmodium, highlighted in red, along with other apicomplexans and non-parasitic free living organisms. Number of proteins identified to be involved in each of the pathways is shown in the form of colour gradient ranging from white, indicating identification of all proteins in a pathway, to grey/black, depicting majority of members of the pathway unidentified. Each row represents an organism and columns represent different pathways. Large fractions of fully characterized pathways in free living unicellular eukaryotes could be observed in the figure. However, for pathogenic organisms, the compendium of proteins known to function in pathways is low. Pathways such as purine/pyrimidine biosynthesis and glycolysis are well represented in *P. falciparum* and other closely related apicomplexa. However, pathways involved in amino acid and cofactor metabolism show poor representation in *P. falciparum* and other related organisms. The plausible reasons responsible for the observed discrepancy are briefed in this section.

Within the host red blood cells (RBC) the parasite expresses a highly efficient hexose transporter along with glucose transporter. It induces massive influx of glucose to meet the parasite's high energy requirements during the process of replication and maturation [4,7,8]. Other changes include massive remodelling of infected RBCs by export of parasite proteins [9], utilization of host proteins to carry out specific processes [10, 11], which greatly enhance the permeability of the infected RBC enabling easy uptake of nutrients and purging of toxins. These adaptive changes induced by *P. falciparum* point towards enormous dependency on the host for its metabolic needs. Such an improved adaptation calls for major reorganization both at genomic and proteomic level pertaining to metabolic genes/proteins to compensate growth for reduced metabolic flexibility [4,7,11,12]. The entire process of genome reduction with intracellular pathogenesis can eventually lead to “Muller’s ratchet” scenario, a phenomenon where genes of small asexual populations of organisms accrue deleterious mutations and are ultimately eliminated from the genome [13]. Such bottlenecks in *P. falciparum*

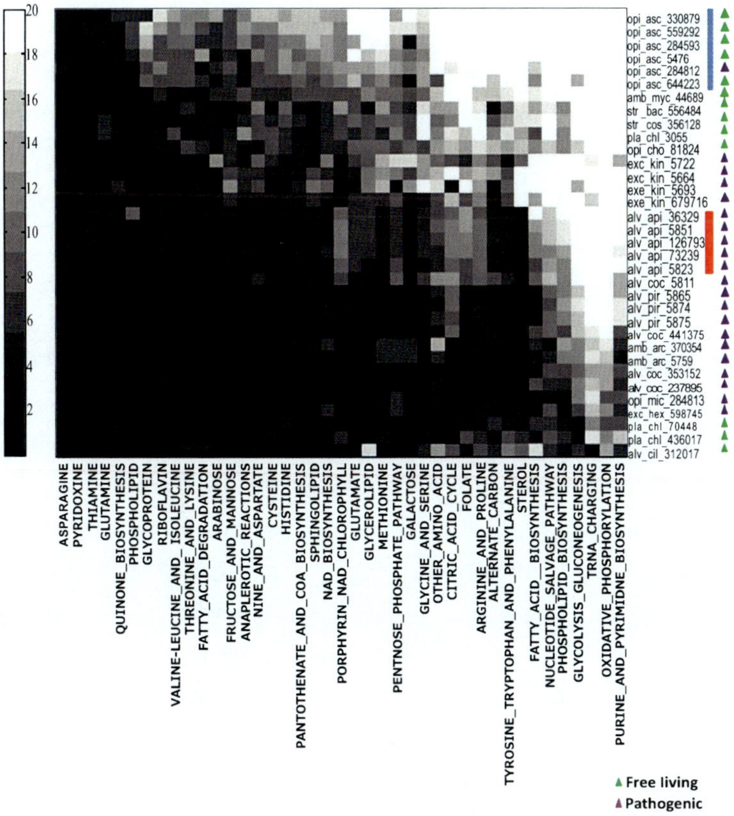

Figure 1: Overview of the extent of coverage of enzymes in pathways across 34 unicellular eukaryotes. The figure is a matrix representing the number of enzymatic steps currently annotated in a pathway in each of the organisms considered with each row indicating the distribution of functionally characterized enzymes across various pathways in an organism. Each column indicates a pathway. Parasitic and free living organisms have been labelled by green and purple triangles, respectively. Amongst the pathways the color ranges from white, indicating almost complete pathways to grey/black cell where majority of the members are not known in a pathway. The various organisms considered and their order of occurrence is as following: opi_asc_330879: *N. fumigata*, opi_asc_559292: *S. cerevisiae*, opi_asc_284593: *C. glabrata*, opi_asc_5476: *C. albicans*, opi_asx_284812: *S. pombe*, opi_asc_644223: *P. pastoris*, amb_myc_44689: *D. discoideum*, str_bac_556484: *P. tricornutum*, str_cos_35128: *T. pseudonana*, pla_chlr_3055: *C. reinhardtii*, opi_cho_81824: *M. brevicollis*, exc_tri_5722: *T. vaginalis*, exc_kin_5664: *L. major,* exc_kin_5693: *T. cruzi*, exc_kin_679716: *T. brucei*, alv_api_36329: *P. falciparum*, alv_api_5851: *P. knowlesi*, alv_api_126793: *P. vivax*, alv_api_73239:, *P. yoelii*, alv_api_ 5823: *P. berghei*, alv_coc_5811: *T. gondii*, alv_pir_5865: *B. bovis*, alv_pir_5874: *T. annulata*, alv_pir_5875: *T. parva*, alv_coc_441375: *C. muris*, amb_arc_370354: *E. dispar*, amb_arc_5759: *E. histolytica*, alv_coc_353152: *C. parvum*, alv_coc_237895:, *C. hominis*, opi_mic_284813: *E. cuniculi*, exc_hex_598745: *G. intestinalis*, pla_chlr_70448: *O. tauri*, pla_chlr_436017: *O. lucimarinus*, alv_coc_312017: *T. thermophile*. Here the abbreviations stand for the major eukaryotic clades containing unicellular eukaryotes, Opi: opihsthokonta, alv: alveolates, amb: amoebozoa, str: stramenopiles and exc:excavates.

may arise, thus facilitating accumulation of mutations and eventual loss of complete pathways. Absence of pathways, such as amino acid metabolism, is observed in the parasite compared to free-living organisms since the parasite fulfils its amino acid requirements by digesting haemoglobin. Along with *de novo* amino acid synthesis, the parasite also lacks *de novo* purine biosynthesis and extensively relies on nucleotide salvage pathway [7,14].

Apart from reductive evolution, the apparent incompleteness of the pathways in *P. falciparum* could be due to synthesis of metabolites through unconventional routes because of non-orthologous displacement, *i.e.*, non-homologous proteins carrying out analogous functions [15,16]. For instance, phosphofructokinase B in *P. furiousus* differs completely in terms of sequence from phosphofructokinase A. Incompleteness of some of the pathways may also be due to their non-canonical roles in the parasite. The tricarboxylic-acid (TCA) cycle is one such example; the pathway is completely characterized in *P. falciparum* with active expression of all the genes involved and along with metabolites [11]. However, biochemical studies indicate a lack of dependency of the parasite on oxidative phosphorylation for its survival [7,11,17]. Non canonical role of TCA is further supported by the fact that the only pyruvate dehydrogenase encoded by *P. falciparum* genome localizes to the apicoplast, where it is important for fatty acid biosynthesis, instead of producing acetyl-CoA for TCA cycle [17,18]. However, TCA cycle is still retained by the parasite because of the potential role of this pathway in mosquito stages of *P. falciparum* and also because of many metabolites produced that can act as intermediates in many other pathways. Thus, the retention of this pathway in *P. falciparum* and in other closely related organisms could be due to the adaptive significance of its non-canonical roles.

Lastly, the incomplete characterization of the metabolic pathways could be due to enormous sequence divergence of parasite proteins compared to the homologues from other organisms. In most of the organisms, function annotation of proteins is largely derived on the basis of homology to functionally well characterized proteins in model organisms. Benchmark studies on a limited datasets have shown homology-based function annotations to yield accuracy of about 70% [19]. However, homology-based annotation transfer works best amongst closely related organisms and chances of reliable detection of homologues fail for organisms like

P. falciparum which has an atypical genome [14]. *P. falciparum* has an AT-rich genome with preponderance of low complexity regions in protein coding regions thereby making homology-based predictions difficult [14]. In many cases, experimental evidence suggests the metabolic pathway to be active, but the proteins involved in such pathways are not recognized. Obtaining an exhaustive list of metabolic proteins is not only essential to understand the physiology of the parasite, it also paves way for rational drug discovery against *P. falciparum*. Success of pathway proteins as drug targets is further exemplified by the fact that many of the effective antimalarial drugs target various metabolic processes [20-22]. Emergence of drug resistant strains in the recent past, against almost all classes of anti-malarial drugs [23-26], necessitates an urgent need in identification of complete list of metabolic proteins and their enhanced structural and functional characterization to facilitate recognition of potential drug targets which would be useful against newly developed drug resistant strains.

In the present chapter we explore and quantify the extent of divergence that is observed amongst proteins participating in various pathways in *P. falciparum* and use this knowledge to search for the various 'missing' metabolic proteins. We show that there is significant divergence observed in *P. falciparum* pathway proteins, not only due to presence of low complexity regions as reported earlier [14], but also due to domain duplication, divergence and domain splitting. We employ extremely sensitive profile-based search approaches in order to identify remote evolutionary relationships of metabolic proteins to address the apparent difficulty in their annotation. Finally, we conclude with identification of "new" metabolic proteins, few of which would be discussed in-depth.

2. EXPLORING THE EXTENT OF DIVERGENCE OF *P. FALCIPARUM* METABOLIC PROTEINS

This section presents the quantification of adaptive changes that are brought about in the various proteins involved in metabolic pathways in *P. falciparum* by performing an exhaustive comparison of metabolic proteins already characterized in the parasite with functionally equivalent homologous proteins in other unicellular eukaryotes. Diversity in the metabolic proteins of *P. falciparum* is quantified at three levels: 1) deviation in domain organization, 2) deviation in protein/domain lengths and 3) extent of deviation in amino acid sequences. Each aspect is elaborated in the following subsections.

2.1. Deviation in Domain Organization

Domain family assignments for proteins participating in various pathways were obtained from Interpro [27] for *P. falciparum* and 33 other unicellular eukaryotes to facilitate a comparative analysis. Table **1** lists the number of proteins involved in the metabolic processes in each of 34 organisms.

Table 1: Number of metabolic proteins for each of the 34 organisms considered.
Source: MetaPhOrs [28] and genome scale metabolic reconstructed models [29-33]

S. No.	Organism Type	Organism Name (Taxonomic ids)	Numbers of Metabolic Proteins
1	Apicomplexa	*Plasmodium vivax* (126793)	264
2	Apicomplexa	*Plasmodium falciparum* (isolate 3D7) (36329)	301
3	Apicomplexa	*Plasmodium berghei* (strain Anka) (5823)	227
4	Apicomplexa	*Plasmodium knowlesi* (strain H) (5851)	274
5	Apicomplexa	*Plasmodium yoelii yoelii* (73239)	241
6	Stramenophiles	*Tetrahymena thermophila* (strain SB210) (312017)	183
7	Apicomplexa	*Cryptosporidium hominis* (237895)	103
8	Apicomplexa	*Cryptosporidium parvum Iowa* II (353152)	107
9	Apicomplexa	*Cryptosporidium muris* (strain RN66) (441375)	152
10	Apicomplexa	*Toxoplasma gondii* (5811)	276
11	Apicomplexa	*Babesia bovis* (5865)	178
12	Apicomplexa	*Theileria annulata* (5874)	175
13	Apicomplexa	*Theileria parva* (5875)	156
14	Amoebozoa	*Entamoeba dispar* (370354)	170
15	Amoebozoa	*Entamoeba histolytica* (5759)	151
16	Amoebozoa	*Dictyostelium discoideum* (44689)	481
17	Excavates	*Giardia intestinalis* (598745)	116
18	Excavates	*Leishmania major* (5664)	470
19	Excavates	*Trypanosoma cruzi* (5693)	480
20	Excavates	*Trypanosoma brucei gambiense* DAL972 (679716)	348
21	Excavates	*Trichomonas vaginalis* (5722)	384
22	Ophisthokonts	*Candida glabrata* (284593)	661

Table 1: contd...

23	Ophisthokonts	*Schizosaccharomyces pombe* (284812)	578
24	Ophisthokonts	*Aspergillus niger* (330879)	870
25	Ophisthokonts	*Candida albicans* (strain WO-1) (5476)	610
26	Ophisthokonts	*Saccharomyces cerevisiae* (559292)	785
27	Ophisthokonts	*Pichia pastoris* (644223)	493
28	Ophisthokonts	*Monosiga brevicollis* (81824)	378
29	Ophisthokonts	*Encephalitozoon cuniculi* (284813)	104
30	Green algae	*Chlamydomonas reinhardtii* (3055)	440
31	Green algae	*Ostreococcus tauri* (70448)	136
32	Green algae	*Ostreococcus lucimarinus* (strain CCE9901) (436017)	162
33	Stramenophiles	*Phaeodactylum tricornutum* (strain CCAP 1055/1) (556484)	429
34	Stramenophiles	*Thalassiosira pseudonana* (35128)	450

Domain architectures, thus obtained, for each of the proteins were compared with each other using a dynamic programming-based domain alignment algorithm [34,35]. The primary structure of the protein, in this case, is viewed as a sequence of domain families instead of sequence of amino acids and global pairwise alignment is performed based on the formula described below.

$$\text{Score} = [S(i,j)]_{n1 \times n2} = max \begin{cases} S(i-1, j-1) + \delta(i,j) \\ S(i-1, j) + \varepsilon \\ S(i, j-1) + \varepsilon \end{cases}$$

$$\text{Domain Distance (DD)} = (n1 + n2) - 2 \times (\mathit{Score})$$

δ(i, j) = Domain scores (identical match=1, mismatch=0)

n1, n2 = Number of domains

ε = -0.01 (gap penalty)

An example describing the use of the above stated algorithm is shown in Fig. **2**, where domain distances are obtained between proteins. Such an approach has been employed to study the evolution of domain architectures in eukaryotic proteins [34].

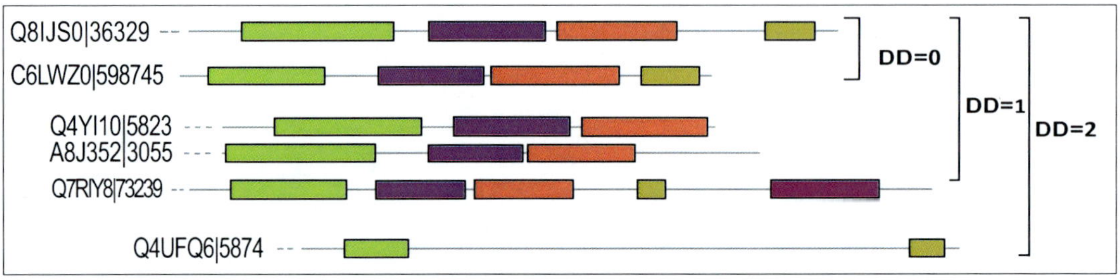

Figure 2: Comparison of domain architectures of few proteins functioning as phosphoacetylglucosamine mutase. Each line represents one protein entry (UniProt ID| NCBI taxonomic ID). Coloured boxes represent various domain assignments. Colour code for the Pfam domains shown here are: Green- PGM_PMM_I; Purple- PGM_PMM_II; Red- PGM_PMM_III; Mustard- PGM_PMM_IV; Violet- Pribosyltran_1.

Domain distance of zero between two proteins indicates completely conserved architecture (both composition and order of appearance of domains). Higher values of domain distances indicate events such as addition/loss of domains, differences in domain architecture (for same composition) and duplication. Infinite distance is obtained between two proteins sharing no common domain family.

Domain organization of every protein in *P. falciparum* catalyzing specific step of a pathway is compared with every protein catalyzing the same step in other organisms. For all the metabolic enzymes in *P. falciparum*, the results obtained are plotted as a bar diagram, shown in Fig. **3**. The figure describes the fraction of protein pairs obtained at various domain distances for each of the enzymatic step that is identified in *P. falciparum*.

In nearly 54% of the enzymatic steps, it is observed that more than 50% of the pairwise domain distances between proteins of *P. falciparum* and other organisms is zero. Out of these enzymatic processes, nearly 67% of the steps involve predominantly single domain proteins. The presence of large fraction of pairwise comparisons at domain distance of zero is indicative of conserved domain organization between *P. falciparum* proteins and proteins of other organisms catalyzing the same reaction. This conservation of domain organization is prevalent across all the major metabolic processes. One of such instance is illustrated in Fig. **4**, showing the domain organizations of proteins functioning as enolase (4.2.1.11) in the glycolytic pathway.

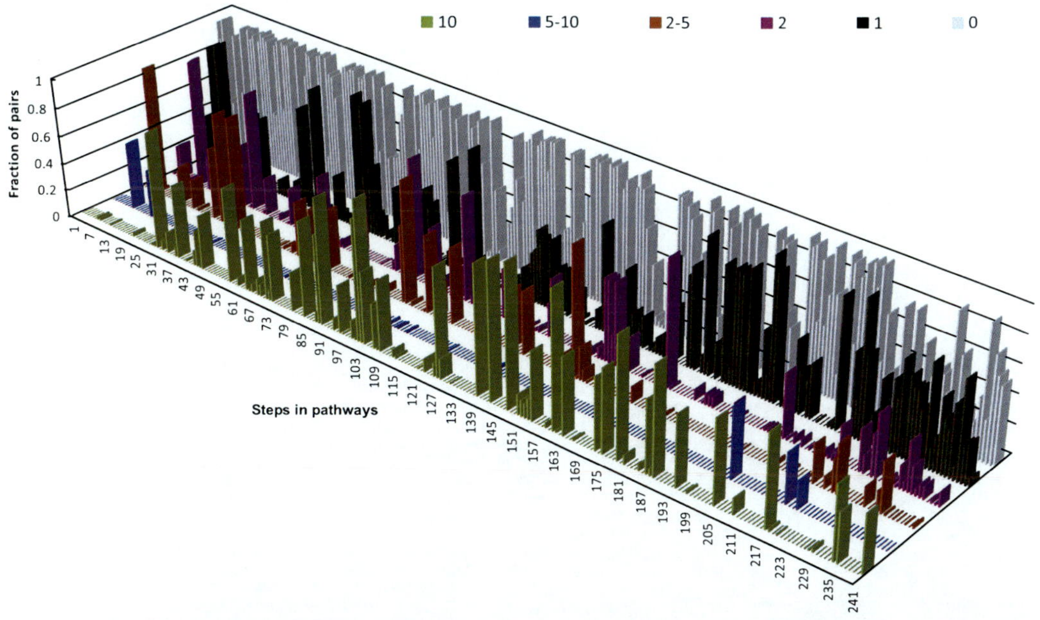

Figure 3: Domain distance distribution across various metabolic steps in the pathways. For every enzymatic step the Plasmodium proteins are compared with all the proteins catalyzing the same reaction in other organisms. Every bar shows the fraction of protein pairs found at a given domain distance for a particular enzymatic step. In the 3D-bar plot each colour represents specific domain distance, thus the height would indicate fraction of pairs found at a particular domain distance. The orders of occurrence of the various metabolic processes in the X-axis are 1-46: Amino acid metabolism, 47-98: Carbohydrate metabolism, 99-142: lipid metabolism, 143-188: Cofactor biosynthesis, 189-219: Nucleotide metabolism and 220-241 t-RNA charging.

Proteins involved in nearly 12% of pathways, differ in their domain organization. Amongst the differences observed, in nearly 68% of the cases there is absence of domain(s) in *P. falciparum* and in about 31% cases an additional domain is observed in the *P. falciparum* proteins.

Fig. **5a** depicts the domain organization of ornithine decarboxylase (4.1.1.17) from various organisms. An additional domain (Pfam ID: PF01536, Adenosylmethionine decarboxylase) is found in the *P. falciparum* homologue (Q8IJ77) and in the homologues in other organisms belonging to the same genus (UniProt IDs: Q7RFF2, A5KDQ6, Q4YHB2 and B3L294). In *P. falciparum,* the presence of adenosylmethionine decarboxylase domain makes this enzyme bi-functional and is implicated in the biosynthesis of polyamines, such as putrescine and spermidine, which are essential for the survival of the organism [36,37]. Both ornithine decarboxylase and S-adenosyl-l-methionine decarboxylase are the rate limiting steps of polyamine synthesis [37]. Presence of these two enzymes as a single bi-functional protein is a feature unique to the Plasmodium genus and is speculated to provide added adaptive advantage in comparison to the mono-functional proteins [38]. Several other bi functional enzymes are present in *P. falciparum* that are entailed as important drug targets [39], such as coexistence of dihydrofolate reductase and thymidylate synthase in a single protein (UniProt ID: Q8I1R6).

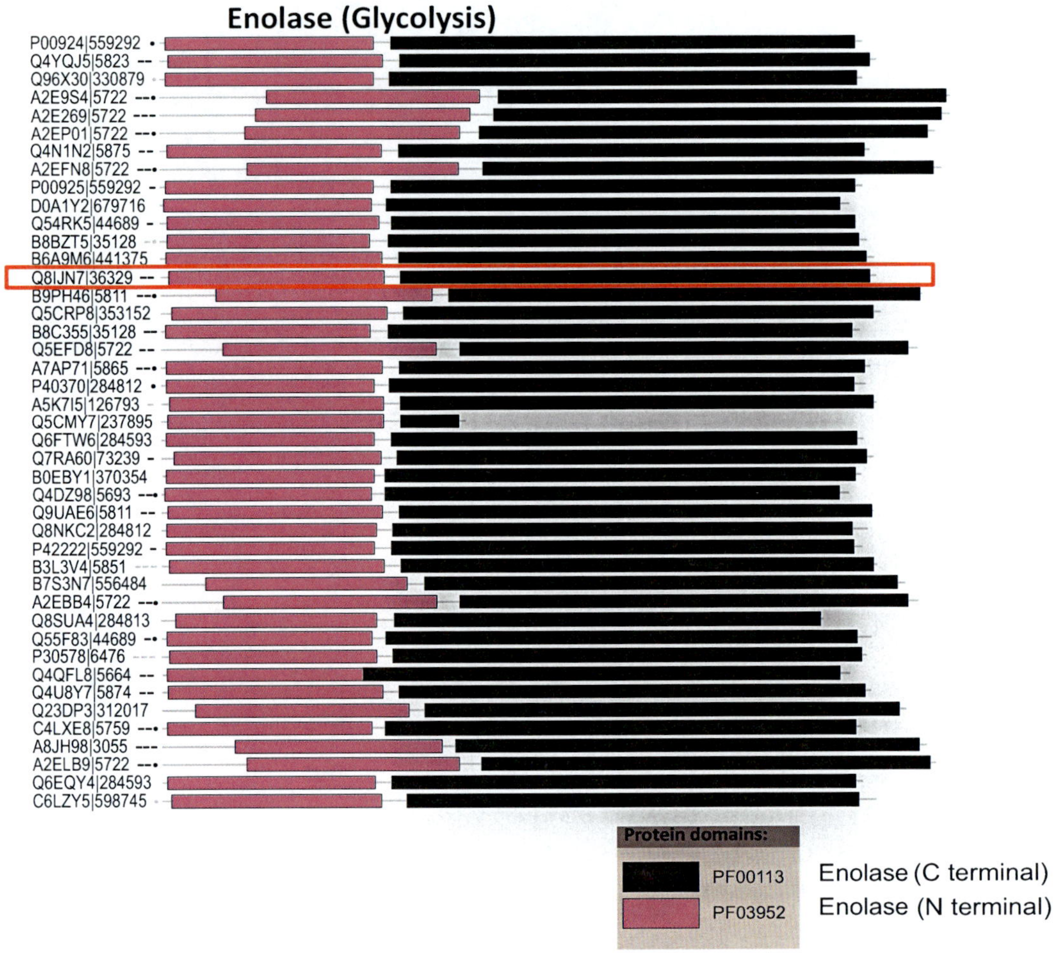

Figure 4: Conserved domain organization. Proteins from different organisms, functioning as enolase with completely conserved domain architecture are shown. *P. falciparum* protein is highlighted in red coloured box and coloured rectangular boxes depict domains assigned to the proteins. The enzymatic step and the key to colour code is provided.

In addition to insertion of additional domains, difference in domain organization in *P. falciparum* is also brought about by loss of domain families as illustrated in Fig. **5**. Phosphoinositide-specific phospholipase C is involved in phospholipid biosynthesis pathway. This enzyme is involved in the generation of IP3 (Inositol trisphosphate) and DAG (Diacylglycerol) and is shown to be membrane associated [40]. Most of the proteins catalyzing this reaction contain C2 domain family which has the characteristic property of locating itself close to the membrane. In *P. falciparum* this domain is absent. However it has a Pleckstrin Homology (PH) domain which is also implicated in targeting proteins to different cellular compartments and might localize the parasite protein to the membrane.

a) **Ornithine Carboxylase(Arginine and Proline metabolism)**

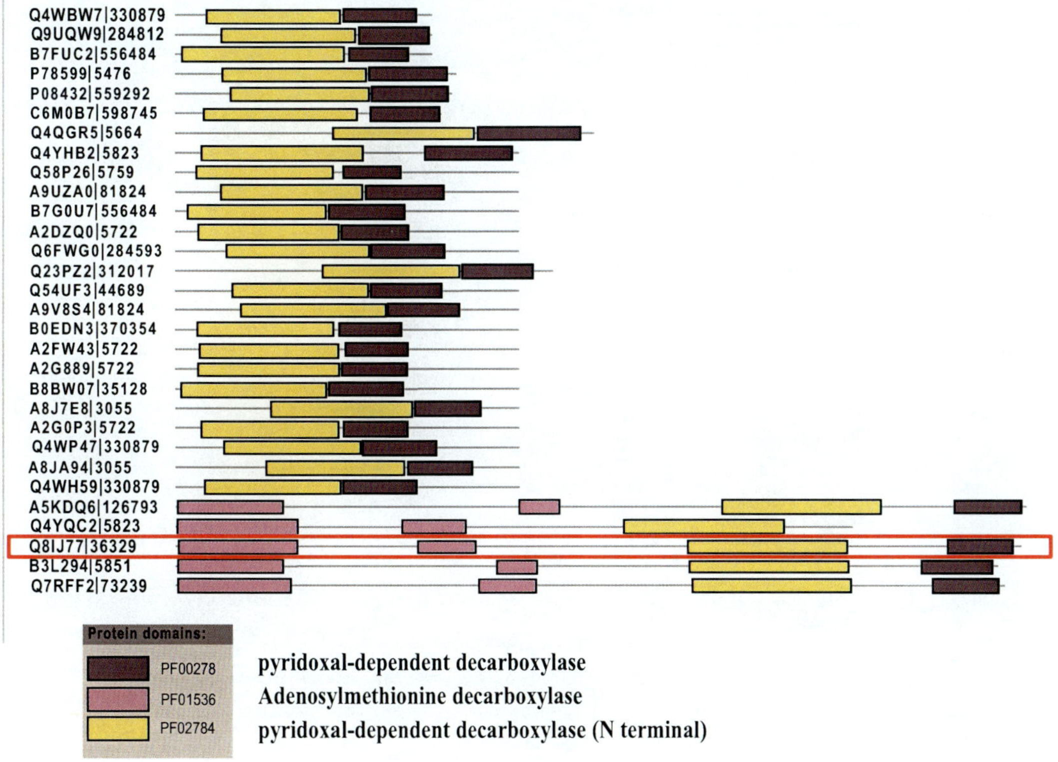

pyridoxal-dependent decarboxylase
Adenosylmethionine decarboxylase
pyridoxal-dependent decarboxylase (N terminal)

Fig. 5: contd….

b)

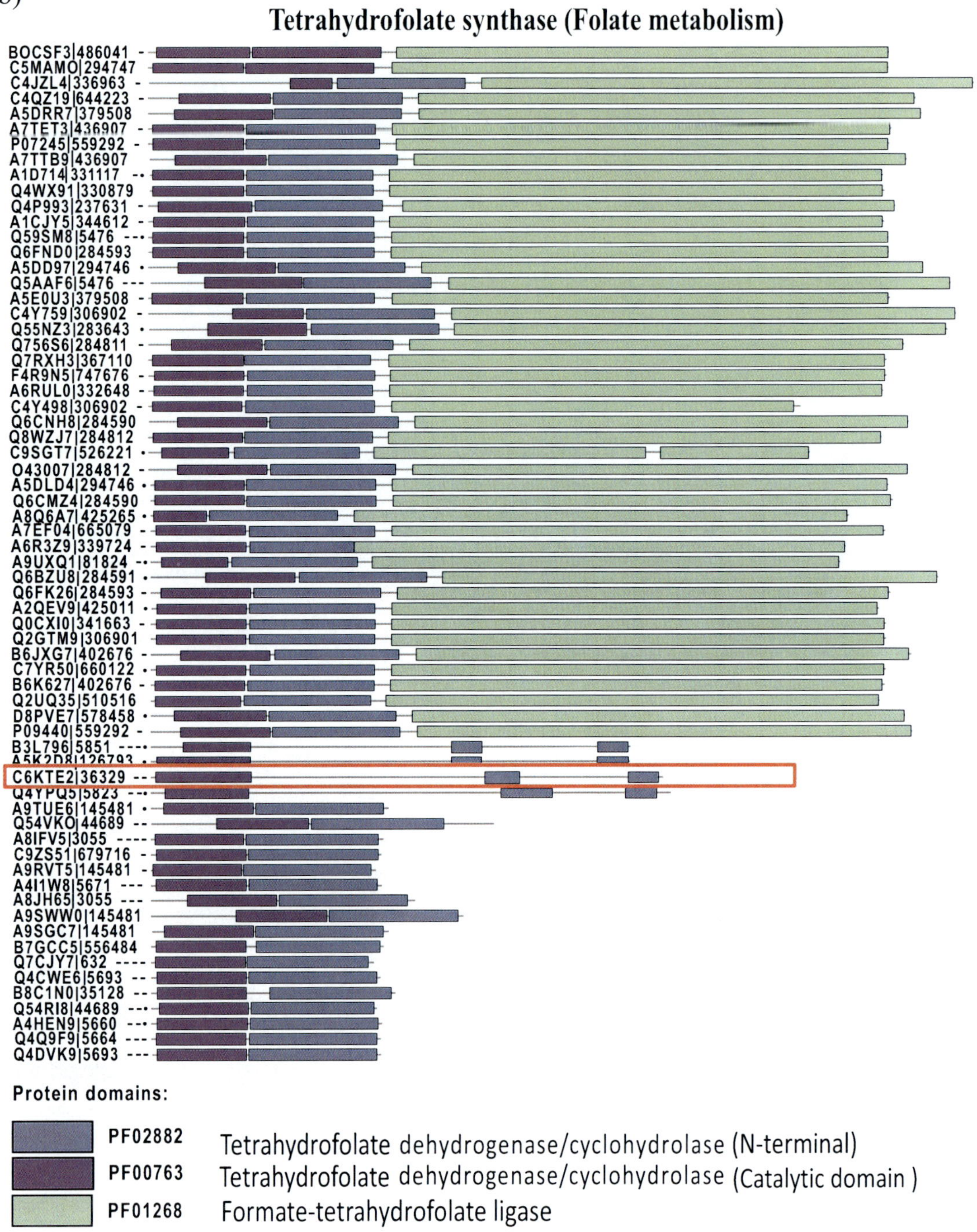
Tetrahydrofolate synthase (Folate metabolism)
BOCSF3|486041
C5MAMO|294747
C4JZL4|336963
C4QZ19|644223
A5DRR7|379508
A7TET3|436907
P07245|559292
A7TTB9|436907
A1D714|331117
Q4WX91|330879
Q4P993|237631
A1CJY5|344612
Q59SM8|5476
Q6FND0|284593
A5DD97|294746
Q5AAF6|5476
A5E0U3|379508
C4Y759|306902
Q55NZ3|283643
Q756S6|284811
Q7RXH3|367110
F4R9N5|747676
A6RUL0|332648
C4Y498|306902
Q6CNH8|284590
Q8WZJ7|284812
C9SGT7|526221
O43007|284812
A5DLD4|294746
Q6CMZ4|284590
A8Q6A7|425265
A7EF04|665079
A6R3Z9|339724
A9UXQ1|81824
Q6BZU8|284591
Q6FK26|284593
A2QEV9|425011
Q0CXI0|341663
Q2GTM9|306901
B6JXG7|402676
C7YR50|660122
B6K627|402676
Q2UQ35|510516
D8PVE7|578458
P09440|559292
B3L796|5851
C6KTE2|36329
A9TUE6|145481
Q54VKO|44689
A8IFV5|3055
C9ZS51|679716
A9RVT5|145481
A4I1W8|5671
A8JH65|3055
A9SWW0|145481
A9SGC7|145481
B7GCC5|556484
Q7CJY7|632
Q4CWE6|5693
B8C1N0|35128
Q54RI8|44689
A4HEN9|5660
Q4Q9F9|5664
Q4DVK9|5693
Protein domains:
PF02882 Tetrahydrofolate dehydrogenase/cyclohydrolase (N-terminal)
PF00763 Tetrahydrofolate dehydrogenase/cyclohydrolase (Catalytic domain)
PF01268 Formate-tetrahydrofolate ligase

c) **Ethanolamine-phosphate cytidylyltransferase (Phospho lipid biosynth)**

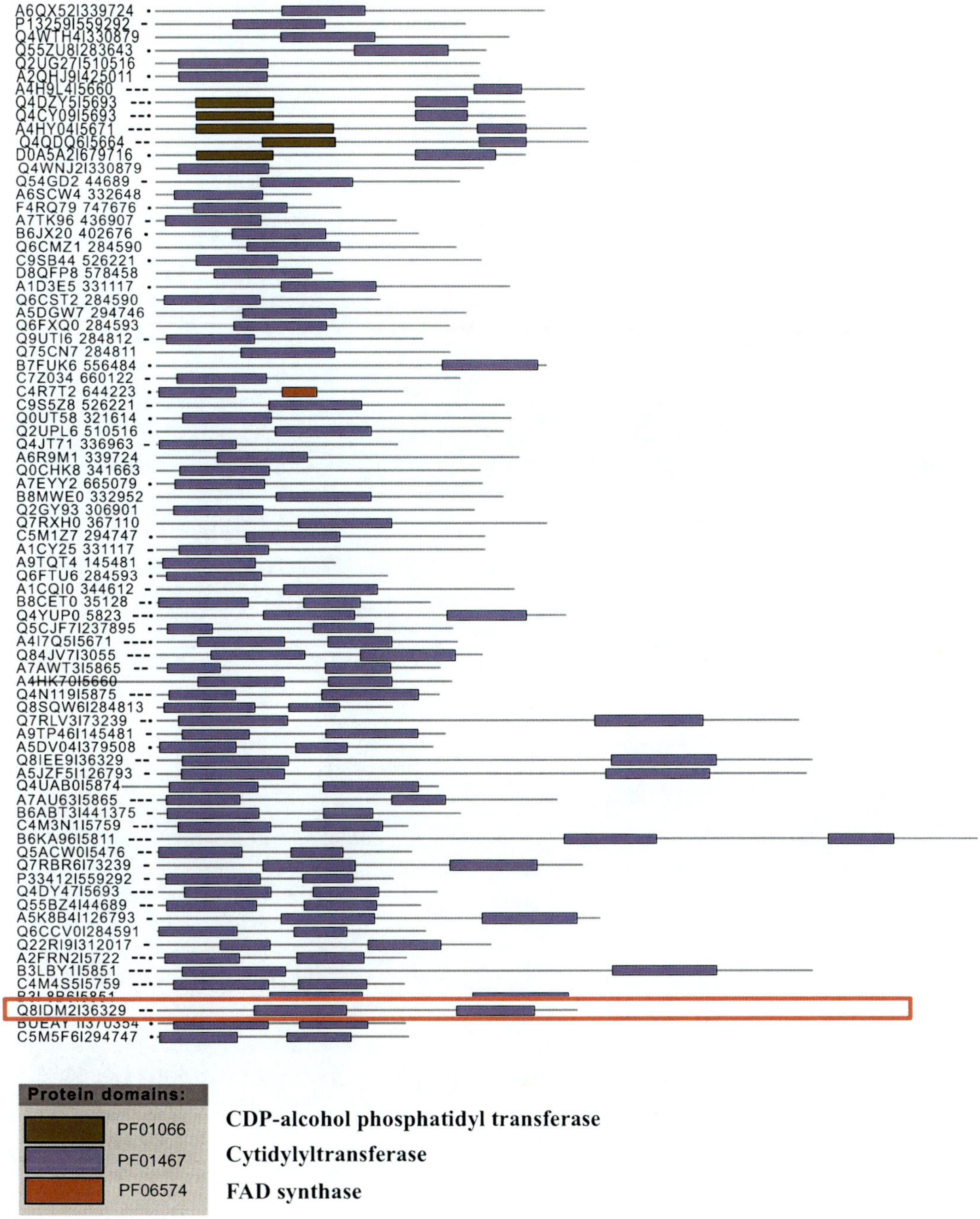

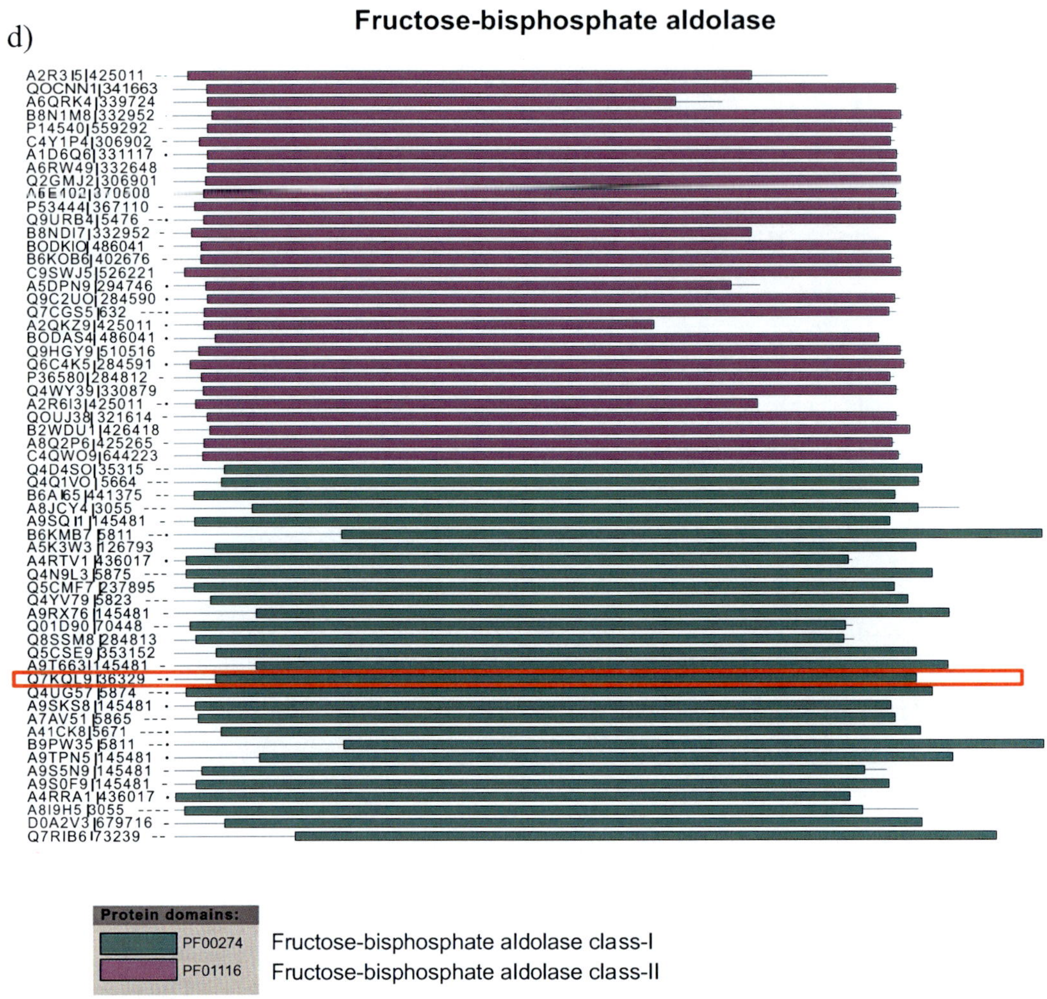

Figure 5: Differences in domain organization. Each example depicts the domain organization of *P. falciparum* proteins (highlighted in red box) and the functional counterparts in other organisms. Each line represents a protein (Uniprot id | taxonomic id) and colored rectangular boxes depict the domains assigned to the proteins. The enzymatic step and the key to the color code are present alongside each example. a) Represents gain of novel domain in *P. falciparum*, b)denote loss of domains c) duplications of domain in the parasite and d) recruitment of different domain in *P. falciparum* in comparison to other functional counterparts.

Fig. **5b** shows yet another example of domain loss. Tetrahydrofolate ligase (THF ligase) exists typically as a trifunctional enzyme in eukaryotes [41] and contains 3 distinct domain families as depicted in the figure. It is involved in catalyzing the formation of 10-formyl THF which is essential for purine biosynthesis. However, *P. falciparum* lacks formate-tetrahydrofolate ligase domain. Experimental studies suggest the presence of purine salvage pathway rather than *de novo* purine biosynthesis [7] in *P. falciparum*. Thus, the exact role of this protein in the parasite is not clear. Other pathways where domain losses are observed include pentose phosphate pathway, thiamine biosynthesis and folate biosynthesis to name a few. All these pathways are necessary for the growth of the parasite. Therefore loss of domains and, in some cases, replacement of domains may offer novel adaptive changes.

Involvement of homologous repeats of domains is also observed in a few of the enzymatic processes. Fig. **5c** shows an example of a protein with domain repeats. Typically, eukaryotic organisms show numbers of repeats in the proteins and such repeats are often involved in protein-protein interactions [42]. Presence of such domain repeats may imply their importance in adaptation.

The most extreme cases of domain architecture differences are cases with large domain distance due to recruitment of completely different domain families. Such families typically show significant structural similarity and, sequence variation is largely manifested due to mechanistic variations. One such example is depicted in the Fig. **5d** which shows an example of Fructose bis-phosphate aldolase. The two domain families represent class I and class II enzymes and they vary from each other in their mechanisms [43]. Class I Fructose bis-phosphate aldolase is prevalent in plants and animals [44] and is also observed in *P. falciparum*. Despite the prevalence of such domain families with distinct sequences and similar functions, infinite domain distance between proteins could also indicate errors in annotation and thus require a careful scrutiny to rationalize the enzymatic roles of these domain families.

Table **2** lists some of the other examples of pathways with different types of variations observed in domain architectures for *P. falciparum* proteins characterized based on domain distances. However, it must be pointed out that the distance calculation does not explicitly account for variation in domain length, variation in lengths due to unassigned regions or sequence diversity. In the following subsections further quantification of diversity of *P. falciparum* proteins involved in metabolic processes is made by taking into account various factors such as:

1) Variation in protein lengths.

2) Sequence diversity for the domain assigned regions.

Table 2: Diversity in domain organization observed in *P. falciparum* proteins involved in different pathways

Pathway	**Enzymatic Steps**	**Type of Variation**	**Plasmodium Protein (UniProt id)**	**Domain Organization in Plasmodium Protein (Pfam ids Separated by "-")**
Citric acid cycle	Dihydrolipoyl dehydrogenase	Completely conserved	Q8IAZ6, Q8I5A0	PF07992-PF02852
Pyrimidine biosynthesis	Thymidylate synthase	Domain gain	Q8I1R6	PF00186-PF00303
Pentose phosphate pathway	Transketolase 1	Domain loss	C6KSV3	PF00456-PF02779
Glutamine metabolism	Glucosamine-fructose-6-phosphate aminotransferase	Domain Duplication	Q8IJF3	PF13522-PF01380-PF01380
Pyrimidine biosynthesis	Dihydroorotase	Completely different	Q8IKA9	PF04909

2.2. Deviation in Protein Lengths

In the previous section, the extents of divergence observed between *P. falciparum* proteins and their functional counterparts in other organisms were quantified by considering domain distances. A significant fraction of pathways shows involvement of proteins with conserved domain architectures and many of these pathways involve single domain proteins. However, despite similarities at the level of domains, these proteins exhibit significant differences due to the presence of domain unassigned regions. Analysis of orthologous proteins, in the past, has revealed large deviations in protein lengths, with eukaryotic proteins being larger than prokaryotes [45,46]. Such variations in length are speculated to be non-random with evolutionary significance [45,46].

Typically, *P. falciparum* proteins are longer and have an average length of 644.6 (±450.6) amino acids, while proteins in organisms belonging to the group alveolates and the rest of the eukaryotes have average protein lengths of 559.3

(± 345.3) and 538.3 (± 102.5) amino acids respectively. Fig. **6** is a scatter plot which shows the difference in lengths of proteins catalyzing various enzymatic reactions in *P. falciparum* with respect to the proteins from other organisms.

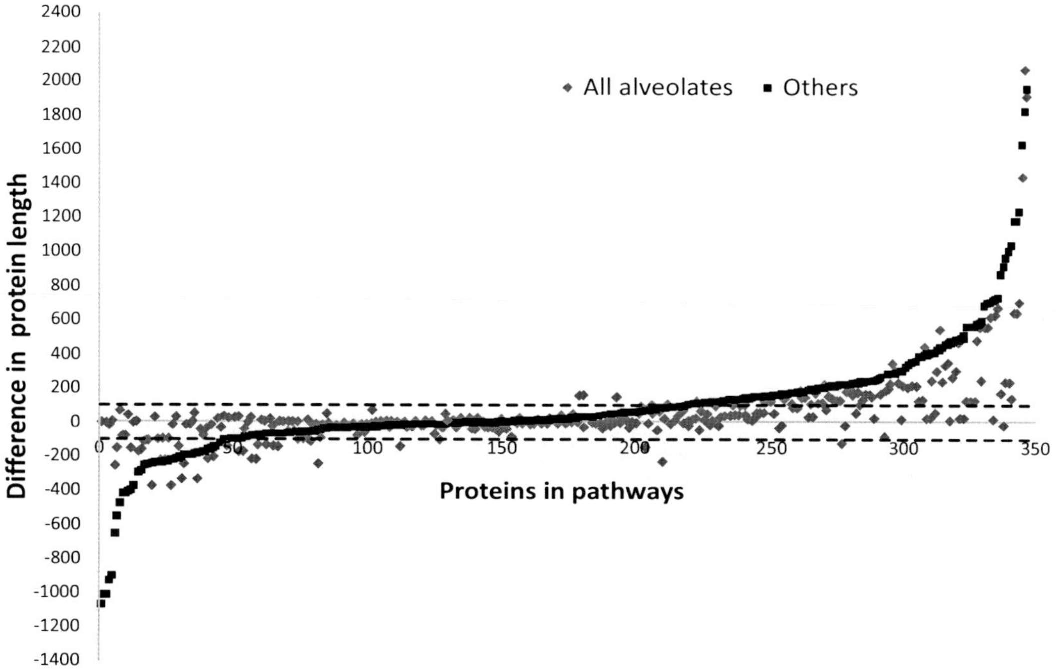

Figure 6: Differences between the lengths of *P. falciparum* proteins and their homologues in other organisms. Scatter plot shows the distribution of differences in the lengths of proteins from *P. falciparum* in comparison to the functional counterparts in other organisms. The Y-axis represents the average of the difference in lengths obtained for *P. falciparum* proteins and proteins belonging to alveolates (grey dots) and the rest of the organisms (black dots) for every pathway.

Each dot represents the average difference in the length of *P. falciparum* protein participating in a pathway with respect to proteins in other organisms involved in the same pathway step. Grey dots represent the average differences obtained between *P. falciparum* protein and other proteins belonging to different alveolates in a particular pathway step. Black dots indicate the average differences in the length of *P. falciparum* proteins in comparison to proteins in the rest of the organisms (non-alveolates). The extent of deviation in lengths of *P. falciparum* proteins is further illustrated by considering as an example of phosphatidyl inositol phosphatase (Fig. **7**). It can be clearly seen that, in general, *P. falciparum* proteins are longer than the other proteins catalyzing the same reaction.

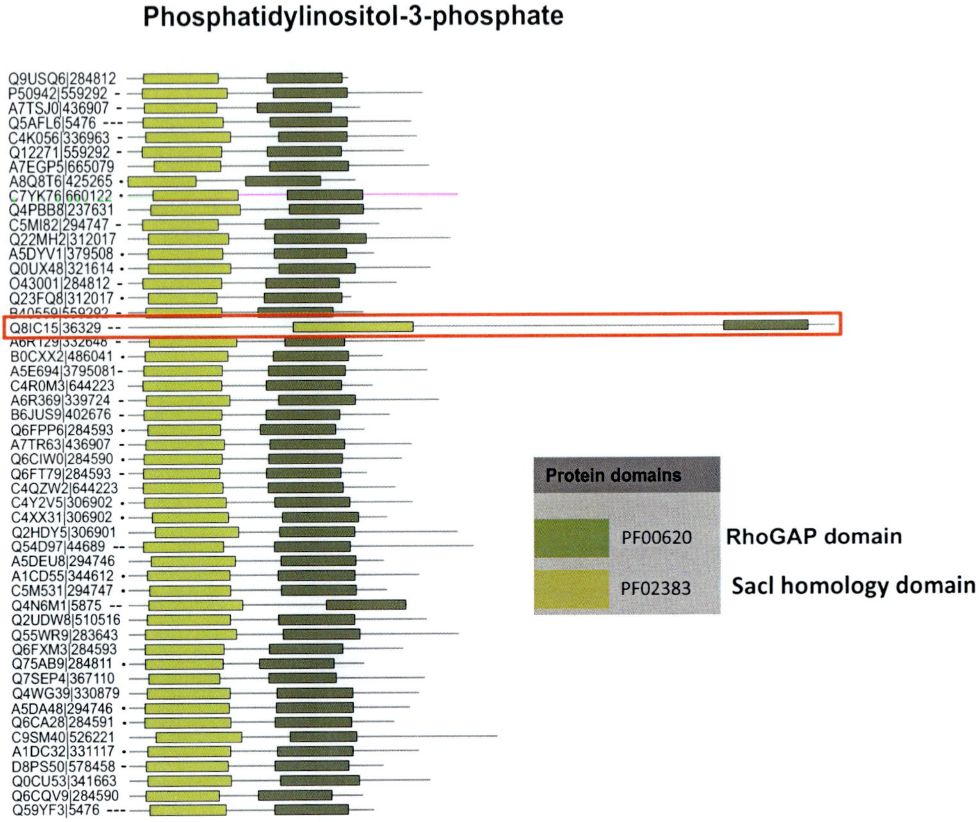

Figure 7: Deviation in protein lengths. An example of a protein showing large deviation in protein length in *P. falciparum* in comparison to other proteins catalyzing the same reaction. *P. falciparum* protein is highlighted within red rectangle.

Table **3** lists examples of a few other proteins of the parasite showing large deviations in their lengths.

Fig. **6** also shows examples where *P. falciparum* proteins are significantly shorter than their functional counterparts, which are represented as black and grey dots below the dashed line along the direction of negative Y-axis in the figure. Number of shorter *P. falciparum* proteins is significantly lower than the fraction of longer proteins (which are more than twice as prevalent). Careful examination of these shorter proteins reveals few examples where *P. falciparum* protein has lost domain(s) as exemplified previously in the case of tetrahydrofolate synthase (Fig. **5b**). The exact functional significance of such length deviations observed in *P. falciparum* is not clear and experimental studies would aid in understanding their adaptive roles.

Table 3: Deviation in lengths of proteins participating in pathways

S. No.	Enzymatic Step	Avg. Length in *P. Falciparum* (Residues)	Avg. Length in Alveolates (Residues)	Avg. Length in Rest of the Organisms (Residues)
1.	**Ornithine decarboxylase**	1434	1204.25	467.21
2.	**6-phosphofructokinase**	1494	1403	701.92
3.	**Mannose-6-phosphate isomerase**	1011	459.25	427.20
4.	**Nucleoside diphosphate kinase**	988.5	398.43	199.3

Overall differences in the lengths of proteins in *P. falciparum* could be manifested at two different levels: 1) variation in domain lengths and 2) variation in the unassigned regions typified by the inter-domain regions and the N/C terminal overhangs. In the following subsections a detailed analysis is presented that illustrates the contribution of both of these factors in differences in protein lengths.

2.2.1. Deviation in Domain Lengths

Analysis of differences in lengths of homologous domains, between *P. falciparum* and other organisms, present in proteins catalyzing specific steps of pathways is illustrated in Fig. **8**. In order to consider the importance of a domain catalyzing a specific step in pathway, only those domain families were taken into account that are present in more than 50% of the organisms, known to catalyze a particular enzymatic step. All the comparisons were made by considering *P. falciparum* proteins and comparing it with other putative functional counterparts. These comparisons have been carried out in a set of three categories: 1) *P. falciparum* proteins *vs.* proteins belonging to organisms in the genus *Plasmodium*, 2) *P. falciparum* proteins *vs.* proteins from other apicomplexans and 3) *P. falciparum* proteins *vs.* the rest of the non-apicomplexans (others). Length of a domain in a protein is quantified by normalizing it with respect to the length of an HMM (Hidden Markov Model), *i.e.*, number of consensus positions in a family.

Fig. **8a** clearly shows limited deviation in lengths (± 50 residues) of domains in substantial fraction of proteins. The results are however consistent with a recent study carried out with structural domain families in proteins across various organisms, where no significant differences in domain lengths were observed [46]. Fig. **8** also shows a few examples where *P. falciparum* proteins contain domains with significant differences in the domain lengths with respect to the functional counterparts from other organisms. However, there are only 9 such cases where domains in *P. falciparum* proteins are on an average 50 amino acids longer than their functional counterparts.

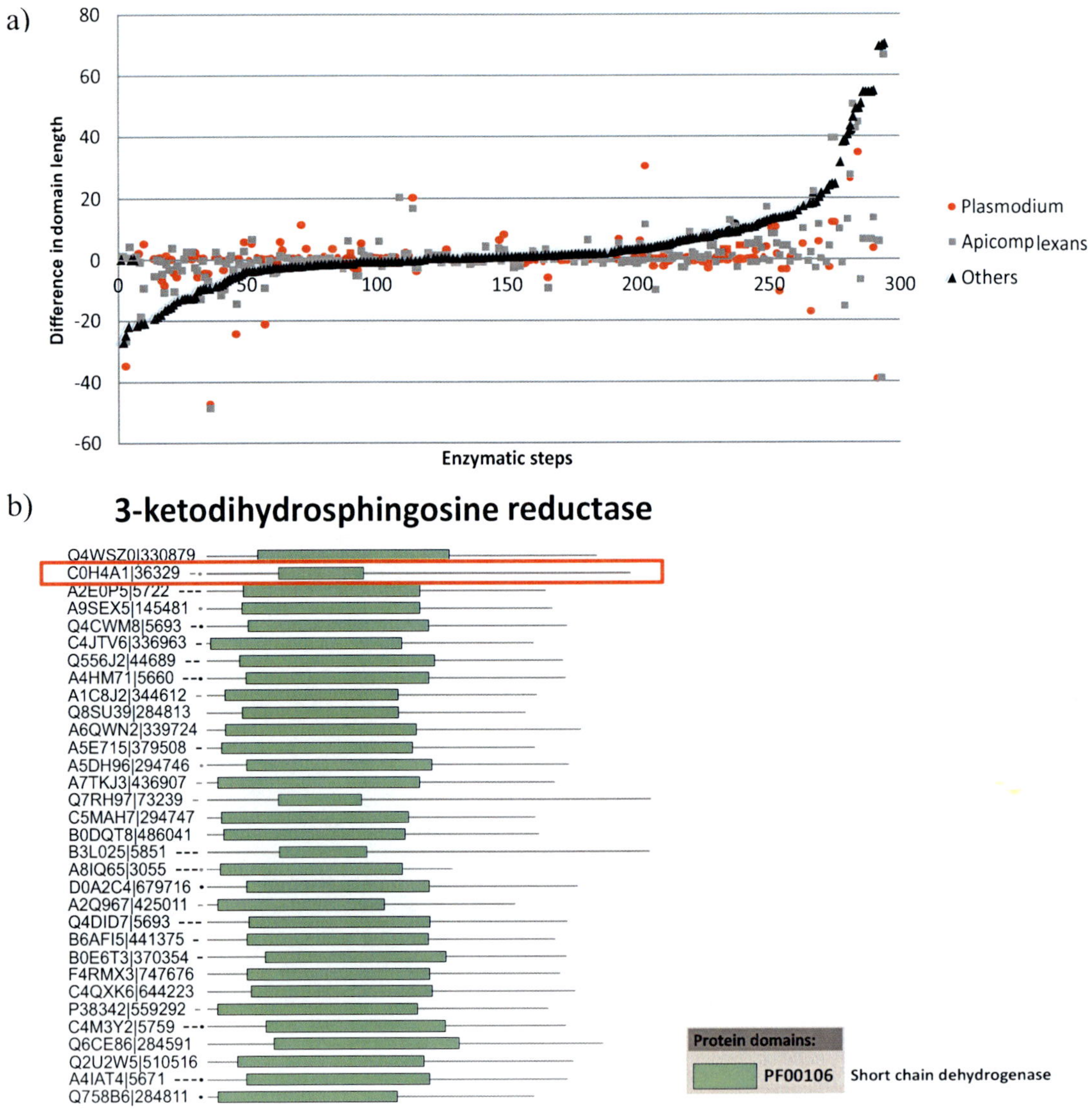

Fig. 8: contd...

c)

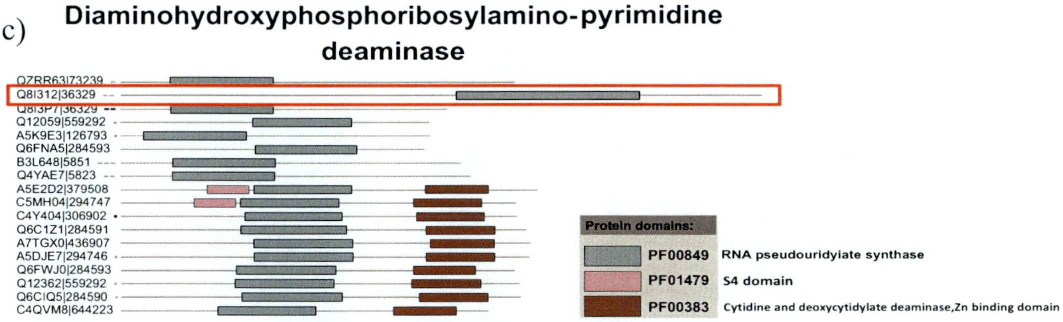

Figure 8: Difference in normalized domain lengths. a) Scatter plot showing the distribution of the differences in the lengths of domain families obtained upon comparing *P. falciparum* domains with domains in organisms belonging to genus *Plasmodium*, apicomplexans and rest of the organisms (others). Domain lengths are quantified as normalized lengths; subsequently differences are calculated and averaged across all the domains present in proteins involved in every enzymatic step. b) Example of *P. falciparum* protein showing a shorter domain and c) example of a *P. falciparum* protein with longer domain.

Even though domain lengths do not exhibit significant differences, variation in protein length can be manifested by the presence of non-conserved inserts within domains (intra-domain inserts). During the process of domain assignments of proteins it was observed that significant fraction of domain family assignments were obtained as fragments. A split domain assignment is made when all the fragments obtained show less than 10% overlap amongst each other and together they add up to more than 70% of the domain family profile. Split domain families are cases where domain family assignments are split across small fragments of non-contiguous statistically significant aligned regions interspersed with long non-conserved inserts. One such split domain family is shown in Fig. **9**.

The presence of a non-conserved insert amongst these domain families is not only present in *P. falciparum* but is also prevalent in other organisms belonging to the same genus indicating potential biological significance of such inserts. Fig. **10** shows the percentage of proteins containing split and ambiguous cases across various organisms.

In *P. falciparum* nearly 16% of the proteins contain split domains and 45% of the proteins contain ambiguously fragmented domain family assignments due to the presence of non-conserved inserts. Amongst the ambiguous cases many domain families could be cases of split domains and several cases correspond to duplicated domains with significant sequence divergence. Many of the inserts observed within the domains are specifically found in the parasite, along with inserts conserved across closely related organisms. Presence of these inserts within domains has been shown to be critical for the functioning of the enzymes. For instance, the presence of parasite specific inserts in bi-functional enzyme S-adenosylmethionine decarboxylase and ornithine decarboxylase in *P. falciparum* are implicated in mediating essential protein-protein interactions [38]. Deletion of these inserts diminishes the overall activity of the enzyme [38]. Similar traits are also exhibited by inserts observed in yet another bi-functional enzyme glucose-6-phosphate dehydrogenase-6-phosphogluconolactonase in *P. falciparum*. These inserts are found conserved across other organisms belonging to the same genus and are critical for the functioning of the protein [47]. Apart from functional importance of these inserts, the uniqueness of these inserts also makes them attractive targets for drug design [48].

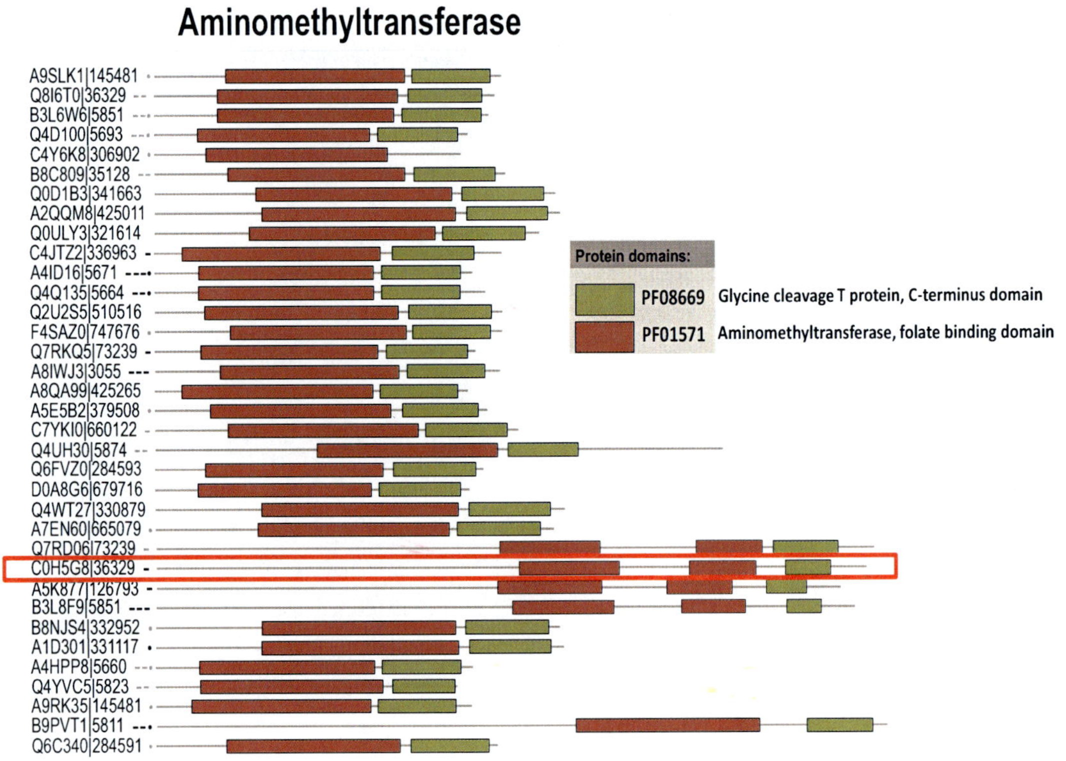

Figure 9: Presence of split domains. *P. falciparum* protein (highlighted in red coloured box) showing the fragmented domain of aminomethyltransferase.

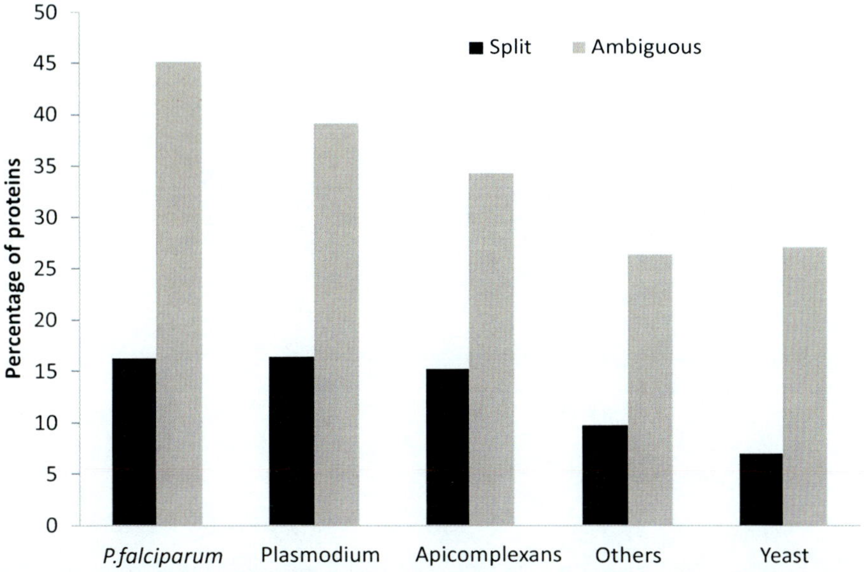

Figure 10: Distribution of fragmented domain families. Percentage of proteins containing split domains in *P. falciparum*, organisms in genus *Plasmodium*, apicomplexans, rest of the organisms (others) and also in yeast.

2.2.2. N and C-Terminal Overhangs and Inter-Domain Separation

Variation in protein lengths are also brought about due to the presence of unassigned N and C-terminal regions along with inter-domain regions. Fig. **11 (a-c)** shows the difference in lengths for all the three regions in *P. falciparum* in comparison to other organisms. Fig. **11a** shows the difference in N-terminal overhang lengths for *P. falciparum* proteins and proteins from organisms belonging to same genus, organisms belonging to apicomplexa and the rest of the organisms (others). Fig. **11b** shows differences for the C-terminal overhangs. On an average N -terminal and C-terminal regions of *P. falciparum* proteins have 133.67 (±178.9) and 103.53 (±197.8) amino acids, respectively. These values are higher than the average values for the N and C-terminal regions in the rest of the organisms (N-terminus: 89.81 and C-terminus average 74.6). It could be inferred that *P. falciparum* proteins are characterized by long and, possibly parasite specific, unassigned regions. It is very possible that many of these regions correspond to domains which could not be assigned due to high evolutionary divergence. Fractions of cases with large differences are, however, lesser when compared with proteins from apicomplexa and organisms from Plasmodium genus.

a)

b)

c)

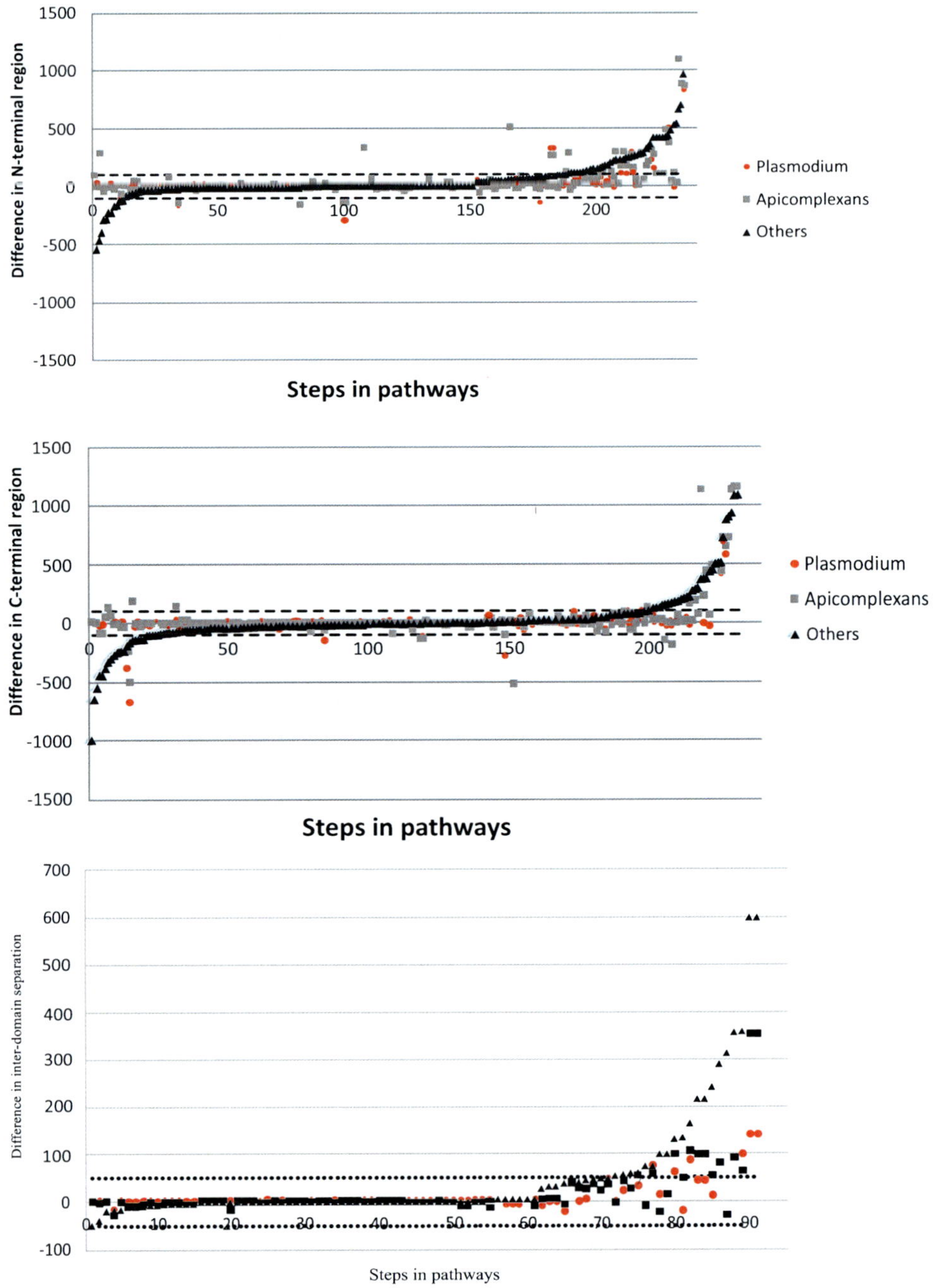

Figure 11: Variation in N and C-terminal overhangs and the inter-domain separation. In each of the scatter plots dots represent average difference obtained between specific regions of *P. falciparum* protein (N, C-termini and inter-domain regions) and proteins from functional counterparts of other organisms. a) Difference in N-terminal unassigned regions. b) Difference C-terminal unassigned regions and c) Difference in unassigned region separating assigned domains.

Apart from inserts within the domains, non-conserved unassigned regions are also found between domains. Fig. **11c** shows the difference in such inter-domain unassigned regions in proteins of *P. falciparum* and the functional counterparts from other organisms. For this analysis, only contiguous domain pairs were considered which are found in more than 50% of the organism in a particular metabolic step. The average inter-domain separation for *P. falciparum* is 89.8 (±131.5) residues and it is more than twice the separation observed on an average in non-apicomplexans (41.3 ± 45). Such non-conserved unassigned regions are either low complexity in nature as has been noted earlier [14] or they correspond to no clear functional roles. In spite of a finite possibility that such domain assignment lacking regions may get family assignment in the future with increase in the size of sequence data, many such regions can also act as potential linkers between the flanking domains and may therefore profoundly influence the function of the protein. An analysis in understanding the length variations in proteins from the extant organisms reveals large variation in these linker regions connecting domains with potential effect on functions [46]. Fig. **12** depicts examples of proteins showing large N and C-terminal overhangs and inter-domain separating regions.

a)

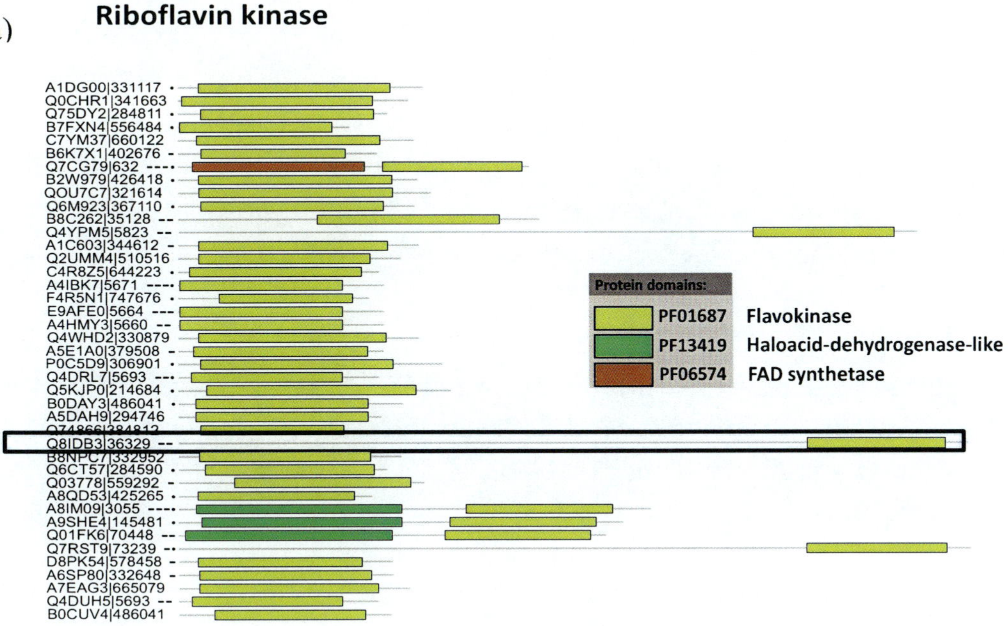

Fig. 12: contd….

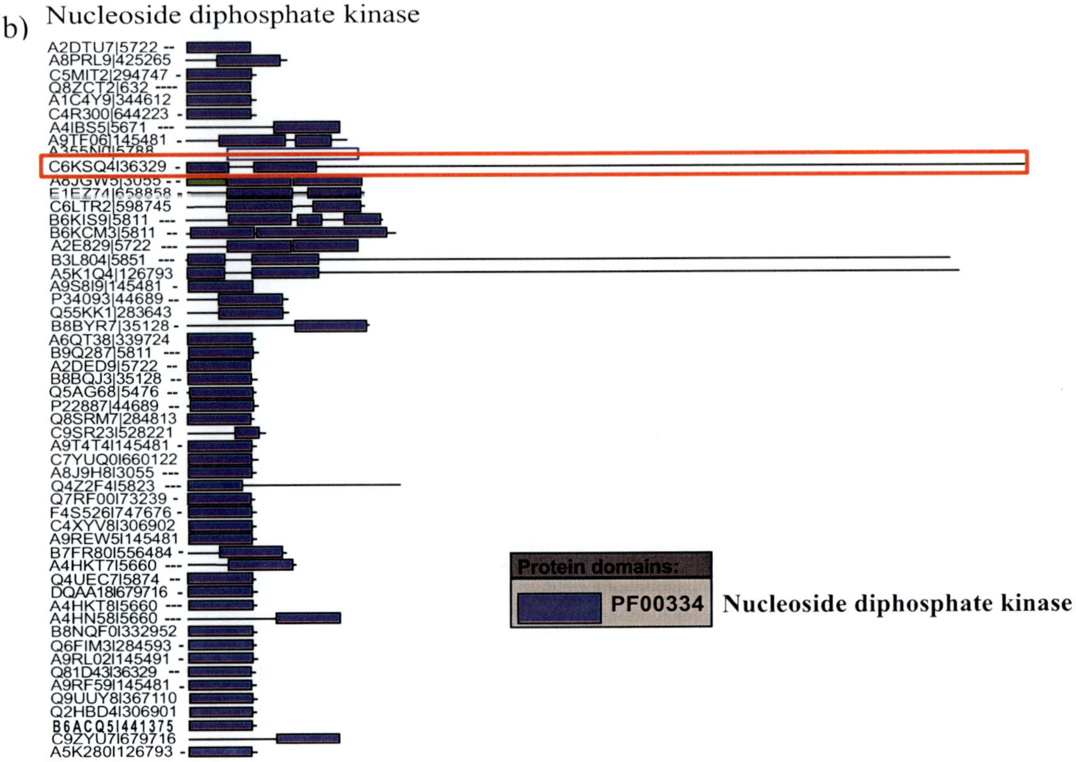

Figure 12: Families of homologous proteins showing variation in N and C-terminal overhangs and inter-domain separating regions. a) and b) proteins with large unassigned N and C-termini respectively. (*P. falciparum* proteins highlighted in red coloured box).

In all these examples it is observed that extra regions are also found in proteins belonging to other closely related organisms which are indicative of possible but as yet unknown adaptive roles in pathogenesis and parasitism. The functional roles of such regions, however, are still unclear.

2.3. Sequence Diversity

The previous section discussed the variation observed in the lengths of proteins catalyzing various enzymatic steps in pathways present in *P. falciparum*. Primary factors causing variation in lengths include 1) presence of non-conserved regions within domains and 2) presence of long, domain-unassigned N and C-terminal overhangs and 3) long domain-unassigned regions between two contiguous domains. The final measure that has been used to quantify the extent of divergence observed between *P. falciparum* proteins and their functional counterparts from other organisms is the sequence similarity. A local pairwise sequence alignment algorithm MAFFT [49] was employed for each *P. falciparum* protein and its putative functional counterparts. Local pairwise sequence alignment with appropriate gap opening and gap extension penalties (MAFFT einsi option) ensured reasonable quality of optimal alignment despite the presence of large non-conserved inserts.

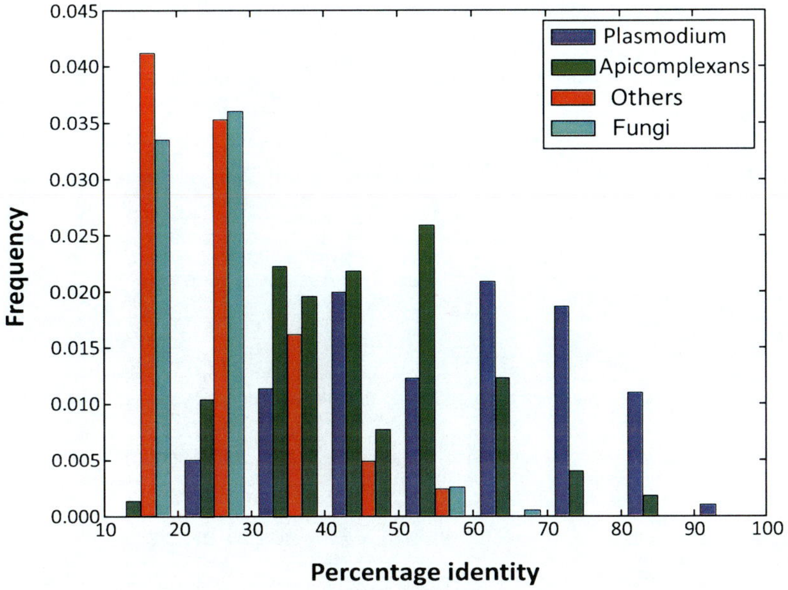

Figure 13: Histogram showing the frequency distribution of pairwise sequence identity obtained between *P. falciparum* proteins and functional counterparts from other organisms. Each coloured bar represents the distribution of pairwise sequence identity obtained between *P. falciparum* proteins and functional counterparts from other organisms for proteins involved in various pathways.

Fig. **13** shows the distribution of values of sequence identity obtained for local pairwise sequence alignments between a *P. falciparum* protein and proteins from other organisms catalyzing the same enzymatic step as the *P. falciparum* protein. Each coloured bar shows the distribution of pairwise sequence identity obtained for *P. falciparum* proteins with respect to the functional counterparts in specific organism groups. It is not surprising to note from the peaks in the histogram that the extent of sequence similarity is high for proteins belonging to the same genus as compared to others, the average being 48.3% and sequence identity being 23.8% when compared to yeast (*Saccharomyces cerevisiae*). In more than 40% of the enzymes the pairwise sequence identities obtained between *P. falciparum* proteins and their functional counterparts from other non-apicomplexans are less than 20%. Despite weak similarity such proteins carry out same or highly similar functions due to the conservation of functionally critical residues, which is exemplified in Fig. **14** which shows the alignments of phosphoglycerate mutase in yeast, *P. falciparum* and a few other organisms.

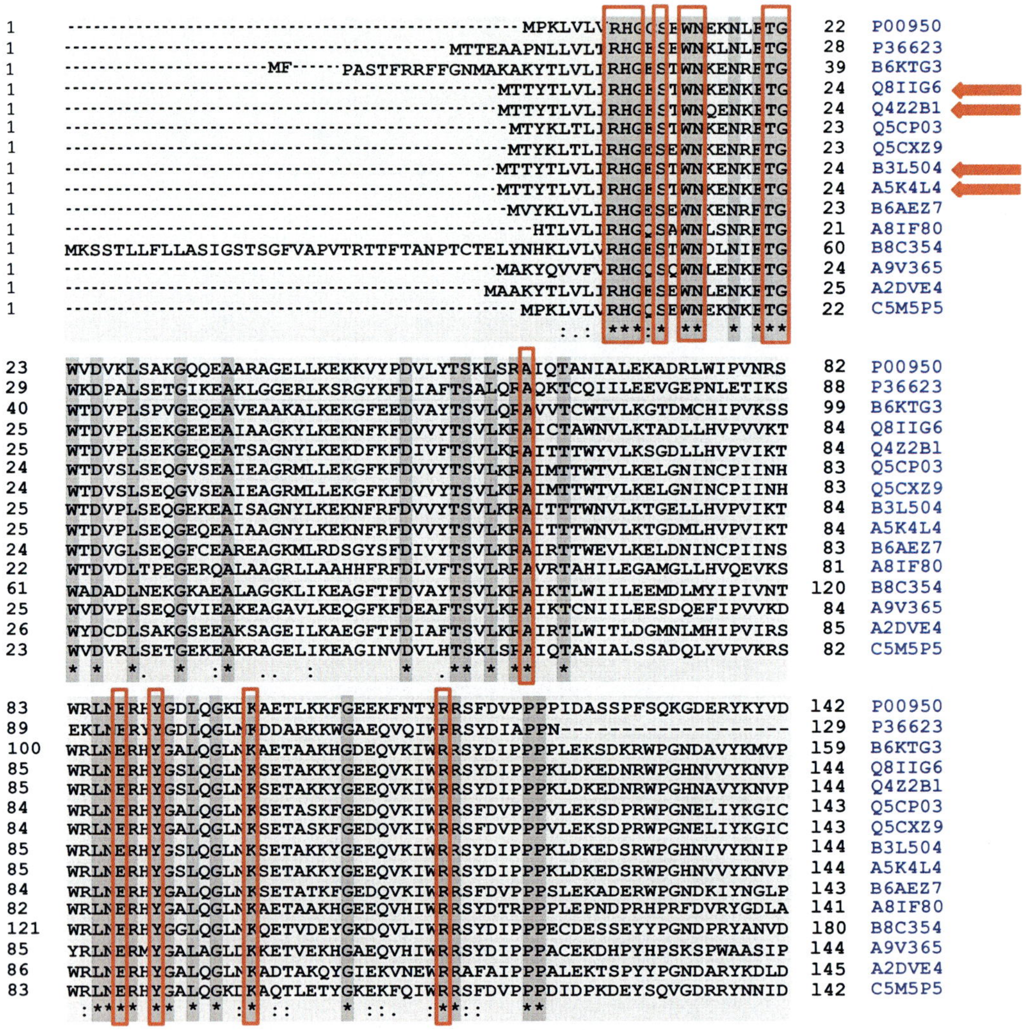

Figure 14: Multiple sequence alignment of *P. falciparum* phosphoglycerate mutases showing conservation of functionally critical residues in plasmodium proteins and in the homologues from other organisms. Red arrows indicate *P. falciparum* proteins and proteins from organisms belonging to the same genus. All the functionally critical residues are encased by red boxes and all the identical positions are highlighted by a light grey background.

It can be seen that all the critical residues necessary for function are conserved despite overall poor sequence similarity. Functionally critical residues were derived from PDB entry 1bq3, the structure of phosphoglycerate mutase from yeast [50].

The difficulty posed by the divergent proteins in *P. falciparum* for function determination is addressed in the next section which explores structural and functional features of the metabolic proteins with the help of extremely sensitive remote homology detection approaches. The results presented are an update to the analysis previously made [83].

3. CHARACTERIZATION OF METABOLIC PROTEINS IN *P. FALCIPARUM* AND IDENTIFICATION OF "NEW" ONES: A REMOTE HOMOLOGY DRIVEN ANALYSIS

Exploring closely related homologues for protein function annotation and detection of distant homologues in order to explore evolutionary relationships have played a pivotal role in understanding evolution of protein structure, function and interactions. Sensitive search procedures based on the use of profiles such as Hidden Markov Models (HMMs) and Position Specific Scoring Matrices (PSSMs) have effectively improved the identification of well-characterized homologues. Employment of such techniques for detection of remote homologous *P. falciparum* proteins forms the basis of this section.

An enhanced structural and functional characterization of metabolic proteins in *P. falciparum* was brought about with the help of sensitive profile-based search algorithms employed by HMMER [53,54], MulPSSM [55,56], and HHblits (A highly sensitive algorithm employing iterative searches using HMMs) [57]. MulPSSM database, developed by our group earlier (http://mulpssm.mbu.iisc.ernet.in/), comprises of about 385,258 profiles for 13672 domain families from Pfam database (Version 26) [58]. Benchmark studies [59-61] have shown a reasonably good coverage of proteins for function annotation achieved with the use of MulPSSM. Measures were taken to ensure reliability of the domain assignment made in the domain unassigned regions of the metabolic proteins, a few of which are listed out in Table **4**.

Table 4: New domain assignments. List of metabolic proteins in *P. falciparum* with domain assignments made

S. No.	Protein Code	Domain Region(s)	Pfam Domain Family	Nature of the Assignment Made
1.	PFI1310w	14-196	CN_hydrolase	N-terminal region
		476-507, 544-761	NAD_synthase	Split domain
2.	PFF0230c	52-176, 184-302	Glyoxalase	Domain duplication
3.	PFF0530w	8-341	Transketolase_N	Already assigned
		356-530	Transket_pyr	Already assigned
		549-659	Transketolase_C	C-terminal region
4.	PF11_0311	200-262	PGM_PMM_I	Already assigned
		482-601	PGM_PMM_II	Inter-domain region
		869-931	PGM_PMM_IV	Already assigned

Enhanced residue coverage for 13% of the metabolic proteins in terms of function recognition was aided by associating domain unassigned regions with structural/functional domain families.

Pfam database, a collection of protein domain families, also comprises of DUFs or Domains of Unknown Function that are a large set of functionally uncharacterized protein families [67]. Numerous attempts in the past [72-76] have been successful in identifying function for DUFs. Based on our analysis, 2 of 3 such DUFs, associated with two metabolic proteins in *P. falciparum* as briefed in Table **5**, could be related to families of known structure. Such structural domain associations enable identification of residue determinants of function of the protein. SUPFAM database [65,66], developed by our group (http://supfam.mbu.iisc.ernet.in/), was employed for this purpose. SUPFAM explores the evolutionary relationships between sequence and structural families, thereby identifying potentially related protein domain families, with the use of highly sensitive profile-profile alignment algorithm AlignHUSH [77], also developed by our group.

PFD0980w, involved in pantothenate and CoA biosynthesis pathway, comprises of an uncharacterized protein family UPF0054 along with a 4'-phosphopantetheinyl transferase domain. Apart from our predicted association, structural elucidations [68,69] suggest a reliable homology of UPF0054 to metal-dependent proteinases such as collagenase and gelatinases. However, experimental characterization failed to show any detectable activity [68]. Thus, UPF0054 remains an uncharacterized protein family of putative metal-dependent proteases.

Table 5: Two DUFs which could be associated with a known structural family based on SUPFAM database

Protein Code	Domain Region	Pfam Domain Family	SCOP Family ID	SCOP Family Description
PFD0980w	185-247	UPF0054	d.92.1.15	Predicted metal-dependent hydrolase
PFF1350c	929-972	DUF4009	e.23.1.1	Acetyl-CoA synthetase-like
	52-136	DUF3448	-	-

PFF1350c, participating in pyruvate metabolism, is annotated to be an acetyl-CoA synthetase [71] which consists of AMP-binding domain associated with DUF4009 at its C-terminus and DUF3448 at N-terminus. As shown in Table **5**, DUF4009 could be associated with a structural family acetyl-CoA synthetase-like, while no relationship could be obtained for DUF3448. The presence of DUF4009 at the C-terminal of AMP-binding enzymes has been previously noted in other organisms [70], where this enzyme family includes luciferase, long chain fatty acid CoA ligase, acetyl-CoA synthetase and various other closely related synthetases. The structure adopted by the domain family DUF4009, was predicted and compared with the associated structural family to check for consistency in the prediction made. Fig. **15a** shows the predicted fold of the domain family DUF4009 obtained with the use of fold recognition algorithm employed by PHYRE2 [62]. The fold predicted for most of the members within DUF4009 family was acetyl-CoA synthetase-like, with >20% sequence identity and >90% confidence, which is consistent with the prediction made. Based on the benchmark studies [62,81], given a high confidence match, the overall fold is most certainly correct and the central core of the model tends to be accurate even at low sequence identities *i.e.* <20%.

a)

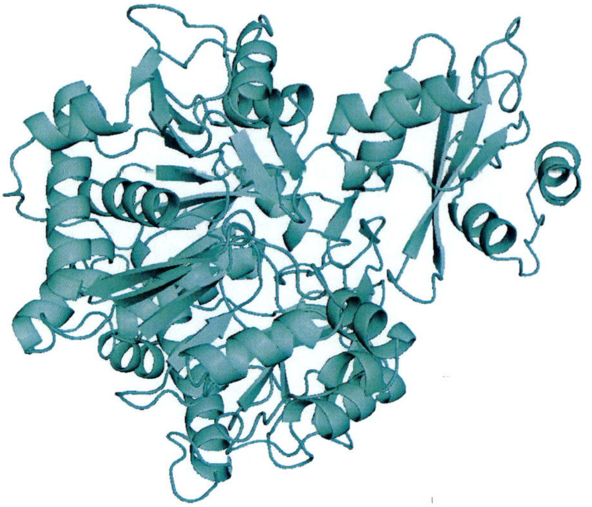

b)

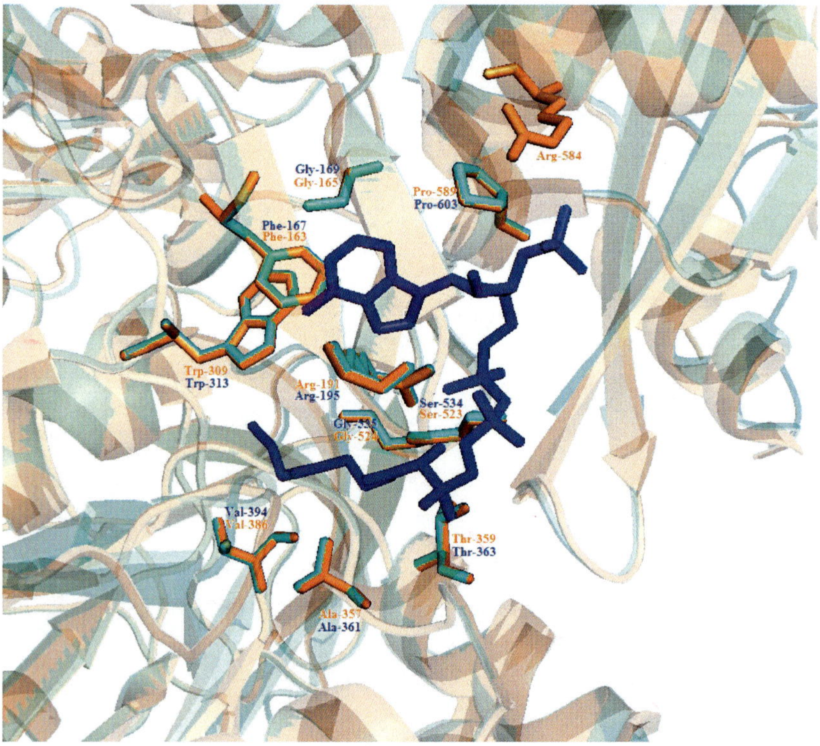

Figure 15: Extending structural/functional information across DUFs. a) Predicted fold adopted by one of the members of DUF4009 family. b) Potential active site residues identified (labelled in cyan) based on inferences from structural alignment with respect to crystal structure of acetyl-CoA synthetase (PDB code: 1PG4; residues labelled in orange). The ligand acetyl-CoA is highlighted in dark blue. Images are rendered using PyMOL (http://pymol.org/).

Structural alignment with crystal structure of acetyl-CoA synthetase aided in the scrutiny of active site conservation in the proteins housing DUF4009, which were identified to conserved as illustrated in Fig. **15b**. These elucidations add up to the credibility of the association made in order to extend structural/functional information across domains of unknown function.

Furthermore, the highly sensitive profile-based approaches employed and their implications were used as guiding tools to identify "missing" metabolic proteins in *P. falciparum*. A list of enzymatic steps corresponding to enzymes not yet identified in *P. falciparum* was created by collating information from KEGG [51], PlasmoCyc [52] and a dataset of malarial parasite metabolic pathways [20]. These steps were referred as "pathway holes". This list comprises of intermediate steps operating in the pathways for which the *P. falciparum* proteins have not yet been identified. A few of such pathway holes identified are listed in Table **6** along with the reactions carried out and the corresponding Pfam domain family.

Table 6: List of pathways in *P. falciparum* with enzymes, involved in pathways, not identified in the parasite

S. No.	Reaction Catalyzed	Pathway Involved	E.C. Number of the Expected Enzyme	Pfam Domain Family Assignment of the Expected Enzyme
1.	dihydro-neo-pterin => glycolaldehyde + 2-amino-4-hydroxy-6-hydroxymethyl-7,8-dihydropteridine	Folic acid biosynthesis, folate biosynthesis	4.1.2.25	FolB
2.	phosphatidylcholine + H_2O => a fatty acid + L-1-lysolecithin	Glycerolipid metabolism	3.1.1.4	PLA2G12
3.	phosphate + D-myo-inositol-4-phosphate => 1D-myo-Inositol 1,4-bisphosphate + H_2O	Inositol phosphate metabolism	3.1.3.57	Inositol_P
4.	ATP + NAD(+) <=> ADP + NADP(+)	Nicotinate and nicotinamide metabolism	2.7.1.23	NAD_kinase
5.	hydroxymethylbilane = >H_2O + uroporphyrinogen-III	Porphyrin metabolism, heme biosynthesis II, biosynthesis of proto- and siro-heme	4.2.1.75	HEM4

In order to identify the missing metabolic proteins of *P. falciparum*, the protein sequences from the organism were associated with structural/functional domain families using the highly sensitive approaches as described previously in this section. To manually ascertain the reliability of the domain assignments made to the 'pathway holes', a fold recognition algorithm [62] was employed. In retrospect, all the remote homology relationships identified using multiple PSSM search approach corresponded to a protein family with at least one member of known three dimensional structure. Another level of validation was adopted by integrating microarray [63,64] data where use of stage-specific gene expression profile information allowed identification of pairs of proteins that are co-expressed. Proteins that are co-expressed in specific stages in *P. falciparum* are expected to participate in same pathway. Thus, co-expression of the newly annotated proteins in pathways with already annotated members can be used as an additional measure to ascertain the validity of the predictions made. A total of 112 proteins were identified with potential roles in metabolic pathways. Table **7** lists a few proteins corresponding to 'pathway holes' newly detected. This analysis was initially made with previous release of Pfam database (version 18) and PlasmoDB [71] version 4.4, results of which successfully coincided with the current releases, confirming the predictions made. Such entries are indicated by an asterix in the Table.

Table 7: *P. falciparum* proteins annotated as potential enzymes using MulPSSM approach

Protein Code	**Domain Assignment Made (Pfam18)**	**Enzyme Classification**	**Pathway**
*MAL8P1.81	Flavoprotein	6.3.2.5	Pantothenate and coenzyme-A biosynthesis
*PF14_0573	BPL_LplA_LipB	6.4.1.1	Pyruvate metabolism
*PF11_0276	Abhydrolase_1	3.1.1.23	Glycerol biosynthesis
*PFF1405c	Acetyltransf_1	2.3.1.4	Aminosugars metabolism
*PF11_0453	Carb_kinase	2.7.1.49	Thiamine metabolism
*PFL0575w	Amino_oxidase	1.4.3.4	Arginine and proline metabolism
*PFL0230w	dCMP_cyt_deam_1	3.5.4.1	Pyrimidine ribonucleotide/ribonucleoside metabolism
*PFF1210w	PAP2	3.6.1.43	Dolichol metabolism
PFL2285c	HEM4	4.2.1.75	Porphyrin metabolism, heme biosynthesis II, biosynthesis of proto- and siroheme
MAL8P1.38	Abhydrolase_5	3.1.1.23	Glycerol biosynthesis
MAL13P1.19	PI3_PI4_kinase	2.7.1.137 2.7.1.67	Phospholipid biosynthesis

* Predictions consistent with current releases of Pfam and PlasmoDB.

One of the many proteins identified to be potentially involved in metabolic pathways in *P. falciparum* is discussed further in depth.

a) Phosphatidylinositol 3- and 4- kinase (PI3_PI4_kinase)

Phosphatidylinositol 3-kinase (E.C.: 2.7.1.137) phosphorylates phosphoinositides on the 3-hydroxyl group of the inositol ring which generates products essential as secondary messengers in cell signalling [78]. Phosphatidylinositol 4-kinase (E.C: 2.7.1.67) acts on phosphatidylinositol (PI) in the production of secondary messenger inositol-1`4`5`-triphosphate [79]. This domain is present in a wide range of enzymes, involved in diverse cellular functions, such as control of cell growth, regulation of cell cycle progression, a DNA damage checkpoint, recombination and maintenance of telomere length [80]. The catalytic domain of the kinase has a typical bilobal structure with the ATP cofactor binding in the crevice formed by the N-and C-terminal lobes. The *P. falciparum* protein MAL13P1.19 could be associated with PI3_PI4_kinase domain family with the help of highly sensitive profile-based approaches, particularly multiple PSSM [55,56]. For the malarial protein sequence of length of 9271 a.a., the region of domain assignment obtained was 8765-8915 with 24.5% sequence identity, at an e-value of 7e-07. Assessment of the fold adopted by this region of the protein with the use of fold recognition algorithm, suggested consistency with the domain assignment made (Fig. **16a**). To assess the structurally equivalent residues and the active site conservation in the predicted structure of the domain region obtained for P. falciparum protein, a structure-based alignment between the malarial protein MAL13P1.19 and crystal structure of PI3 kinase was of immense aid. Fig. **16b** highlights the residues in PI3 kinase (dark blue) involved in binding with the ligand, 3-(4-morpholin-4-ylthieno [3,2-d]pyrimidin-2-yl)phenol, (red) which are partially conserved in the predicted structure of MAL13P1.19 domain region (yellow).

The *P. falciparum* protein, MAL13P1.19, is currently known to be a conserved *Plasmodium* protein of unknown function [71, 82] predicted to be a peptidase/kinase [80,84]. Sequence alignment obtained for this protein with its potential homologues showed predominance of large N-terminal unassigned region which is apparently unique to Plasmodium genus. The poor sequence identity shared by the *P. falciparum* protein with its homologues accounts for the difficulty in recognition by remote homology detection approaches. For MAL13P1.19, apart from PI3_PI4_kinase at the C-terminal region, no other domain families could be assigned. The presence of such long N-terminus region unique to *Plasmodium* could possibly play some role in regulation of the protein function, pertaining specifically to *P. falciparum* physiology.

a)

b)

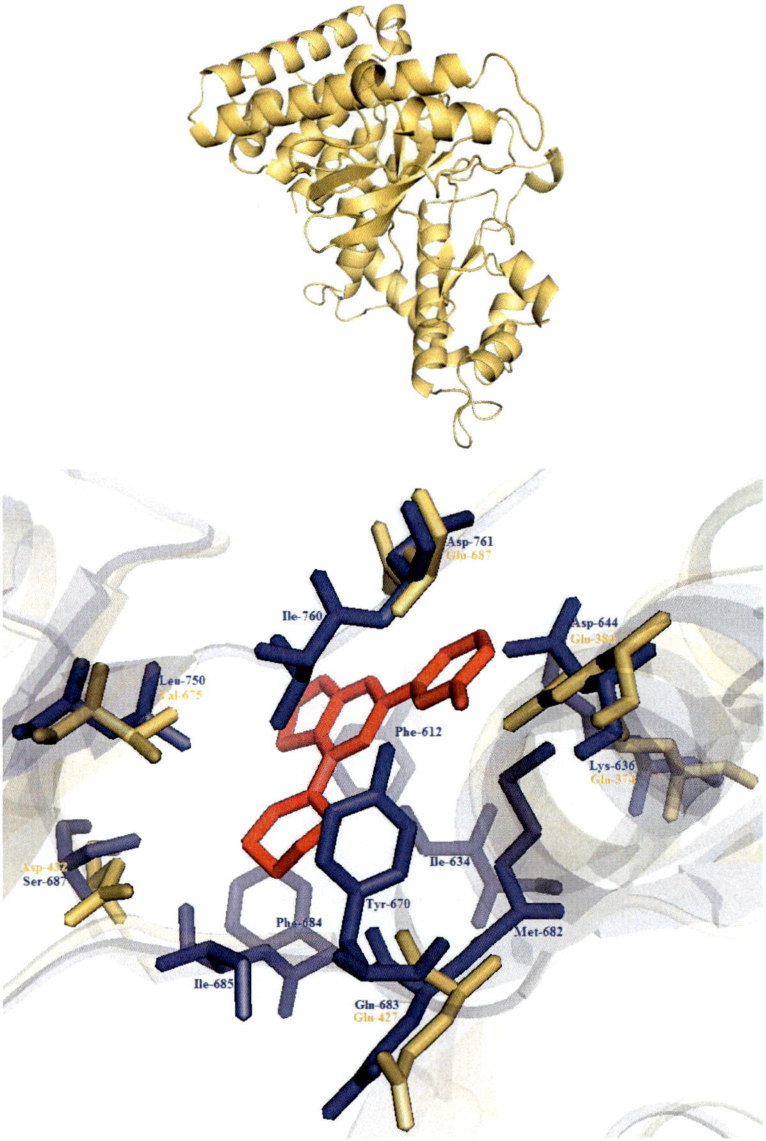

Figure 16: Identifying "new" metabolic proteins in *P. falciparum*. a) Predicted fold adopted by the domain region in MAL13P1.19. b) Ligand binding site residues are highlighted in dark blue for the crystal structure of PI3 kinase (PDB code: 3LS8), complexed with its ligand shown in red, and the structurally equivalent residues obtained in the malarial protein are highlighted in yellow. Images are rendered using PyMOL (http://pymol.org/).

Comparison of expression profiles with other proteins (PFE0485w, PFE0765w) participating in phospholipid biosynthesis indicated consistent expression in trophozoite, schizont and late schizont stages in the life cycle of the parasite [64,71]. Co-expression of MAL13P1.19 with other proteins participating in the same pathway further adds the credibility of the domain assessment made.

4. CONCLUSIONS

Quantification of the extent of divergence in *P. falciparum* proteins involved in various pathways has provided the basis to understand the significance of the observed divergence not only due to presence of low complexity regions but also due to the observed deviation in domain organization, length of domains, length of the proteins and sequence similarity. Exhaustive comparison of domain organization reveals that in nearly 54% of the pathways, domain organizations are conserved. Many of these conserved architectures arise due to single domain proteins. Amongst the proteins showing variation in domain organization, features like loss of domains (63%), gain of domains along with cases of proteins constituting domain repeats are prevalent. Domain distance based approach however is insensitive to variations that are brought about due to variations in lengths of proteins especially in the domain unassigned regions. It is seen that *P. falciparum* proteins show prevalence of domain families which are intervened with non-conserved inserts (15%). Many of these inserts are found to be conserved across closely related organisms and may have functional roles, as is also indicated in some cases through experimental studies [38,47,48]. *P. falciparum* proteins also show abundance of large domain unassigned N and C-terminal overhangs (23% and 9% respectively) in comparison to other model organisms. Finally it is also observed that *P. falciparum* proteins typically show poor sequence similarity in comparison to homologous proteins carrying out similar functions in other organisms. Factors contributing to the observed divergence in the metabolic proteins make detection of function of *P. falciparum* proteins using homology a cumbersome exercise.

The later part of the analysis presented in this chapter focuses on addressing the problem of identification of functions of divergent *P. falciparum* proteins, wherein extremely sensitive profile-based approaches have been employed. Typically, the employment of multiple profiles of domain families for domain family assignments along with an iterative profile-profile search algorithm has aided in enhanced structural and functional characterization of the proteins. Associating a domain family of unknown function with known structural family has provided useful insights in understanding the role of protein domain family concerned. Identification of "new" metabolic proteins was facilitated with the help of multiple profile search approach. Totally 112 proteins were identified with potential roles in metabolic pathways, providing a basis to comprehend the metabolic pathways concerned. In most of these cases, there is prevalence of domain unassigned regions accompanied with poor sequence similarity thereby making them tough targets for function annotations.

AUTHOR CONTRIBUTIONS

N.S. and S.M. conceptualized and designed the study. S.M. made substantial contributions in data analysis and interpretation, along with considerable inputs from G.R. and P.D. S.M. and G.R. drafted the manuscript.

ACKNOWLEDGEMENTS

Authors thank Ms. Richa Mudgal for her valuable inputs on remote homology detection. This research was supported by Mathematical Biology programme, Department of Science and Technology and by the Department of Biotechnology, New Delhi. S.M. is supported by Council of Scientific and Industrial Research, New Delhi, India.

CONFLICT OF INTEREST

The authors confirm that this chapter contents have no conflict of interest.

DISCLOSURE

The chapter "Analysis of Sequence Divergence in Metabolic Proteins of Plasmodium falciparum: Implications for Remote Homology Detection" submitted for eBook series titled "Frontiers in Protein and Peptide Science" Volume 1" is an update of our article "Identification of "missing" metabolic proteins of Plasmodium falciparum: a bioinformatics approach" published in Protein and Peptide Letter (Research article), Volume 16, Number 8, Pages 961-968, 2009 (PMID: 19689423).

ABBREVIATIONS

Abhydrolase	=	alpha/beta hydrolase fold (Pfam domain family)
Acetyltransf_1	=	Acetyltransferase (Pfam domain family)
AlignHUSH	=	Alignment of hidden Markov models using structural and hydrophobicity information
Amino_oxidase	=	Flavin containing amine oxidoreductase (Pfam domain family)
AMP-binding	=	Adenosine monophosphate binding enzyme (Pfam domain family)
BPL_LplA_LipB	=	Biotin/lipoate A/B protein ligase family (Pfam domain family)
C2	=	C2 domain (Pfam domain family)
Carb_kinase	=	Carbohydrate kinase (Pfam domain family)
CN_hydrolase	=	Carbon-nitrogen hydrolase (Pfam domain family)
DAG	=	Diacylglycerol
dCMP_cyt_deam_1	=	Cytidine and deoxycytidylate deaminase zinc-binding region (Pfam domain family)
DD	=	Domain distance
DUF	=	Domain of unknown function

HEM4	=	Uroporphyrinogen-III synthase HemD (Pfam domain family)
HMM	=	Hidden Markov Model
IP3	=	Inositol triphosphate
KEGG	=	Kyoto encyclopedia of genes and genomes
MAFFT	=	Mutiple alignment using fast Fourier transform
MetaPhOrs	=	Database of phylogeny-based orthology and paralogy predictions
MulPSSM	=	Multiple position-specific scoring matrices
NAD_synthase	=	Nicotinamide adenine dinucleotide(+) synthase (Pfam domain family)
PAP2	=	Type 2 phosphatidic acid phosphatase (Pfam domain family)
PGM_PMM	=	Phosphoglucomutase/phosphomannomutase (Pfam domain family)
PH	=	Pleckstrin homology domain (Pfam domain family)
PHYRE2	=	Protein homology recognition engine
PI3_PI4_kinase	=	Phosphatidylinositol 3- and 4-kinase (Pfam domain family)
PlasmoDB	=	Plasmodium genomics resource
Plasmocyc	=	Pathway/genome database
PSSM	=	Position-specific scoring matrix
SUPFAM	=	Superfamily database

THF	=	Tetrahydrofolate
Transketolase_C	=	Transketolase, C-terminal domain (Pfam domain family)
Transketolase_N	=	Transketolase, thiamine diphosphate binding domain (Pfam domain family)
Transket_pyr	=	Transketolase, pyrimidine binding domain (Pfam domain family)
UPF	=	Uncharacterized protein family

REFERENCES

[1] World Global Malaria Programme. *World Malaria Report 2011,* World health Organization: Switzerland, **2011.**

[2] Kantele, A; Jokiranta, T.S. Review of cases with the emerging fifth human malaria parasite, *Plasmodium knowlesi. Clin. Infect. Dis.*, **2011**, *52(11),* 1356-1362.

[3] Casadevall, A. Evolution of intracellular pathogens. *Annu. Rev. Microbiol.,* **2008**, *62,* 19-33.

[4] Plattner, F; Soldati-Favre, D. Hijacking of host cellular functions by the Apicomplexa. *Annu. Rev. Microbiol.*, **2008**, *62,* 471-487.

[5] Mackinnon, MJ; Marsh, K. The selection landscape of malaria parasites. *Science,* **2010**, *328(5980),* 866-871.

[6] Coban, C; Ishii, KJ; Horii, T; Akira, S. Manipulation of host innate immune responses by the malaria parasite. *Trends in Microbiol.*, **2007**, *15(6)*, 271-278.

[7] Lang-Unnasch, N; Murphy, A.D. Metabolic changes of the malaria parasite during the transition from the human to the mosquito host. *Annu. Rev. Microbiol.*, **1998**, *52*, 561-590.

[8] Kutner, S; Ginsburg, H; Cabantchik, Z.I. Permselectivity changes in malaria (*Plasmodium falciparum*) infected human red blood cell membranes. *J. Cell. Physiol.,* **1983**, *114(2)*, 245-251.

[9] Kirk, K; Martin, R.E.; Broer, S; Howitt, S.M.; Saliba, K.J. Plasmodium permeomics: membrane transport proteins in the malaria parasite. *Curr. Top Microbiol. Immunol.,* **2005**, *295,* 325-356.

[10] Koncarevic, S; Rohrbach, P; Deponte, M; Krohne, G; Prieto, J.H.; Yates III, J.; Rahlfs, S.; Becker, K. The malarial parasite *Plasmodium falciparum* imports the human protein peroxiredoxin 2 for peroxide detoxification. *Proc. Natl. Acad. Sci. U. S. A.,* **2009**, *106(32)*, 13323-13328.

[11] Olszewski, K.L.; Morrisey, J.M.; Wilinski, D.; Burns, J.M.; Vaidya, A.B.; Rabinowitz, J.D.; Llinas, M. Host-parasite interactions revealed by *Plasmodium falciparum* metabolomics. *Cell Host Microbe,* **2009**, *5(2)*, 191-199.

[12] Polonais, V; Soldati-Favre, D. Versatility in the acquisition of energy and carbon sources by the Apicomplexa. *Biol. Cell.,* **2010**, *102(8)*, 435-445.

[13] Andersson, S.G.; Kurland, C.G. Reductive evolution of resident genomes. *Trends in Microbiol.,* **1998**, *6(7)*, 263-268.

[14] Gardner, M.J.; Hall, N.; Fung, E.; White, O.; Berriman, M.; Hyman, R.W.; Carlton, J.M.; Pain, A.; Nelson, K.E.; Bowman, S.; Paulsen, I.T.; James, K.; Eisen, J.A.; Rutherford, K.; Salzberg, S.L.; Craig, A.; Kyes, S.; Chan, M.S.; Nene, V.; Shallom, S.J.; Suh, B.; Peterson, J.; Angiuoli, S.; Pertea, M.; Allen, J.; Selengut, J.; Haft, D.; Mather, M.W.; Vaidya, A.B.; Martin, D.M.; Fairlamb, A.H.; Fraunholz, M.J.; Roos, D.S.; Ralph, S.A.; McFadden, G.I.; Cummings, L.M.; Subramanian, G.M.; Mungall, C.; Venter, J.C.; Carucci, D.J.; Hoffman, S.L.; Newbold, C.; Davis, RW.; Fraser, C.M.; Barrell, B. Genome sequence of the human malaria parasite *Plasmodium falciparum. Nature,* **2002**, *419*, 498-511.

[15] Galperin, M.Y.; Walker, D.R.; Koonin, E.V. Analogous enzymes: independent inventions in enzyme evolution. *Genome Res.,* **1998**, *8(8),* 779-790.

[16] Omelchenko, M.V.; Galperin, M.Y.; Wolf, Y.I.; Koonin, E.V. Non-homologous isofunctional enzymes: a systematic analysis of alternative solutions in enzyme evolution. *Biol. Direct.,* **2010**, *5*, 31

[17] Olszewski, K.L.; Mather, M.W.; Morrisey, J.M.; Garcia, B.A.; Vaidya, A.B.; Rabinowitz, J.D.; Llinas, M. Branched tricarboxylic acid metabolism in *Plasmodium falciparum. Nature,* **2010**, *466*, 774-778.

[18] Foth, B.J.; Stimmler, L.M.; Handman E.; Crabb, B.S.; Hodder, A.N.; McFadden, G.I. The malaria parasite *Plasmodium falciparum* has only one pyruvate dehydrogenase complex, which is located in the apicoplast. *Mol. Microbiol.,* **2005**, *55(1)*, 39-53.

[19] Watson, J.D.; Sanderson, S.; Ezersky, A.; Savchenko, A.; Edwards, A.; Orengo, C.; Joachimiak, A.; Laskowski, R.A.; Thornton, J.M. Towards fully automated structure-based function prediction in structural genomics: a case study. *J. Mol. Biol.,* 2007, *367(5)*, 1511-1522.

[20] Ginsburg, H. Progress in *in silico* functional genomics: the malaria Metabolic Pathways database. *Trends Parasitol.,* **2006**, *22(6)*, 238-240.

[21] Ginsburg, H. Caveat emptor: limitations of the automated reconstruction of metabolic pathways in *Plasmodium. Trends Parasitol.,* **2009**, *25(1)*, 37-43.

[22] Yeh, I.; Hanekamp, T.; Tsoka, S.; Karp, P.D.; Altman, R.B. Computational analysis of *Plasmodium falciparum* metabolism: organizing genomic information to facilitate drug discovery. *Genome Res.,* **2004**, *14*, 917-924.

[23] Dondorp, A.M.; Nosten, F.; Yi, P.; Das, D.; Phyo, AP.; Tarning, J.; Lwin, K.M.; Ariey, F.; Hanpithakpong, W.; Lee, S. J.; Ringwald, P.; Silamut, K.; Imwong, M.; Chotivanich, K.; Lim, P.; Herdman, T.; Sen Sam An.; Yeung, S.; Singhasivanon, P.; Nicholas, P.J.; Day, D.M.; Lindegardh N.; Socheat, D.; and White N.J.;, F.R.S. Artemisinin resistance in *Plasmodium falciparum* malaria. *N. Engl. J. Med.,* **2009**, *361,* 455-467.

[24] Mu, J.; Myers, R.A.; Jiang, H.; Liu, S.; Ricklefs, S.; Waisberg, M.; Chotivanich, K.; Wilairatana, P.; Krudsood, S.; White, N.J.; Udomsangpetch, R.; Cui, L.; May, Ho.; Ou, F.; Li, H.; Song, J.; Li, G.; Wang, X.; Seila, S.; Sokunthea, S.; Socheat, D.; Sturdevant, D.E.; Porcella, S.F.; Fairhurst, R.M.; Wellems, T.E.; Awadella, P.; Su, X. *Plasmodium falciparum* genome-wide scans for positive selection, recombination hot spots and resistance to antimalarial drugs. *Nat. Genet.,* **2010**, *42,* 268-271.

[25] Noedl, H.; Se, Y.; Schaecher, K.; Smith, BL.; Socheat, D.; Fukuda, M. Evidence of artemisinin-resistant malaria in western Cambodia. *N. Engl. J. Med.,* **2008**, *359*, 2619-2620.

[26] Zalis, M.G.; Pang, L.; Silveira, M.S.; Milhous, W.K.; Wirth, D.F. Characterization of *Plasmodium falciparum* isolated from the Amazon region of Brazil: evidence for quinine resistance. *Am. J. Trop. Med. Hyg.*, **1998**, *58(5)*, 630-637.

[27] Hunter,S.; Apweiler, R.; Attwood, T.K.; Bairoch, A.; Bateman, A.; Binns, D.; Bork, P.; Das U.; Daugherty, L.; Dugenne, L.; Finn, R.D.; Gough, J.; Haft, D.; Hulo, N.; Kahn, D.; Kelly, E.; Laugraud, A.; Letunic, I.; Lonsdale, D.; Lopez, R.; Madera, M.; Maslen, J.; McAnulla, C.; McDowall, J.; Mistry, J.; Mitchell, A.; Mulder, N.; Natale, D.; Orengo, C.; Quinn, A.F.; Selengut, J.D.; Sigrist, C.J.; Thimma, M.; Thomas, P.D.; Valentin, F.; Wilson, D.; Wu, C.H.; Yeats, C. InterPro: the integrative protein signature database. *Nucleic Acids Res.*, **2009**, *37(D)*, 211-215.

[28] Pryszcz, L.P.; Huerta-Cepas, J.; Gabaldon, T. MetaPhOrs: orthology and paralogy predictions from multiple phylogenetic evidence using a consistency-based confidence score. *Nucleic Acids Res.*, **2011**, *39(5)*, e32.

[29] Andersen, M.R.; Nielsen, M.L.; Nielsen, J. Metabolic model integration of the bibliome, genome, metabolome and reactome of Aspergillus niger. *Mol. Syst. Biol.*, **2008**, *4*, 178.

[30] Chavali, A.K.; Whittemore, J.D.; Eddy, J.A.; Williams, K.T.; Papin, J.A. Systems analysis of metabolism in the pathogenic trypanosomatid Leishmania major. *Mol. Syst. Biol.*, **2008**, *4*, 177.

[31] Mo, M.L.; Palsson, B.O.; Herrgard, M.J. Connecting extracellular metabolomic measurements to intracellular flux states in yeast. *BMC Syst. Biol.*, **2009**, *3*, 37.

[32] Roberts, S.B.; Robichaux, J.L.; Chavali, A.K.; Manque, P.A.; Lee, V.; Lara, M.A.; Papin, J.A.; Buck, G.A. Proteomic and network analysis characterize stage-specific metabolism in *Trypanosoma cruzi. BMC Syst. Biol.*, **2009**, *3*, 52.

[33] Plata, G.; Hsiao, T.L.; Olszewski, K.L.; Llinas, M.; Vitkup, D. Reconstruction and flux-balance analysis of the *Plasmodium falciparum* metabolic network. *Mol. Syst. Biol.*, **2010**, *6*, 408.

[34] Bjorklund, A.K.; Ekman, D.; Light, S.; Frey-Skott, J.; Elofsson, A. Domain rearrangements in protein evolution. *J. Mol. Biol.*, **2005**, *353(4)*, 911-923.

[35] Krishnadev, O.; Rekha, N.; Pandit, S.B.; Abhiman, S.; Mohanty, S.; Swapna, L.S.; Gore, S.; Srinivasan, N. PRODOC: a resource for the comparison of tethered protein domain architectures with in-built information on remotely related domain families. *Nucleic Acids Res.*, **2005**, *33(W)*, 126-129.

[36] Muller, I.B.; Das, Gupta, R.; Luersen, K.; Wrenger, C.; Walter, R.D. Assessing the polyamine metabolism of *Plasmodium falciparum* as chemotherapeutic target. *Mol. Biochem. Parasitol.*, **2008**, *160(1)*, 1-7.

[37] Wrenger, C.; Luersen, K.; Krause, T.; Muller, S.; Walter, R.D. The *Plasmodium falciparum* bifunctional ornithine decarboxylase, S-adenosyl-L-methionine decarboxylase, enables a well balanced polyamine synthesis without domain-domain interaction. *J. Biol. Chem.*, **2001**, *276(32)*, 29651-29656.

[38] Birkholtz, L.M.; Wrenger, C.; Joubert, F.; Wells, G.A.; Walter, R.D.; Louw, A.I. Parasite-specific inserts in the bifunctional S-adenosylmethionine decarboxylase/ornithine decarboxylase of Plasmodium falciparum modulate catalytic activities and domain interactions. *Biochem. J.*, **2004**, *377(Pt 2)*, 439-448.

[39] Yuvaniyama, J.; Chitnumsub, P.; Kamchonwongpaisan, S.; Vanichtanankul, J.; Sirawaraporn, W.; Taylor, P.; Walkinshaw, M.D.; Yuthavong, Y. Insights into antifolate resistance from malarial DHFR-TS structures. *Nat. Struct. Biol.*, **2003**, *10*, 357-365.

[40] Wakui, H.; Komatsuda, A.; Ishino, T.; Imai, H.; Kobayashi, R.; Nakamoto, Y.; Miura, A.B. Localization of phosphoinositide-specific phospholipase C-alpha in porcine kidney. *Kidney Int.,* **1992**, *42(4)*, 888-895.

[41] Song, J.M.; Rabinowitz, J.C. Function of yeast cytoplasmic C1-tetrahydrofolate synthase. *Proc. Natl. Acad. Sci. U. S. A.,* **1993**, *90(7)*, 2636-2640.

[42] Bjorklund, A.K.; Ekman, D.; Elofsson, A. Expansion of protein domain repeats. *PLoS Comput. Biol.,* **2006**, *2(8)*, e114.

[43] Marsh, J.J.; Lebherz, H.G. Fructose-bisphosphate aldolases: an evolutionary history. *Trends Biochem. Sci.,* **1992**, *17(3)*, 110-113.

[44] Patron, N.J.; Rogers, M.B.; Keeling, P.J. Gene replacement of fructose-1,6-bisphosphate aldolase supports the hypothesis of a single photosynthetic ancestor of chromalveolates. *Eukaryot. Cell,* **2004**, *3(5)*, 1169-1175.

[45] Kurland, C.G.; Canback, B.; Berg, O.G. The origins of modern proteomes. *Biochimie.,* **2007**, *89(12)*, 1454-1463.

[46] Wang, M.; Kurland, C.G.; Caetano-Anolles, G. Reductive evolution of proteomes and protein structures. *Proc. Natl. Acad. Sci. U. S. A.*, **2011**, *108(29)*, 11954-11958.

[47] Clarke, J.L.; Sodeinde, O.; Mason, P.J. A unique insertion in *Plasmodium berghei* glucose-6-phosphate dehydrogenase-6-phosphogluconolactonase: evolutionary and functional studies. *Mol. Biochem. Parasitol.,* **2003**, *127(1)*, 1-8.

[48] Gilberger, T.W.; Schirmer, R.H.; Walter, R.D.; Muller, S. Deletion of the parasite-specific insertions and mutation of the catalytic triad in glutathione reductase from chloroquine-sensitive *Plasmodium falciparum* 3D7. *Mol. Biochem. Parasitol.,* **2000**, *107(2)*, 169-179.

[49] Katoh, K.; Misawa, K.; Kuma, K.; Miyata, T. MAFFT: a novel method for rapid multiple sequence alignment based on fast Fourier transform. *Nucleic Acids Res.,* **2002**, *30(14)*, 3059-3066.

[50] Rigden, D.J.; Walter, R.A.; Phillips, S.E.; Fothergill-Gilmore, L.A. Polyanionic inhibitors of phosphoglycerate mutase: combined structural and biochemical analysis. *J. Mol. Biol.,* **1999**, *289(4)*, 691-699.

[51] Kanehisa, M.; Araki, M.; Goto, S.; Hattori, M.; Hirakawa, M.; Itoh, M.; Katayama, T.; Kawashima, S.; Okuda, S.; Tokimatsu, T.; Yamanishi, Y. KEGG for linking genomes to life and the environment. *Nucleic Acids Res.,* **2008**, *36(D)*, 480-484.

[52] Caspi, R.; Foerster, H.; Fulcher, C.A.; Kaipa, P.; Krummenacker, M.; Latendresse, M.; Paley, S.; Rhee, S.Y.; Shearer, A.G.; Tissier, C.; Walk, T.C.; Zhang, P.; Karp, P.D. The MetaCyc Database of metabolic pathways and enzymes and the BioCyc collection of Pathway/Genome Databases. *Nucleic Acids Res.,* **2008**, *36(D),* 623-631.

[53] Eddy, S.R. Profile hidden Markov models. *Bioinformatics*, **1998**, *14(9),* 755-763.

[54] Eddy, S.R. Accelerated profile HMM searches. *PLoS Comput. Bio.,* **2011**, *7(10),* e1002195.

[55] Gowri, V.S.; Krishnadev, O.; Swamy, C.S.; Srinivasan, N. MulPSSM: a database of multiple position-specific scoring matrices of protein domain families. *Nucleic Acids Res.,* **2006**, *34(D)*, 243-246.

[56] Gowri, V.S.; Tina, K.G.; Krishnadev, O.; Srinivasan, N. Strategies for the effective identification of remotely related sequences in multiple PSSM search approach. *Proteins,* **2007**, *67*, 789-794.

[57] Remmert, M.; Biegert, A.; Hauser, A.; Soding, J. HHblits: lightning-fast iterative protein sequence searching by HMM-HMM alignment. *Nat. Methods,* **2012**, *9(2)*, 173-175.

[58] Punta, M.; Coggill, P.C.; Eberhardt, R.Y.; Mistry, J.; Tate, J.; Boursnell, C.; Pang, N.; Forslund, K.; Ceric, G.; Clements, J.; Heger, A.; Holm, L.; Sonnhammer, E.L.L.; Eddy,

S.R.; Bateman, A.; Finn, R.D. The Pfam protein families database. *Nucleic Acids Res.*, **2012**, *40(D)*, 290-301.

[59] Mohanty, S. *Functionally interacting proteins: analyses and predictions*. PhD Thesis, Indian Institute of Science: Bangalore, November **2012.**

[60] Gowri, V.S. *Analysis of protein evolution and its implications in remote homology detection and function recognition*. PhD Thesis, Indian Institute of Science: Bangalore, October **2007**.

[61] Krishnadev, O. *Development and applications of computational methods to aid recognition of protein functions and interactions*. PhD Thesis, Indian Institute of Science: Bangalore, March **2010.**

[62] Bennett-Lovsey, R.M.; Herbert, A.D.; Sternberg, M.J.; Kelley, L.A. Exploring the extremes of sequence/structure space with ensemble fold recognition in the program Phyre. *Proteins*, **2008**, *70(3)*, 611-625.

[63] Ben Mamoun, C.; Gluzman, I.Y.; Hott, C.; MacMillan, S.K.; Amarakone, A.S.; Anderson D.L.; Carlton, J.M.; Dame, J.B.; Chakrabarti, D.; Marti, R.K.; Brownstein, B.H.; Goldberg, D.E. Co-ordinated programme of gene expression during asexual intraerythrocytic development of the human malaria parasite *Plasmodium falciparum* revealed by microarray analysis. *Mol. Microbiol.*, **2001**, *39(1)*, 26-36.

[64] De Risi, J.L.; Iyer, V.R.; Brown, P.O. Exploring the metabolic and genetic control of gene expression on a genomic scale. *Science*, **1997**, *278(5338)*, 680-686.

[65] Pandit S.B., Gosar D., Abhiman S., Sujatha S., Dixit S.S., Mhatre N.S., Sowdhamini R., Srinivasan N. SUPFAM-a database of potential protein superfamily relationships derived by comparing sequence-based and structure-based families: implications for structural genomics and function annotation in genomes. *Nucleic Acids Res.*, **2002**, *30(1)*, 289-293.

[66] Pandit, S. B., Bhadra, R., Gowri, V. S., Balaji, S., Anand, B. and Srinivasan, N. SUPFAM: A database of sequence superfamilies of protein domains. *BMC Bioinformatics*, **2004**, *5(28)*.

[67] Bateman, A.; Coggill, P.; Finn, R.D. DUFs: families in search of function. *Acta. Crystallogr. Sect. F Struct. Biol. Cryst. Commun.*, **2010**, *66(Pt 10)*, 1148-1152.

[68] Oganesyan, V.; Busso, D.; Brandsen, J.; Chen, S.; Jancarik, J.; Kim, R.; Kim, S.-H. Structure of the hypothetical protein AQ_1354 from *Aquifex aeolicus.Acta. Crystallogr. D. Biol. Crystallogr.* **2003**, *59(Pt 7)*, 1219-1223.

[69] Penhoat, C.H.; Li, Z.; Atreya, H.S.; Kim, S.; Yee, A.; Xiao, R.; Murray, D. ;Arrowsmith, C.H.; Szyperski, T. NMR solution structure of *Thermatoga maritima* protein reveals a Zn-metalloprotease-like tertiary structure. *J. Struct. Funct. Genomics.*, **2005**, *6(1)*, 51-62.

[70] Wellcome Trust Sanger Institute. Pfam. http://pfam.sanger.ac.uk/ (Accessed 22 Oct, **2012**).

[71] Aurrecoechea, C.; Brestelli, J.; Brunk, B.P.; Dommer, J.; Fischer, S.; Gajria, B.; Gao, X.; Gingle, A.; Grant, G.; Harb, O.S.; Heiges, M.; Innamorato, F.; Iodice, J.; Kissinger, J.C.; Kraemer, E.; Li, W.; Miller, J.A.; Nayak, V.; Pennington, C.; Pinney, D.F.; Roos, D.S.; Ross, C.; Stoeckert, C.J.Jr.; Treatman, C.; Wang, N. PlasmoDB: a functional genomic database for malarial parasites. *Nucleic Acids Res.*, **2008**, *37(D)*, 539-543.

[72] Dlakic, M. DUF283 domain of Dicer proteins has a double-stranded RNA-binding fold. *Bioinformatics*, **2006**, *22(22)*, 2711–2714.

[73] Martzen, M.R.; McCraith, S.M.; Spinelli, S.L.; Torres, F.M.; Fields, S.; Grayhack, E.J.; Phizicky, E.M. A biochemical genomics approach for identifying genes by the activity of their products. *Science*, **1999**, *286(5442)*, 1153–1155.

[74] Karras, G.I.; Kustatscher, G.; Buhecha, H.R.; Allen, M.D.; Pugieux, C.; Sait, F.; Bycroft, M.; Ladurner, A.G. The macro domain is an ADP-ribose binding module. *EMBO J.,* **2005**, *24(11)*, 1911–1920.

[75] Schulze-Gahmen, U.; Pelaschier, J.; Yokota, H.; Kim, R.; Kim, S.-H. Crystal structure of a hypothetical protein, TM841 of *Thermatoga maritima,* reveals its function as a fatty acid binding protein. *Proteins,* **2003**, *50(4)*, 526–530.

[76] Krishna, S.S.; Tautz, L.; Xu, Q.; McMullan, D.; Miller, M.D.; Abdubek, P.; Ambing, E.; Astakhova, T.; Axelrod, H.L.; Carlton, D.; Chiu, H.-J.; Clayton, T.; DiDonato, M.; Duan, L.; Elsliger, M.-A.; Grzechnik, J.H.; Hampton, E.; Han, G.W.; Haugen, J.; Jaroszewski, L.; Jin, K.K.; Klock, H.E.; Knuth, M.W.; Koesema, E.; Morse, A.T.; Mustelin, T.; Nigoghossian, E.; Oomamachen, S.; Reyes, R.; Rife, C.L.; van den Bedem, H.; Weekes, D.; White, A.; Hodgson, K.O.; Wooley, J.; Deacon, A.M.; Godzik, A.; Lesley, S.A.; Wilson, I.A. Crystal structure of NMA1982 from *Neisseria meningitidis* at 1.5 Å resolution provides a structural scaffold for nonclassical, eukaryotic-like phosphatases. *Proteins,* **2007**, *69(2),* 415–421.

[77] Krishnadev, O.; Srinivasan, N. AlignHUSH: Alignment of HMMs using structure and hydrophobicity information. *BMC Bioinformatics,* **2011**, *12(275).*

[78] Hiles, I.D.; Otsu, M.; Volinia, S.; Fry, M.J.; Gout, I.; Dhand, R.; Panayotou, G.; Ruiz-Larrea, F.; Thompson, A.; Totty, N.F. Hsuan, J.J.; Courtneidge, S.A.; Parker, P.J.; Waterfield, M.D. Phosphatidylinositol 3-kinase: structure and expression of the 110 kd catalytic subunit. *Cell,* **1992**, *70(3),* 419-429.

[79] Garcia-Bustos, J.F.; Marini, F.; Stevenson, I.; Frei C.; Hall, M.N. PIK1, an essential phosphatidylinositol 4-kinase with the yeast nucleus. *EMBO J.,* **1994**, *13(10),* 2352-2361.

[80] European Bioinformatics Institute. InterPro. http://www.ebi.ac.uk/interpro/ (Accessed 22 Oct. **2012**).

[81] Kelly, L.A.; Sternberg, M.J.E. Protein structure prediction of the web: a case study using the Phyre server. *Nat. Protocols.***2009**, *4(3),* 363-371.

[82] Ortholog Groups of Protein Sequences. OrthoMCL DB. http://orthomcl.cbil.upenn.edu (Accessed 22 Oct **2012**).

[83] Mohanty, S.; Srinivasan, N. Identification of "missing" metabolic proteins of *Plasmodium falciparum*: a bioinformatics approach. *Protein Pept. Lett.* **2009**, *16,* 961-968.

[84] UniProt Consortium. UniProt. http://www.uniprot.org/ (Accessed 22 Oct **2012**).

Index

A

Ben M. Dunn (Ed)

B

C

D

I

J

K

L

M

N

O

P

R

S

T

V

W